The Virial Coefficients of Pure Gases and Mixtures
A Critical Compilation

J. H. Dymond

and

E. B. Smith

CLARENDON PRESS · OXFORD · 1980

Oxford University Press, Walton Street, Oxford OX2 6DP

OXFORD   LONDON   GLASGOW
NEW YORK   TORONTO   MELBOURNE   WELLINGTON
KUALA LUMPUR   SINGAPORE   JAKARTA   HONG KONG   TOKYO
DELHI   BOMBAY   CALCUTTA   MADRAS   KARACHI
NAIROBI   DAR ES SALAAM   CAPE TOWN

TO
OUR MENTOR

Joel H. Hildebrand

British Library Cataloguing in Publication Data

Dymond, J H
    The second virial coefficients of pure gases
    and mixtures.—(Oxford science research papers).
    1. Virial coefficients—Tables
    I. Title   II. Smith, Eric Brian
    536'.412   QD458   79-40667

    ISBN 0-19-855361-7

Printed in Great Britain by
Thomson Litho Ltd, East Kilbride, Scotland

PREFACE

Tables of the virial coefficients of pure gases in substantially the present form were prepared by Professor M. L. McGlashan in 1955. In 1964 these tables were revised and extended to almost double the original length at the Physical Chemistry Laboratory, Oxford University. Both these sets of tables were produced in very limited quantities and privately circulated. As it proved impossible to satisfy the demand by this method it was decided in 1969 to publish a new and enlarged edition in co-operation with Oxford University Press. The latest version is again much enlarged; it is some three times longer than the 1969 edition and includes a comprehensive collection of material published up to early 1979 not only for pure substances but for the virial coefficient data on binary mixtures.

Again we would acknowledge the numerous scientists who have communicated with us and provided their data often prior to publication. The previous compilation was remarkably free of errors but as the present work is three times as long the problem of eliminating errors has become increasingly difficult. We would be most grateful if errors and omissions could be brought to our attention.

We would also appreciate it if those workers engaged in the study of gas imperfection would send reprints of their work to J. H. D. at the Department of Chemistry, University of Glasgow or E. B. S. at the Physical Chemistry Laboratory, South Parks Road, Oxford to ensure their inclusion in any subsequent editions.

J. H. D.
E. B. S.

CONTENTS

INTRODUCTION

THE imperfection of real gases is a subject that has concerned physicists and chemists for over a century. Some of this interest has arisen from the importance of the study of gas imperfections in the elucidation of the forces between molecules. But many of those involved with the behaviour of real gases have been concerned with the resolution of practical problems that occur in many diverse aspects of thermodynamics.

Gas Imperfections

The perfect gas is characterized by the equation of state

$$Z = \frac{P\bar{V}}{RT} = 1,$$

where P is the pressure, $\bar{V}$ the molar volume, T the absolute temperature, and R the gas constant. Real gases may show significant deviations from this equation of state, even at low pressure. At low temperatures and pressures the compressibility factor Z is usually than unity, whereas at high temperatures and pressures the converse is true. Typical behaviour of Z is illustrated in Fig. 1 for a series of temperatures such that $T_3 > T_2 > T_1$. The temperature T_3, at which the density dependence of Z as $\frac{1}{\bar{V}} \to 0$ is zero, is termed the Boyle temperature.

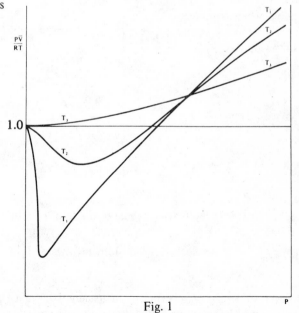

Fig. 1

The equation of state of real gases is best represented, at all but the highest pressures, by the series

$$\frac{P\bar{V}}{RT} = 1 + \frac{B(T)}{\bar{V}} + \frac{C(T)}{\bar{V}^2} + \frac{D(T)}{\bar{V}^3} + \ldots,$$

where $B(T)$, $C(T)$, and $D(T)$ are respectively termed the second, third, and fourth virial coefficients. $B(T)$ is defined as follows:

$$B(T) = \lim_{1/\bar{V} \to 0} \left(\frac{P\bar{V}}{RT} - 1 \right) \bar{V} \equiv \lim_{1/\bar{V} \to 0} \mathscr{A}$$

and $B(T)$ is zero at the Boyle temperature. The general manner of the variation of $B(T)$ with temperature is indicated in Fig. 2. At low temperatures $B(T)$ is large and negative, whereas at high temperatures it has small positive values.

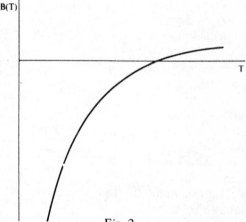

Fig. 2

The third virial coefficient is equal to the slope of $\mathscr{A}$ at zero pressure.

$$C(T) = \lim_{1/\bar{V} \to 0} (\mathscr{A} - B)\,\bar{V}.$$

However, values for this coefficient are usually determined experimentally from gas-compressibility data by fitting the results at a given temperature by a polynomial in reciprocal volume. The coefficients of this polynomial are then identified with the coefficients of the infinite virial series. Values for the virial coefficients obtained in this way depend on the degree of polynomial used and on the density range of the compressibility data. The resulting uncertainties in the second virial coefficient are small (usually less than 1 per cent) but are much larger for the third virial coefficient.[1] The general dependence of $C(T)$ on temperature is illustrated in Fig 3.

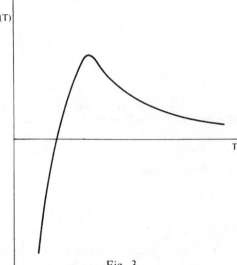

Fig. 3

Virial Coefficients and Intermolecular Energy

Besides providing a convenient method of describing gas imperfection, the virial form of the equation of state is important because the coefficients are related in a fairly simple manner to the intermolecular potential energy function, U(R), of the molecules concerned.[2] Thus it can be shown that, for molecules with a spherically symmetric potential energy function, such as that illustrated in Fig. 4,

$$B(T) = -2\pi N \int_0^\infty (e^{-U(R)/kT} - 1) . R^2 . dR.$$

In principle it is not possible to determine the potential energy function $U(R)$ uniquely from second virial coefficient data above, (except for a purely monotonic repulsion function). This can be seen when the second virial coefficient is expressed.[3]

$$B(T) = \frac{2\pi N}{3kT} e^{\epsilon/kT} \int_0^\infty \Delta \exp\left[-\frac{\phi}{kT}\right] d\phi$$

where $\Delta = r_L^3 - r_R^3$, ϵ is the maximum depth of the potential energy well, and $\phi = U + \epsilon$ is the potential energy measured from the bottom of the well. r_L and r_R are the inner and outer coordinates of the potential energy function at ϕ. In the repulsive region $\Delta = r_L$. A formal inversion of this expression should be possible as

$$B(T)T/\tfrac{2}{3}\,\pi N e^{\epsilon/kT}$$

is the Laplace transform of Δ. The inversion of B can, in principle, give the repulsive branch of $U(R)$ and the well width as a function of the depth. This has only proved useful in the case of helium at high reduced temperatures.

Despite the formal limitations, inversion methods have been devised which enable the pair potential energy function, $U(R)$ to be determined from high accuracy second virial coefficient data.[4] In brief, the methods define a characteristic length, $\tilde{r}$, defined by

$$\tilde{r} = \left\{ \frac{B + T\,(\mathrm{d}B/\mathrm{d}T)}{2/3\,\pi\,N} \right\}^{1/3}.$$

The intermolecular potential energy at this separation $U(\tilde{r})$, can be related to kT where T is the temperature at which B was determined by

$$U(\tilde{r}) = \mathcal{G}(T)\,kT.$$

The function $\mathcal{G}(T)$ is called the inversion function. It is comparatively insensitive to the detailed form of $U(R)$ and can be calculated using a very crude model potential such as the Lennard-Jones (12-6) function. Thus $U(R)$ can be directly determined from $B(T)$. With data of very high quality, this approach can be made the basis of an iterative procedure.

The third virial coefficient can also be related to $U(R)$ if the intermolecular energies are considered pair-wise additive. However, this assumption introduces considerable error.

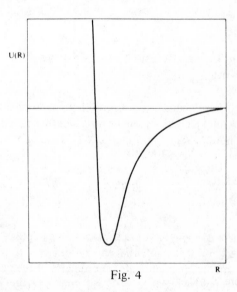

Fig. 4

Properties of Gases in terms of virial coefficients

The thermodynamic properties of gases may readily be deduced from a knowledge of the virial coefficients and their dependence on temperature.[5] For calculations at pressures not much greater than 1 atm, a knowledge of the second virial coefficient is usually sufficient. At high pressures, the contribution of the third virial coefficient may become significant.

In the following paragraphs, expressions for the thermodynamic properties are given, generally, in terms of the equation of state and, specifically, in terms of virial coefficients ($\bar{\ }$ indicates a quantity per mol and ^{0} refers to the perfect gas state).

(i) *Fugacity, f*

Fugacity may be defined in terms of the chemical potential, μ, by the relation

$$\partial \ln f = \partial \mu / RT.$$

From this the following expression may be derived:

$$RT \ln \frac{f}{P} = - \int_0^P \left(\frac{RT}{P} - \bar{V} \right) dP.$$

Expressing $\bar{V}$, the molar volume of the real gas, in terms of the virial coefficients leads to the equation:

$$\ln \frac{f}{P} = \frac{B}{\bar{V}} + \frac{C + B^2}{2(\bar{V})^2} + \ldots.$$

(ii) *Internal Energy, U*

The departure of the molar internal energy from the perfect gas value may be written

$$\bar{U} - \bar{U}^0 = - \int_{\bar{V}}^{\infty} \left(T \left(\frac{\partial P}{\partial T} \right)_{\bar{V}} - P \right) d\bar{V}.$$

Using the virial equation of state, this becomes

$$\bar{U} - \bar{U}^0 = - RT \left(\frac{B_1}{\bar{V}} + \frac{C_1}{2(\bar{V})^2} + \ldots \right),$$

where

$$B_1 = T \left(\frac{dB}{dT} \right); \quad C_1 = T \left(\frac{dC}{dT} \right).$$

(iii) *Enthalpy, H*

The difference between the enthalpy of a real gas and that of a perfect gas under the same conditions is simply related to the corresponding difference in internal energy

$$\bar{H} - \bar{H}^0 = \bar{U} - \bar{U}^0 + PV - RT.$$

Thus, using the virial equation of state,

$$\bar{H} - \bar{H}^0 = RT\left(\frac{B - B_1}{\bar{V}} + \frac{2C - C_1}{2(\bar{V})^2} + \cdots\right).$$

(iv) *Heat capacity at constant volume, C_V*

This function for a real gas is given in terms of the standard state value by the expression

$$\bar{C}_V - \bar{C}_V{}^0 = -T \int_{\bar{V}}^{\infty} \left(\frac{\partial^2 P}{\partial T^2}\right)_{\bar{V}} \cdot d\bar{V},$$

which, in terms of the virial equation of state, becomes

$$\bar{C}_V - \bar{C}_V{}^0 = -R\left(\frac{2B_1 + B_2}{\bar{V}} + \frac{2C_1 + C_2}{2(\bar{V})^2} + \cdots\right),$$

where

$$B_2 = T^2\left(\frac{d^2 B}{dT^2}\right); \quad C_2 = T^2\left(\frac{d^2 C}{dT^2}\right).$$

(v) *Heat capacity at constant pressure, C_p*

This function is given by the equation:

$$\bar{C}_p - \bar{C}_p{}^0 = -R - T\left(\frac{\partial P}{\partial T}\right)_{\bar{V}}^2 \bigg/ \left(\frac{\partial P}{\partial \bar{V}}\right)_T - T\int_{\bar{V}}^{\infty}\left(\frac{\partial^2 P}{\partial T^2}\right)_{\bar{V}} d\bar{V}.$$

Substituting the virial equation of state leads to the result

$$\bar{C}_p - \bar{C}_p{}^0 = -R\left\{\frac{B_2`}{\bar{V}} - \frac{(B - B_1)^2 - (C - C_1) - C_2/2}{\bar{V}^2} + \cdots\right\}.$$

(vi) *Entropy, S*

The departure of the entropy from the perfect gas value is given by

$$\bar{S} - \bar{S}^0 = -R\ln P + R\ln\frac{P\bar{V}}{RT} - \int_{\bar{V}}^{\infty}\left\{\left(\frac{\partial P}{\partial T}\right)_{\bar{V}} - \frac{R}{\bar{V}}\right\}d\bar{V},$$

which leads to the following result

$$\bar{S} - \bar{S}^0 = -R\left\{\ln P + \frac{B_1}{V} + \frac{(B^2 - C + C_1)}{2(\bar{V})^2} + \cdots\right\}.$$

(vii) *Joule–Thomson coefficient, χ.*

The Joule–Thomson coefficient is given by the relation:

$$\chi = \frac{1}{\bar{C}_p}\left(T\left(\frac{\partial \bar{V}}{\partial T}\right)_p - \bar{V}\right),$$

which takes the following form in terms of virial coefficients:

$$\chi = \frac{1}{\bar{C}_p{}^0}\left\{(B_1 - B) + \frac{2B^2 - 2B_1 B - 2C + C_1}{\bar{V}} + \frac{R}{\bar{C}_p{}^0}\left(\frac{B_2(B_1 - B)}{\bar{V}}\right) + \ldots\right\},$$

where $\bar{C}_p{}^0$ is the zero pressure value of the molar heat capacity.

The virial coefficients of mixtures

The virial coefficient of a binary gas mixture may be expressed

$$B_M(T) = B_{11}(T)x_1^2 + 2B_{12}(T)x_1 x_2 + B_{22}(T)x_2^2 .$$

Here x_1 and x_2 are the mole fractions of species 1 and 2 in the mixture. B_{11} and B_{22} are the second virial coefficients of pure component 1 and component 2, respectively. B_{12} the so-called interaction virial coefficient is the second virial coefficient which corresponds to the potential energy function $U_{12}(R)$ which describes the interaction of molecules of species 1 with those of species 2. B_{12} is also referred to as the cross virial coefficient, the cross-term virial coefficient, and the mixed virial coefficient.

The excess virial coefficient E is given by

$$E = B_{12}(T) - 0.5\,[B_{11}(T) + B_{22}(T)] .$$

The third virial coefficient of a binary gas mixture may be written

$$C_M(T) = C_{111}x_1^3 + 3C_{112}\,x_1^2\,x_2 + 3C_{122}\,x_1 x_2^2 + C_{222}\,x_2^3 .$$

C_{111} and C_{222} are the third virial coefficients of the pure components, C_{112} is the contribution that arises from two molecules of species 1 interacting with one molecule of species 2, and C_{122} arises from the interaction of one molecule of species 1 with two of species 2.

The second virial coefficients of multicomponent mixtures may be calculated if the second virial coefficients of the pure components and the interaction virial coefficients of all pairs of molecules present in the mixture are known.

Thus for an n component mixture

$$B_M(T) = \sum_{a=1}^{n} \sum_{\beta=1}^{n} B_{a\beta}(T) x_a x_\beta$$

which written in full for a three component mixture is

$$B_M(T) = B_{11} x_1^2 + 2B_{12} x_1 x_2 + 2B_{13} x_1 x_3 + 2B_{23} x_2 x_3 + B_{22} x_2^3 + B_{33} x_3^3 .$$

The Tables

The revised tables include material published up to early 1979. For the most part, substances are listed by the name used by the authors who reported the virial coefficients. However, when more than one set of data is available for a compound this procedure cannot always be followed. The proper entry to the tables is by way of the formula index, which is based on the system used in the *Chemical abstracts* formula index. A name index is also provided for convenience in finding data, especially for complex organic substances, but in view of the diversity of names in use for many substances, the formula index should always be checked if the compound does not appear to be present in the name index.

An attempt has been made to give some assessment of the quality of the second virial coefficients quoted. However, in view of the variety of techniques used, it has proved virtually impossible to give completely reliable independent estimates of the accuracy of the data in these tables. The following procedures have been adopted.

(i)   Where the authors make general statements about the supposed accuracy or precision of their work these are quoted.

(ii)   Where estimates of the reproducibility or precision of individual determinations are given these are reproduced.

(iii)   If information about the quality of the results is not given in the original paper the results have been classified, where possible, into one of three groups, according to the precision to be expected from the technique used.

The groups are:

Class I.   Estimated precision < 2 per cent or <1 cm^3 mol^{-1}, whichever is the greater.

Class II.   Estimated precision < 10 per cent or <15 cm^3 mol^{-1}, whichever is the greater.

Class III. Estimated precision greater than 10 per cent or 15 cm^3 mol^{-1}, whichever is the greater.

Results obtained by the analysis of thermodynamics data (such as heats of vaporization) and determinations made at a single temperature have generally not been classified. All estimates of accuracy must be regarded with caution, and those for whom an assessment of this factor is of prime importance should consult the original papers.

Where sets of data for a single substance of apparently high precision are in obvious conflict, or in cases where there is a profusion of data, an attempt has been made to indicate with a few critical comments the relative value of the different sets of data. These comments appear at the head of the section dealing with that particular substance.

For certain important substances we have tabulated 'recommended' values of the second virial coefficients obtained from a graphical analysis of what we regard as the best data. These tables, which appear after the critical comments, may prove useful to those who require reasonably reliable estimates of second virial coefficients but do not wish to undertake a full analysis of all the available data. The second virial coefficients have all been converted to the units $cm^3 \ mol^{-1}$, which is the unit in most common use at the present time. The S I units $m^3 \ k \ mol^{-1}$ should attain more importance in the next few years, but the conversion is trivial. For third virial coefficients the units $cm^6 \ mol^{-2}$ are used. Further details of the conversion procedures and the lay-out of the material are given in the notes that follow.

1.   The symbol (*) indicates that experimental data from these references have been converted to values of B and C in the series:

$$P\bar{V} = RT\left(1 + \frac{B}{\bar{V}} + \frac{C}{\bar{V}^2} + \ldots\right)$$

and have been expressed in units of $cm^3 \ mol^{-1}$ B and $cm^6 \ mol^{-2}$ C.

2.   Certain unpublished experimental results have been communicated and are denoted in the tables by (†)

3.   For the calculations, the following formulae and values of RT_0 and T_0 have been used:

$$P\bar{V} = A\left(1 + \frac{B}{\bar{V}} + \frac{C}{\bar{V}^2} + \ldots\right),$$

$$P\bar{V} = A^* + B^* \, P + C^* \, P^2 + \ldots,$$

$$P\bar{V} = A' + \frac{B'}{\bar{V}} + \frac{C'}{\bar{V}^2} + \ldots.$$

For conversion from one series to another: (strictly true for infinite series only)

$$A = A^* = A',$$

$$B = B^* = \frac{B'}{A'},$$

$$C = B^{*2} + A^* C^* = \frac{C'}{A'}.$$

$$R T_0 = 22\,414 \, \text{cm}^3 \text{atm. mol}^{-1}.$$

$$T_0 = 273 \cdot 15 \text{ K.}$$

1 Amagat unit of volume = volume occupied by 1 mol of gas at 0°C and 1 atm

1 Berlin unit of volume = volume occupied by 1 mol of gas at 0°C and 1 m mercury.

4.   Where possible the entries indicate the pressure range (or the maximum pressure) utilized. The degree of the polynomial used in analysing the data is given where this is greater than 2.

5.   The conversion factors from the CGS units used in this work to S I units are:

$$B(\text{m}^3 \text{ k mol}^{-1}) = B\,(\text{cm}^3 \text{ mol}^{-1}) \times 10^{-3},$$

$$C(\text{m}^6 \text{ k mol}^{-2}) = C\,(\text{cm}^6 \text{ mol}^{-2}) \times 10^{-6}.$$

References

1 Michels, A., Abels, J. C., Ten Seldam, C. A., and de Graaff, W. *Physica* **26** 381 (1960).
2 Hirschfelder, J. O., Curtiss, C. F., and Bird, R. B. *Molecular theory of gases and liquids*, Chapter 3. Wiley, New York (1954).
3 Frisch, H. L. and Helfand, E. *J. chem Phys.* **32**, 269 (1960).
4 Crawford, F. W., Harris, E. J. R, and Smith, E. B. *Molec. Phys.* **37**, 1323 (1979).
5 See ref. 2. p: 230.

PERIODICALS CITED

The abbreviations used are those of the *World list of scientific periodicals,* 4th
   edition (1963-65).
Advances in Cryogenic Engineering
American Institute of Chemical Engineers Journal
American Journal of Science
Anales de la Reál Sociedad espanol a de fisica y quimica (Madrid)
Annalen der Physik
Applied Scientific Research (Section A)
Archives neerlandaises des sciences exactes et naturelles.
Australian Journal of Chemistry
Bulletin de l'Academie polonaise des sciences. Seriedes sciences chimiques.
Canadian Journal of Chemistry
Canadian Journal of Research
Chemical Engineering Progress
Chemical Engineering Progress Symposium Series.
Communications from the Kamerlingh Onnes Laboratory, University of Leiden
Compte rendue hebdomadaire des seances de l'Academie des Sciences.
Helvetica chimica acta
High Temperature (English translation of Teplofizica vysokikh temperature)
Industrial and Engineering Chemistry
Industrial and Engineering Chemistry (Fundamentals)
Journal of the American Chemical Society
Journal of Applied Chemistry (London)
Journal of Chemical and Engineering Data
Journal of Chemical Physics
Journal of the Chemical Society (London)
Journal de chimie physique
Journal of Physical Chemistry
Journal of Research of the National Bureau of Standards. Section A.
Molecular Physics
Nature
Philosophical Transactions of the Royal Society. Series A.
Physica
Physical Review
Physics of Fluids
Physikalische Zeitschrift (Leipzig)
Proceedings of the American Academy of Arts and Sciences.
Proceedings of the Physical Society. (London)
Proceedings of the Royal Irish Academy.
Proceedings of the Royal Society. (London) Series A.
Refrigerating Engineering
Russian Journal of Physical Chemistry (English translation of Zhurnal
   fizicheskoi khimii)
Science Reports of the Research Institutes, Tohoku University. Series A.

Teploenergetika (see also Thermal Engineering)
Teplofizika vysokikh temperature (see also High Temperature)
Thermal Engineering (English translation of Teploenergetika)
Transactions of the American Institute of Mining and Metallurgical Engineers.
Transactions of the American Society of Mechanical Engineers
Transactions of the Faraday Society
Zeitschrift für angewandte Physik
Zeitschrift für Elektrochemie
Zeitschrift für die gesamte Kalteindustrie
Zeitschrift für Physik
Zeitschrift für physikalische Chemie (Frankfurter Ausgabe)
Zeitschrift für physikalische Chemie (Leipzig)
Zhurnal fizicheskoi khimii (see also Russian Journal of Physical Chemistry)

PURE GASES

ARGON Ar

The second virial coefficients above 130K are in good agreement except for the results of Onnes and Crommelin (1) above 200K which are less negative by up to 5 cm^3mol^{-1}, and above 700K where the values given by Lecocq (8) lie up to 3 cm^3mol^{-1} below the results of Osborne and Saville (23). Below 130K recent results and theoretical calculations support the values reported by Fender and Halsey (11). These have been used in preference to the results of Weir, Wynn Jones, Rowlinson and Saville (14) and Byrne, Jones and Staveley (16) in obtaining the smoothed values given below.

T	B	T	B
81	-276 ± 5	250	-27.9 ± 1
85	-251 ± 3	300	-15.5 ± 0.5
90	-225 ± 3	400	-1.0 ± 0.5
95	-202.5 ± 2	500	+7.0 ± 0.5
100	-183.5 ± 1	600	12.0 ± 0.5
110	-154.5 ± 1	700	15.0 ± 1
125	-123.0 ± 1	800	17.7 ± 1
150	-86.2 ± 1	900	20.0 ± 1
200	-47.4 ± 1	1000	22.0 ± 1

1. H. Kamerlingh Onnes and C.A. Crommelin, Communs phys. Lab. Univ. Leiden 118b (1910) (*).

    6-term fit of PV data (V series; terms in V^0, V^{-1}, V^{-2}, V^{-4}, V^{-6}, and V^{-8}). (Values above 200 K are probably high). Maximum pressure 60 atm. Class I.

T	B	C
151.94	-82.53	2 140
152.91	-82.17	2 240
153.96	-81.19	2 060
156.53	-79.01	2 240
157.29	-79.26	2 390
159.35	-75.70	2 030
163.27	-72.23	2 160

T	B	C
170.64	-65.09	1 830
186.10	-53.85	1 570
215.43	-37.03	1 040
273.15	-16.56	(3)
293.54	-12.55	310

2. L. Holborn and J. Otto, Z. Phys. <u>33</u> 1 (1925)(*).
   5-term fit of PV data (P series; terms in P^0, P^1, P^2, P^4, and P^6).
   Maximum pressure 100 atm.
   Class I.

T	B	C
173.15	-64.32	890
223.15	-37.78	1 750
273.15	-22.08	1 680
323.15	-11.02	1 190
373.15	- 4.29	1 120
423.15	+ 1.16	970
473.15	4.67	990
573.15	11.22	610
673.15	15.29	

3. C.C. Tanner and I. Masson, Proc. R. Soc. <u>A126</u> 268 (1930)(*).
   3-term fit of PV data (P series). Maximum pressure 126 atm.
   Class I.

T	B	C
298.15	-16.35	1 470
323.15	-11.49	1 270
348.15	- 7.48	1 140
373.15	- 4.10	1 020
398.15	- 0.72	820
423.15	+ 2.17	705
447.15	3.72	835

4. A. Michels, Hub. Wijker, and Hk. Wijker, Physica, 's Grav. <u>15</u> 627
   (1949)(*).
   Class I.
   (a) 7-term fit of PV data (P→2900 atm).

T	B	C
273.15	-21.13	1 054
298.15	-15.49	991

T	B	C
323.15	-11.06	1 017
348.15	- 7.14	959
373.15	- 3.89	918
398.15	- 1.08	877
423.15	+ 1.42	833

(b) 3-term fit of PV data (P→80 atm).

T	B	C
273.15	-21.45	1 271
298.15	-15.76	1 158
323.15	-11.24	1 130
348.15	- 7.25	1 040
373.15	- 4.00	1 004
398.15	- 1.18	968
423.15	+ 1.38	884

5. E. Whalley, Y. Lupien, and W.G. Schneider, Can. J. Chem. <u>31</u> 722 (1953) (*).

3-term fit of PV data (P series) up to 873.15 K, 2-term fit at higher temperatures. Standard deviation: 0.2 for B, 30 for C.

T	B	C
273.15	-22.41	1 690
323.15	-11.20	1 230
373.15	- 4.34	1 110
423.15	+ 1.01	990
473.15	5.28	865
573.15	10.77	760
673.15	15.74	
773.15	17.76	
873.15	19.48	

6. T.L. Cottrell, R.A. Hamilton, and R.P. Taubinger, Trans. Faraday Soc. <u>52</u> 1310 (1956).

T	B
303.2	-13.2 ± 1.8
333.2	- 9.3 ± 1.3
363.2	- 4.3 ± 2.2

4

7. A. Michels, J.M. Levelt, and W. de Graaff, Physica,'s Grav. <u>24</u> 659 (1958)(*).

Class I.

(a) 7-term fit of PV data (P→1000 atm).

T	B	C
248.15	-28.25	1 234
223.15	-37.09	1 340
203.15	-45.99	1 369
188.15	-54.27	1 416
173.15	-64.24	1 379
163.15	-71.87	1 163

(b) 4-term fit of PV data (P→80 atm).

T	B	C
150.65	-85.58	2 233
148.15	-88.29	2 224

(c) 8-term fit of PV data (P→350 atm).

T	B	C
153.15	-82.54	2 129

(d) 3-term fit of PV data (P→50 atm).

T	B	C
248.15	-28.57	1 415
223.15	-37.43	1 541
203.15	-46.52	1 702
188.15	-54.83	1 791
173.15	-65.21	2 015
163.15	-73.25	2 104
153.15	-82.97	2 278
150.65	-85.63	2 313
148.15	-88.45	2 357
143.15	-94.42	2 417
138.15	-100.88	2 418
133.15	-107.98	2 356

8. A. Lecocq, J. Rech. Cent. Nat. Rech. Scient. $\underline{50}$ 55 (1960) (*).
    3-term fit to PV data.

T	B	C
573.16	9.79	948
673.16	13.10	903
773.16	16.01	852
923.16	18.85	827
1073.16	19.93	777
1223.16	21.05	752

9. I.A. Rogavaya and M.G. Kaganer, Zh. fiz. Khim. $\underline{35}$ 2135 (1961); Russ.
    J. phys. Chem. $\underline{35}$ 1049 (1961).

    PVT data given: P range → 200 atm, T 90.14-248.18 K. Accuracy in com-
    pressibility data given as ±0.05 per cent.

10. R.A.H. Pool, G. Saville, T.M. Herrington, B.D.C. Shields, and L.A.K.
    Staveley, Trans. Faraday Soc. $\underline{58}$ 1692 (1962).

T	B
90	-231 ± 2

11. B.E.F. Fender and G.D. Halsey, Jr., J. chem. Phys. $\underline{36}$ 1881 (1962).
    3-term fit of PV data (P less than 1 atm). Maximum error in B ± 1.5
    per cent.

T	B
84.79	-249.34
88.34	-229.89
92.30	-211.79
95.06	-200.87
101.40	-178.73
102.01	-177.65
105.51	-166.06
108.15	-160.27
113.32	-149.58
117.50	-140.58
123.99	-127.99

12. J.H. Dymond and E.B. Smith (1962) (†).
    Estimated accuracy of B ± 2.

T	B
273.15	-23.1
298.15	-16.5
323.15	-10.9

13. R.W. Crain, Jr. and R.E. Sonntag, Adv. cryogen. Engng 11 379 (1966).
    Class I.

T	B	C
143.15	-94.69	2 462
163.15	-73.20	2 098
203.15	-46.35	1 610
273.15	-21.18	1 173

Burnett method. Maximum pressure 500 atm.

14. R.D. Weir, I. Wynn Jones, J.S. Rowlinson, and G. Saville, Trans. Faraday Soc. 63 1320 (1967).

    3-term fit of PV data. Errors: ± 10 at 80 K decreasing to ±3 for temperatures 82-85 K and ± 1 at high temperatures.

T	B	T	B
80.43	-285.1	105.89	-167.8
81.95	-272.9	108.07	-161.5
84.23	-268.5	120.00	-134.3
87.12	-245.9	129.56	-114.9
88.85	-235.9	114.60	- 93.18
92.78	-216.9	146.05	- 92.08
94.75	-207.3	157.41	- 80.34
97.65	-194.8	190.52	- 52.85
102.08	-179.1		

15. N.K. Kalfoglou and J.G. Miller, J. phys. Chem., Ithaca 71 1256 (1967).

T	B	C
303.2	-15.05	1283
373.2	- 4.10	1040
473.2	5.05	860
573.2	10.77	765
673.2	14.25	790
773.2	17.07	665

16. M.A. Byrne, M.R. Jones, and L.A.K. Staveley, Trans. Faraday Soc. 64 1747 (1968).

Class I.

T	B	T	B
84.03	-264.3	122.38	-127.3
85.96	-250.2	124.70	-123.5
88.94	-233.5	130.96	-112.7
89.57	-228.2	140.04	- 99.1
93.59	-209.7	159.72	- 77.1
97.69	-193.6	179.85	- 60.7
102.79	-175.8	209.94	- 43.7
107.93	-161.1	241.04	- 30.9
113.97	-146.0	271.39	- 21.9

17. R.K. Crawford and W.B. Daniels, J. chem. Phys. 50 3171 (1969).

PVT data given: P range 0.2 to 6.3 k bar,

T range 95 → 210 K.

18. A.L. Blancett, K.R. Hall and F.B. Canfield, Physica, 's Grav. 47 75 (1970).

Burnett method. Maximum pressure 700 atm.

T	B	C
223.15	-37.30 ± 0.25	1401 ± 54
273.15	-20.90 ± 0.09	1029 ± 31
323.15	-10.82 ± 0.07	974 ± 18

19. R.N. Lichtenthaler and K. Schäfer, Ber. (dtsch.) Bunsenges. phys. Chem. 73 42 (1969).

Estimated absolute error in B ± 1.

T	B
288.2	-17.95
296.0	-16.07
303.2	-14.69
313.2	-12.82
323.1	-11.10

20. S.L. Robertson, S.E. Babb, Jr., and G.J. Scott, J. chem. Phys. <u>50</u> 2160 (1969).

PVT data given: P range 1500-10000 atm., T range 308-673 K.

21. T.K. Bose and R.H. Cole, J. chem. Phys. <u>52</u> 140 (1970).

T	B
322.85	-15.8 ± 1.0

22. J.A. Provine and F.B. Canfield, Physica, 's Grav. <u>52</u> 79 (1971).

T	B	C
143.15	-94.04 ± .56	2363 ± 149
158.15	-77.87 ± .52	2230 ± 194
183.15	-56.48 ± .18	1287 ± 53

23. J. Osborne and G. Saville (unpublished results). See also J. Osborne, Ph.D. thesis, University of London (1972).

Burnett method, P → 800 bar. Maximum estimated error is ± 1 in B and 200 in C.

T	B	C
300	-15.8	1080
443	4.2	580
478	6.7	590
533	10.2	620
585	11.6	690
635	14.4	570
684	16.1	550
731	16.0	710
777	17.5	660
831	19.1	570
876	19.8	530
924	21.0	690
975	22.3	520
1024	23.2	330

24. G.A. Pope, P.S. Chappelear and R. Kobayashi, J. chem. Phys. $\underline{59}$ 423 (1973).

Burnett method.

T	B	C
101.202	-176.03 ± 0.65	-6949 ± 1371
116.421	-138.28 ± 0.05	1611 ± 44
138.224	-101.05 ± 0.09	2467 ± 20

25. J.Bellm, W. Reineke, K. Schäfer, and B. Schramm, Ber. (dtsch.) Bunsenges. phys. Chem. $\underline{78}$ 282 (1974).

Estimated accuracy of B is ± 2.

T	B	T	B
300	-15.4	430	+ 1.9
320	-12.1	460	4.2
340	- 8.9	490	5.8
370	- 4.6	520	7.0
400	- 1.0	550	7.8

26. R. Hahn, K. Schäfer, and B. Schramm, Ber. (dtsch.) Bunsenges. phys. Chem. $\underline{78}$ 287 (1974).

B values determined assuming B(296 K) = -16.2

Quoted accuracy of B is ± 2.

T	B	T	B
200.5	-47.7	251.5	-27.4
210.9	-42.4	273.2	-21.6
231.5	-34.1		

27. B. Schramm and U. Hebgen, Chem. phys. Letters $\underline{29}$ 137 (1974).

Values measured relative to B(T) for neon and B(296) for neon and argon.

T	B
77.3	-303
87.2	-239
90.2	-224

28. J. Santafe, J.S. Urieta and C. Gutierrez, Revta Acad. Cienc. exact. fis. - quim. nat. Zaragoza <u>31</u> 63 (1976).

B values also given by J. Santafe, J.S. Urieta and C.G. Losa, Chem. phys. <u>18</u> 341 (1976).

Compressibility measurements. Accuracy of B estimated to be ± 3.

T	B	T	B
273.2	-21.6	303.2	-14.3
283.2	-18.6	313.2	-12.5
293.2	-16.4	323.2	-11.0

29. H.-P. Rentschler and B. Schramm, Ber. (dtsch.) Bunsenges. phys. Chem. <u>81</u> 319 (1977).

Maximum error in B estimated to be ± 4.

T	B	T	B
326	-10.9	553	10.4
416	+ 0.6	620	12.5
485	7.0	713	15.9

30. B. Schramm, H. Schmiedel, R. Gehrmann, and R. Bartl, Ber. (dtsch.) Bunsenges. phys. Chem. <u>81</u> 316 (1977).

Maximum error in B estimated to be ± 4.

T	B	T	B
202.5	-47.3	367.0	- 5.3
217.9	-39.0	401.9	- 0.8
233.1	-32.9	431.3	+ 1.8
264.2	-22.7	466.2	4.0
295.2	-15.8	499.9	6.0
332.8	- 8.8		

BORON TRIFLUORIDE BF_3

1. C.J.G. Raw, J. chem. Phys. <u>34</u> 1452 (1961).

T	B	T	B
293.2	-105.1 ± 5.4	323.2	- 82.0 ± 5.5
303.2	- 98.0 ± 6.1	343.2	- 71.2 ± 5.8
313.2	- 89.7 ± 5.6		

2. D.S. Viswanath, J. chem. Engng Data <u>11</u> 453 (1966).
    5-term fit of PV data.
    Class II.

T	B	T	B
273.15	-135.05	313.15	- 97.50
283.15	-117.90	323.15	- 78.55
293.15	-110.67	333.15	- 65.00
298.15	- 98.10		

3. M. Waxman, J. Hilsenrath and W.T. Chen, J. chem. Phys. <u>58</u> 3692 (1973).
    Burnett method. P → 250 atm. Standard deviation in B < 1, but
    chemisorption errors uncertain. Values read from graph.

T	B	T	B
273.15	-135	373.15	- 52
283.16	-123	398.15	- 40
298.15	-106	423.15	- 30
310.65	- 94	448.15	- 22
323.15	- 83	473.16	- 16
335.84	- 75	498.15	- 10
348.15	- 67		

4. B. Schramm and R. Gehrmann (1979) (†).
    Estimated error in B is ± 10.

T	B	T	B
193.6	-368.7	240.7	-197.1
203.6	-324.9	243.9	-197.7
211.6	-288.1	253.2	-172.5
214.9	-274.9	261.5	-166.5
224.4	-247.6	268.6	-141.5
225.8	-237.6	295.5	-110.0

5. B. Schramm and H. Schmiedel (1979)(†).
    Estimated error in B is ± 15.

T	B	T	B
295	-110.0	400	- 34.0
330	- 75.0	425	- 24.0
365	- 52.0	450	- 15.0

CHLOROTRIFLUOROMETHANE $CClF_3$

1. A. Michels, T. Wassenaar, G.J. Wolkers, Chr. Prins, and L.v.d. Klundert, J. chem. Engng Data 11 449 (1966) (*).

    Class I.

    (a) 4-term fit to PV data (P→40 atm).

T	B	C
298.15	-225.5	15 800

    (b) 8-term fit to PV data (P→400 atm).

T	B	C
303.15	-220.4	21 690

    (c) 7-term fit to PV data (P→400 atm).

T	B	C
323.15	-186.5	11 780
348.15	-157.3	11 700
373.15	-132.9	10 820
398.15	-112.8	10 440
423.15	- 95.2	9 400

2. R.G. Kunz and R.S. Kapner, J. chem. Engng Data 14 190 (1969).

    B values calculated from PVT data given by Dupont de Nemours and Co., E.I., Bull. T13 (1959).

T	B	T	B
233.15	-393.8	388.71	-121.3
244.26	-355.1	399.82	-112.2
255.37	-321.3	410.93	-104.0
266.48	-288.2	422.04	- 96.2
277.59	-261.2	433.15	- 90.4
288.71	-238.6	444.26	- 84.4
299.82	-219.8	455.37	- 78.0
310.93	-202.8	466.48	- 71.2
322.04	-183.5	477.59	- 64.8
333.15	-171.0	488.71	- 59.3
344.26	-162.6	499.82	- 54.6
355.37	-152.0	510.93	- 51.5
366.48	-139.8	522.04	- 47.4
377.59	-129.3	533.15	- 40.3

3. H. Sutter and R.H. Cole, J. chem. Phys. 52 132 (1970).

T	B
323.15	-185 ± 5
369.45	-135 ± 3
404.75	-105 ± 4

4. R.F. Hajjar and G.E. MacWood, J. chem. Engng Data 15 3 (1970).

T	B
313.16	-220.9
343.43	-168.1
403.15	-121.5

5. W.S. Haworth and L.E. Sutton, Trans. Faraday Soc. 67 2907 (1971).
    Estimated precision of B ± 10.

T	B
298.2	-222
313.2	-188
328.2	-178

6. R.D. Nelson, Jr. and R.H. Cole, J. chem. Phys. 54 4033 (1971).

T	B
323.2	-202.8

7. B. Schramm and R. Gehrmann (1979)(†)
    Estimated error in B is ± 6.

T	B	T	B
192.2	-539.0	237.5	-353.2
205.6	-468.8	241.5	-336.3
209.4	-451.1	243.4	-334.9
213.8	-422.1	258.2	-296.8
216.1	-418.2	295.2	-235.0

DICHLORODIFLUOROMETHANE CCl_2F_2

1. R.M. Buffington and W.K. Gilkey, Ind. Engng Chem. ind. Edn 23 254 (1931).
    PVT data given.

2. J.O. Hirschfelder, F.T. McClure and I.F. Weeks, J. chem. Phys. <u>10</u> 201 (1942).

B calculated from available data.

T	B	T	B
238.70	-637	255.37	-560
244.26	-604	283.15	-454
249.81	-586	310.93	-347

3. A. Michels, T. Wassenaar, G.J. Wolkers, Chr. Prins, and L.v.d. Klundert, J. chem. Engng Data <u>11</u> 449 (1966) (*).

Class II.

(a) 3-term fit of PV data (P→80 atm).

T	B	C
373.15	-282.0	23 515

(b) 4-term fit of PV data (P→80 atm).

T	B	C
382.65	-264.9	23 325
386.51	-262.1	25 400

(c) 5-term fit of PV data (P→80 atm).

T	B	C
384.70	-262.0	19 480
385.50	-262.6	21 300
390.42	-255.4	21 610
398.15	-244.3	21 445
410.15	-230.5	22 830
423.15	-213.3	21 070

(d) 3-term fit of PV data (P→20 atm).

T	B	C
323.15	-384.2	19 200
348.15	-327.1	22 740
373.15	-281.0	22 120
384.70	-263.0	21 890
398.15	-243.2	20 830
423.15	-211.4	18 690

4. R.G. Kunz and R.S. Kapner, J. chem. Engng Data <u>14</u> 190 (1969).

B values calculated from PVT data given by Dupont de Nemours and Co., Bull. T12 (1956).

T	B	T	B
255.37	-725.0	377.59	-276.3
266.48	-648.7	388.71	-258.3
277.59	-583.7	399.82	-242.0
288.71	-527.3	410.93	-227.0
299.82	-479.3	422.04	-213.4
310.93	-437.9	433.15	-200.9
322.04	-401.8	444.26	-189.3
333.15	-370.3	455.37	-178.6
344.26	-342.3	466.48	-167.9
355.37	-317.7	477.59	-158.5
366.48	-295.8		

5. R.F. Hajjar and G.E. MacWood, J. chem. Engng Data <u>15</u> 3 (1970).

T	B	T	B
313.16	-372.4	373.15	-279.4
343.43	-319.1	403.15	-237.1

6. W.S. Haworth and L.E. Sutton, Trans. Faraday Soc. <u>67</u> 2907 (1971).
Estimated precision of B ± 10.

T	B
298.2	-457
313.2	-415
328.2	-381

7. K. Watanabe, T. Tanaka, and K. Oguchi, Proc. Seventh Symp. Thermophys. Props., Am. Soc. Mech. Engrs., New York, 470 (1977). (*)

Burnett method; P → 108 atm.

Quoted uncertainties are 2 to 4% in B, and vary from 26000 to 73000 in C.

T	B	C
273.15	-473.8	-631770
283.15	-467.8	-320455
293.15	-464.8	-105945
303.15	-430.8	- 67300

313.15	-405.7	- 54820
323.15	-382.8	- 42660
333.15	-385.8	184750
343.15	-336.6	- 4095
353.15	-317.6	+ 4405
363.15	-295.3	- 9275
373.15	-278.6	- 19600
383.15	-262.7	- 5570
393.15	-243.7	- 7100
403.15	-232.6	- 4340

TRICHLOROFLUOROMETHANE CCl_3F

1. J.O. Hirschfelder, F.T. McClure, and I.F. Weeks, J. chem. Phys. <u>10</u> 201 (1942).

   Calculated from available data.

T	B	T	B
238.70	-1 150	366.48	- 523
249.81	-1 070	394.26	- 459
283.15	- 862	422.04	- 381
310.93	- 720	449.81	- 328
338.70	- 617		

2. A.F. Benning and R.C. MacHarness, Ind. Engng Chem. ind. Edn <u>32</u> 698 (1940).

   PVT data.

3. R.G. Kunz and R.S. Kapner, J. chem. Engng Data <u>14</u> 190 (1969).

   B values calculated from PVT data given in Allied Chemical Corp., General Chemical Division, Bulletin "Genetron 11 Refrigerant Tri-chloromonofluoromethane", 1957.

T	B	T	B
405.37	-417.5	444.26	-330.5
410.93	-403.7	449.82	-319.4
416.48	-390.8	455.37	-308.3
422.04	-378.5	460.93	-297.6
427.59	-366.0	466.48	-287.1
433.15	-353.9	472.04	-277.3
438.71	-341.9	477.59	-267.2

4. R.F. Hajjar and G.E. MacWood, J. chem. Engng Data 15 3 (1970).

T	B
343.43	-525.9
373.15	-454.6
403.15	-394.7

5. H. Sutter and R.H. Cole, J. chem. Phys. 52 132 (1970).

T	B
369.45	-543 ± 15

CARBON TETRACHLORIDE (TETRACHLOROMETHANE) CCl_4

The results of Francis and McGlashan (2) support the somewhat scattered values obtained by Lambert et al. (1) for the second virial coefficient of CCl_4. At higher temperatures the results of Masia et al. (4) form a natural continuation. The single determination of Bottomley and Remmington (3) at 295 K appears to be very inaccurate. Smoothed values are given below.

T	B	T	B
320	-1 350 ± 50	380	- 950 ± 40
340	-1 160 ± 40	400	- 870 ± 30
360	-1 040 ± 40	420	- 810 ± 30

1. J.D. Lambert, G.A.H. Roberts, J.S. Rowlinson, and V.J. Wilkinson, Proc. R. Soc. A196 113 (1949) (†).

    Maximum pressure 600 torr. Accuracy ± 50.

T	B	T	B
318.8	-1 390	333.2	-1 200
319.4	-1 480	338.2	-1 150
320.8	-1 485	342.1	-1 150
323.8	-1 170	351.2	-1 060
327.9	-1 280	351.4	-1 030

2. P.G. Francis and M.L. McGlashan, Trans. Faraday Soc. 51 593 (1955).

    Maximum pressure 400 torr. Uncertainty in B less than ± 50.

T	B	T	B
315.7	-1 445	337.2	-1 170
323.2	-1 330	343.2	-1 120
333.7	-1 225		

3. G.A. Bottomley and T.A. Remmington, J. chem. Soc. 3800 (1958).
    Accuracy ±2 per cent.

T	B
295.2	-1 283

4. A. Perez Masia, M. Diaz Pena, and J.A. Burriel Lluna, An. R. Soc.
    esp. Fis. Quim. 60B 229 (1964).

T	B	C
353.29	-1 088 ± 20	46 000
375.39	- 972 ± 22	45 000
398.55	- 876 ± 23	39 000
418.75	- 812 ± 34	36 000

DEUTEROMETHANE CD_4

1. G. Thomaes and R. van Steenwinkel, Molec. Phys. 5 307 (1962).
    Values for B relative to methane are given for the temperature range
    383 - 568 K.

2. I. Gainar, K. Strein, and B. Schramm, Ber. (dtsch.) Bunsenges. phys.
    Chem. 76 1242 (1972).
    Values of ΔB ($B_{CD_4} - B_{CH_4}$) determined on the basis that ΔB (511.1K)
    = 0.6. Repeatability of $\Delta B \pm 0.3$.

T	ΔB	T	ΔB
204.1	3.2	296.2	2.2
213.1	2.9	379.3	1.8
233.1	2.7	425.7	1.4
252.3	2.6	475.6	1.0
273.2	2.4		

3. A.Y. Fang and W.A. Van Hook, J. chem. Phys. 60 3513 (1974).
    B values relative to B_{CH_4} at the same temperature.

T	$B_{CD_4} - B_{CH_4}$	T	$B_{CD_4} - B_{CH_4}$
111.24	5.7 ± 0.2	167.37	4.7 ± 0.3
122.15	5.4 ± 0.1	206.43	4.1 ± 0.3
132.25	5.0 ± 0.2	249.95	3.5 ± 0.2
150.88	4.7 ± 0.2	300.33	3.6 ± 0.4

METHYLAMINE-d_5 CD_3ND_2

1. G. Adam, Kl. Schäfer, and B. Schramm, Ber. (dtsch.) Bunsenges. phys. Chem. <u>80</u> 912 (1976).

   Values of ΔB ($B_{deuterated} - B_{CH_3NH_2}$) determined on the basis that ΔB (539K) = 0. Repeatability of ΔB ± 0.7.

T	ΔB	T	ΔB
296	-2.9	400	-0.1
313	-1.5	439	-0.1
331	-1.4	469	+0.3
361	-0.9	516	+0.5

CARBON TETRAFLUORIDE (TETRAFLUOROMETHANE) CF_4

Excellent agreement is observed between the measurements of Douslin et al. (4) and most other workers. The following values are recommended.

T	B	T	B
225	-172.5 ± 1	400	- 32.0 ± 0.5
250	-137.5 ± 1	450	- 16.0 ± 0.5
275	-109.0 ± 0.5	500	- 4.0 ± 0.5
300	- 87.0 ± 0.5	600	+ 14.0 ± 0.5
325	- 69.0 ± 0.5	700	25.0 ± 0.1
350	- 55.0 ± 0.5	800	33.0 ± 0.1

1. W. Cawood and H.S. Patterson, Phil. Trans. R. Soc. <u>A236</u> 77 (1937).

T	B
294.15	-101.1
294.15	-102.3

2. K.E. MacCormack and W.G. Schneider, J. chem. Phys. <u>19</u> 845, 849 (1951). (*).

   Maximum pressure 50 atm. Accuracy of B: 1 per cent at 273 K and 1.5 per cent at 473 K and above.

T	B	C
273.15	-111.0	7 003
323.15	- 70.4	5 578
373.15	- 43.1	4 259
423.15	- 26.0	5 305
523.15	+ 1.25	3 477

573.15	9.3	3 422
673.15	23.6	2 563

3. S.D. Hamann, J.A. Lambert, and W.J. McManamey, Aust. J. Chem. 7 1 (1954).

Class II.

T	B		T	B
313.15	-83.6		373.15	-43.9
323.15	-72.4		398.15	-30.1
348.15	-56.1		398.15	-31.1

4. D.R. Douslin, R.H. Harrison, R.T. Moore, and J.P. McCullough, J. chem. Phys. 35 1357 (1961).

Maximum pressure 394 atm. 4-term fit. Error in B ± 0.05.

T	B	C
273.16	-111.00	7 100
298.15	- 88.30	6 070
303.15	- 84.40	5 900
323.15	- 70.40	5 380
348.15	- 55.70	4 870
373.16	- 43.50	4 490
398.17	- 33.20	4 210
423.18	- 24.40	3 980
448.20	- 16.80	3 810
473.21	- 10.10	3 660
498.23	- 4.25	3 540
523.25	+ 1.00	3 440
548.26	5.60	3 350
573.27	9.80	3 250
598.28	13.60	3 180
623.29	17.05	3 100

5. N.K. Kalfoglou and J.G. Miller, J. phys. Chem., Ithaca 71 1256 (1967).

T	B	C
303.2	-84.02	5460
373.2	-43.54	5150
473.2	-10.02	4470
573.2	+10.11	4215
673.2	23.53	4210
773.2	32.42	5715

6. H.B. Lange Jr. and F.P. Stein, J. chem. Engng Data <u>15</u> 56 (1970).
   Burnett method.

T	B	C
203.15	-216.75 ± 0.7	9320 ± 500
223.15	-177.71 ± 0.5	9645 ± 300
243.15	-146.36 ± 0.2	8660 ± 100
273.15	-110.87 ± 0.2	7070 ± 100
313.15	- 77.26 ± 0.2	6030 ± 125
368.15	- 45.75 ± 0.2	4890 ± 150

7. J.J. Martin and B.K. Bhada, A. I. Ch. E. Jl <u>17</u> 683 (1971).
   PVT data given: T range 300 - 423 K, P → 100 atm.

8. P.M. Sigmund, I.H. Silberberg and J.J. McKetta, J. chem. Engng Data
   <u>17</u> 168 (1972).
   Burnett method: standard errors given.

T	B	C
271.61	-112.36 ± 0.13	7620 ± 40
308.21	- 81.22 ± 0.56	6440 ± 500
323.55	- 70.93 ± 0.41	5500 ± 240
348.10	- 56.93 ± 0.55	4190 ± 240
373.15	- 44.48 ± 0.30	4580 ± 100
423.15	- 23.71 ± 0.89	

9. T.K. Bose, J.S. Sochanski and R.H. Cole, J. chem. Phys. <u>57</u> 3592 (1972).
   B derived from low pressure dielectric measurements.

T	B
279.8	-104.6 ± 1.7
322.5	- 74.0 ± 1.4
373.4	- 43.0 ± 1.0

CHLORODIFLUOROMETHANE $CHClF_2$

1. M. Zander, Proc. 4th Symp. Thermophys. Props. P.114 (Univ. Maryland
   (1968)). (*)
   Burnett method. PVT data given for P → 350 atm. B, C calculated
   using 3 term fit to low pressure data (maximum 44 atm.).

T	B	C
303.15	-329.2	-55080
323.15	-287.4	-19060
343.15	-247.4	- 9200
373.15	-202.2	- 1830
398.15	-171.6	+ 540
423.15	-147.7	4455
473.15	-113.6	9160

2. R.F. Hajjar and G.E. MacWood, J. chem. Engng Data 15 3 (1970).

T	B	T	B
313.16	-323.5	373.15	-236.1
343.43	-275.0	403.15	-204.2

3. H. Sutter and R.H. Cole, J. chem. Phys. 54 4988 (1971).

T	B
323.05	-274 ± 8
373.35	-214 ± 16
424.95	-142 ± 8

4. W.S. Haworth and L.E. Sutton, Trans. Faraday Soc. 67 2907 (1971).
    Estimated precision of B ± 10.

T	B
298.2	-356
313.2	-318
328.2	-292

DICHLOROFLUOROMETHANE $CHCl_2F$

1. J.O. Hirschfelder, F.T. McClure, and I.F. Weeks, J. chem. Phys. 10 201
    (1942).
    Calculated from available data.

T	B	T	B
238.70	-766	366.48	-403
249.81	-734	394.26	-354
283.15	-616	422.04	-310
310.93	-528	449.81	-271
338.70	-446		

2. R.F. Hajjar and G.E. MacWood, J. chem. Engng Data $\underline{15}$ 3 (1970).

T	B	T	B
313.16	-526.2	373.15	-388.3
343.43	-451.5	403.15	-332.7

CHLOROFORM (TRICHLOROMETHANE) $CHCl_3$

The determinations of the second virial coefficient of chloroform by
Lambert et al. (1) and Zaalishvili and co-workers (3,4) are in good mutual
agreement and are consistent with the results obtained by Francis and
McGlashlan (2) below 350 K. Above this temperature, however, the values
reported by Francis and McGlashlan (2) are about 100 cm^3 mol^{-1} more negative than those of the other workers. These results have been neglected in
obtaining the smoothed values given below.

T	B	T	B
320	-1 000 ± 50	380	- 640 ± 30
340	- 860 ± 30	400	- 560 ± 30
360	- 740 ± 30		

1. J.D. Lambert, G.A.H. Roberts, J.S. Rowlinson, and V.J. Wilkinson, Proc.
    R. Soc. $\underline{A196}$ 113 (1949) (†).

    Maximum pressure 600 torr. Accuracy ± 50.

T	B	T	B
319.2	- 954	343.2	- 840
319.4	- 948	351.8	- 780
326.0	-1 000	374.6	- 690
329.6	- 956	383.2	- 650
335.2	- 905	394.7	- 570

2. P.G. Francis and M.L. McGlashan, Trans. Faraday Soc. $\underline{51}$ 593 (1955).

    Maximum pressure 400 torr. Uncertainty in B less than ± 50.

T	B	T	B
315.7	-1 010	349.3	- 930
323.2	-1 005	363.2	- 835
337.2	- 920	382.1	- 740
343.2	- 860	397.4	- 680
349.3	- 850		

3. Sh. D. Zaalishvili and L.E. Kolysko, Zh. fiz. Khim. $\underline{34}$ 2596 (1960).
    Maximum pressure 1 200 torr. Deviation in B of 1-3 per cent.

T	B
338.15	-880
351.15	-760
358.15	-740

4. Sh. D. Zaalishvili, Z.S. Belousova, and L.E. Kolysko, Russ. J. phys. Chem. <u>39</u> 232 (1965); Zh. fiz. Khim. <u>39</u> 447 (1965).

Accuracy better than 3.5 per cent.

T	B	T	B
353.2	-799	373.2	-681
363.2	-732	383.2	-632

TRIDEUTEROMETHANE CHD_3

1. A.Y. Fang and W.A. Van Hook, J. chem. Phys. <u>60</u> 3513 (1974).

B values relative to B_{CH_4} at the same temperature.

T	$B_{CHD_3} - B_{CH_4}$	T	$B_{CHD_3} - B_{CH_4}$
112.29	4.2 ± 0.1	197.28	2.8 ± 0.3
149.13	3.4 ± 0.1	301.84	2.5 ± 0.3

FLUOROFORM (TRIFLUOROMETHANE) CHF_3

1. P.G.T. Fogg, P.A. Hanks, and J.D. Lambert, Proc. R. Soc. <u>A217</u> 490 (1953).

Class III.

B in T range 288 - 353 K given by:

$\ln(-B) = -14.636 - 2.93 \times 10^{-2}T$.

2. Y.C. Hou and J.J. Martin, A.I. Ch. E. Jl <u>5</u> 125 (1959).

PVT data given: P range 3 - 135 atm, T. range 222-394 K.

3. J.H. Dymond and E.B. Smith, Trans. Faraday Soc. <u>60</u> 1378 (1964).

Class II.

T	B	T	B
273.2	-217	313.2	-159
288.2	-193	323.2	-147
298.2	-177	333.2	-136

4. H. Sutter and R.H. Cole, J. chem. Phys. 46 2014 (1967).

T	B
323.2	-164 ± 5

From dielectric coefficient measurements.

T	B
323.2	-165 ± 8

By series-expansion method.

5. W.S. Haworth and L.E. Sutton (†).
    Class II.

T	B
298	-192
313	-179
328	-156

6. R.F. Hajjar and G.E. MacWood, J. chem. Engng Data 15 3 (1970).

T	B	T	B
313.16	-94.4	373.15	-60.3
343.43	-72.5	403.15	-50.2

7. H. Sutter and R.H. Cole, J. chem. Phys. 52 132 (1970).

T	B
323.15	-155 ± 4
369.45	-109 ± 4
404.75	- 85 ± 3

8. H.B. Lange, Jr., and F.P. Stein, J. chem. Engng Data 15 56 (1970).

T	B	C
243.15	-311.60 ± 1.0	15000 ± 1000
273.15	-233.60 ± 0.4	15700 ± 175
313.15	-165.50 ± 0.2	11630 ± 100
368.15	-109.50 ± 0.2	8410 ± 250

9. W.S. Haworth and L.E. Sutton, Trans. Faraday Soc. 67 2907 (1971).
    Estimated precision of B ± 10.

T	B
298.2	-192
313.2	-179
328.2	-157

10. B. Schramm and R. Gehrmann (1979)(†).

Estimated error in B is ± 6.

T	B	T	B
195.1	-461.5	220.5	-342.4
199.3	-430.6	222.7	-338.2
204.8	-414.9	237.6	-286.2
214.6	-380.4	251.0	-258.1
220.3	-348.8	295.2	-184.0

HYDROGEN CYANIDE HCN

1. T.L. Cottrell, I.M. MacFarlane, and A.W. Read, Trans. Faraday Soc. 61 1632 (1965).

T	B
303.15	-1 332 ± 100
348.15	- 765 ± 75

The following values were calculated by the authors from the PVT data of W.A. Felsing and G.W. Drake, J. Am. chem. Soc. 58 1714 (1936).

T	B
303.15	-1 602
343.15	- 811
383.15	- 507

METHYLENE CHLORIDE (DICHLOROMETHANE) CH_2Cl_2

1. P.G.T. Fogg, P.A. Hanks, and J.D. Lambert, Proc. R. Soc. A219 490 (1953) (†).

Class II.

T	B	T	B
319.4	-733	362.8	-474
330.1	-608	372.2	-435
341.5	-570	382.6	-401
351.2	-497		

2. A. Perez Masia and M. Diaz Pena,An. R. Soc. esp. Fis. Quim. 54B 661 (1958).

Class II.

T	B	T	B
323.15	-676.5	398.15	-389.2
348.15	-544.8	423.15	-348.9
373.15	-467.2		

3. M. Rätzsch, Z. phys. Chem. <u>238</u> 321 (1968).

T	B
303.2	-810
313.1	-777
333.2	-635

DIDEUTEROMETHANE CH_2D_2

1. A.Y. Fang and W.A. Van Hook, J. chem. Phys. <u>60</u> 3513 (1974).
   B values relative to B_{CH_4} at the same temperature.

T	$B_{CH_2D_2} - B_{CH_4}$	T	$B_{CH_2D_2} - B_{CH_4}$
112.48	2.9 ± 0.2	196.96	1.8 ± 0.2
148.91	2.1 ± 0.2	302.01	1.8 ± 0.2

METHYL-d_3-AMINE CD_3NH_2

1. G. Adam, Kl. Schäfer, and B. Schramm, Ber. (dtsch.) Bunsenges. phys. Chem. <u>80</u> 912 (1976).
   Values of ΔB ($B_{deuterated} - B_{CH_3NH_2}$) determined on the basis that ΔB (538K) = 0. Repeatability of ΔB given.

T	ΔB	T	ΔB
297	10.0 ± 0.3	440	1.9 ± 0.8
303	9.4 ± 0.3	463	1.1 ± 0.8
332	6.4 ± 0.5	468	0.6 ± 1.0
362	4.6 ± 0.5	494	0.4 ± 1.0
401	2.7 ± 0.5	514	0 ± 1.0

METHYLENE FLUORIDE (DIFLUOROMETHANE) CH_2F_2

1. P.G.T. Fogg, P.A. Hanks, and J.D. Lambert, Proc. R. Soc. <u>A219</u> 490 (1953) (†).
   Class II.

T	B	T	B
289.2	-346	341.2	-252
327.7	-265	348.8	-240
333.8	-249		

2. J.H. Dymond (†)

Class II.

T	B	T	B
273.2	-397.5	303.2	-311
283.2	-366	313.2	-286.5
293.2	-337	323.2	-264.5

3. P.F. Malbrunot, P.A. Meunier, G.M. Scatena, W.H. Mears, K.P. Murphy, and J.V. Sinka, J. chem. Engng Data 13 16 (1968).

PVT data given: P → 200 atm, T 298 → 473 K.

FORMIC ACID H.COOH

1. J.R. Barton and C.C. Hsu, J. chem. Engng Data 14 184 (1969).

PVT data given; T 323-398 K, P below 1 atm.

METHYL BROMIDE (BROMOMETHANE) CH_3Br

1. P.G.T. Fogg, P.A. Hanks, and J.D. Lambert, Proc. R. Soc. A219 490 (1953) (*).

Class II.

T	B	T	B
291.4	-552	351.8	-345
320.1	-416	361.3	-337
330.7	-407	374.7	-280
341.7	-373	383.2	-298

2. W. Kapallo, N. Lund, and K. Schafer. Z. phys. Chem. Frankf. Ausg. 37 196 (1963).

Class II.

T	B	T	B
244.0	-1 040	297.0	- 567
273.0	- 718	321.0	- 451

3. M. Ratzsch and H.-J. Bittrich, Z. phys. Chem. 228 81 (1965).

Class II.

T	B	T	T
293.2	-478	313.2	-402
305.0	-435	323.2	-372

4. M. Ratzsch, Z. phys. Chem. 238 321 (1968).

T	B	T	B
303.2	-501	333.2	-405
313.2	-483		

5. R.N. Lichtenthaler and K. Schäfer, Ber. (dtsch.) Bunsenges. phys. Chem. 73 42 (1969).

Estimated absolute error in B ± 1, ± 5%.

T	B	T	B
287.8	-622.3	307.7	-541.7
296.0	-593.1	321.1	-484.9
303.2	-561.9		

6. W.S. Haworth and L.E. Sutton, Trans. Faraday Soc. 67 2907 (1971).

Estimated precision of B ± 10.

T	B
298.2	-512
313.2	-482

METHYLCHLORIDE (CHLOROMETHANE) CH_3Cl

There is good agreement above 350 K. Below this temperature the results of Hirschfelder et al. (1) and Mansoonian et al. (9) appear much too negative. The results of these workers have been ignored in estimating the recommended values given below.

T	B	T	B
280	-470 ± 10	400	-212 ± 5
300	-400 ± 10	450	-155 ± 5
320	-345 ± 10	500	-111 ± 3
340	-300 ± 5	550	- 80 ± 2
360	-264 ± 5	600	- 57 ± 2

1. J.O. Hirschfelder, F.T. McClure, and I.F. Weeks, J. chem. Phys. 10 201 (1942).

Calculated from available data.

T	B	T	B
238.70	-764	338.70	-320
249.81	-668	366.48	-265
255.37	-637	394.26	-214
283.15	-500	422.04	-184
310.93	-401	449.81	-155

2. S.D. Hamann and J.F. Pearse, Trans. Faraday Soc. 48 101 (1952).
Initial pressure 0.5 atm. Accuracy ±6.

T	B	T	B
293.15	-426.3	333.15	-314.0
293.15	-431.8	353.15	-273.8
313.15	-364.5		

3. G.A. Bottomley, C.G. Reeves, and R. Whytlaw-Gray, Nature, Lond. 181 1004 (1958).
Maximum pressure 750 torr.

T	B
295.15	-416 ± 1

4. C.C. Hsu and J.J. McKetta. J. chem. Engng Data 9 45 (1960).
PVT data given: T 308 - 498 K, P 6 - 310 atm.

5. G.A. Bottomley and T.H. Spurling, Aust. J. Chem. 20 1789 (1967).
Accuracy ±4.

T	B	T	B
276.16	-480	347.88	-283
292.28	-425	376.04	-244
295.16	-416	399.29	-211
325.68	-330	427.45	-180
343.52	-291		

6. K.W. Suk and T.S. Storvick, A. I. Ch. E. Jl 13 231 (1967).
Burnett method. Maximum pressure 70 atm: Errors in B ± 0.2.

T	B	T	B
473.15	-144.2	573.15	- 68.50
498.15	-113.7	598.15	- 58.71
523.15	- 93.62	623.15	- 47.05
548.15	- 81.19		

7. R.N. Lichtenthaler and K. Schäfer, Ber. (dtsch.) Bunsenges. phys. Chem. <u>73</u> 42 (1969).

Estimated absolute error in B ± 1, ± 1%.

T	B	T	B
288.2	-444.0	307.7	-368.4
296.0	-413.6	313.2	-352.0
303.2	-386.1		

8. H. Sutter and R.H. Cole, J. chem. Phys. <u>52</u> 132 (1970).

T	B	T	B
323.15	-321 ± 5	404.75	-210 ± 3
369.45	-243 ± 3		

9. H. Mansoorian, K.R. Hall and P.T. Eubank, Proc. Seventh Symp. Thermophys. Props., Am. Soc. Mech. Engrs., New York, 456 (1977).

Burnett method; p → 150 atm.

T	B	C
323.15	-415.5 ± 8.0	341000 ± 82000
348.15	-310.3 ± 7.5	97400 ± 23000
373.15	-242.5 ± 7.0	25200 ± 6000
398.15	-207.5 ± 6.0	18200 ± 4400
423.15	-180.6 ± 4.0	14500 ± 3500
448.15	-156.1 ± 3.0	11900 ± 2900
473.15	-135.9 ± 2.5	9800 ± 2400

MONODEUTEROMETHANE CH_3D

1. A.Y. Fang and W.A. Van Hook, J. chem. Phys. <u>60</u> 3513 (1974).

B values relative to B_{CH_4} at the same temperature.

T	$B_{CH_3D} - B_{CH_4}$	T	$B_{CH_3D} - B_{CH_4}$
112.65	1.6 ± 0.2	197.49	1.2 ± 0.2
149.29	1.2 ± 0.1	302.04	0.4 ± 0.3

METHYLAMINE-d_2 CH_3ND_2

1. G. Adam, Kl. Schäfer, and B. Schramm, Ber. (dtsch.) Bunsenges. phys. Chem. <u>80</u> 912 (1976).

Values of ΔB ($B_{deuterated} - B_{CH_3NH_2}$) determined on the basis that ΔB (554K) = 0. Repeatability of ΔB given.

T	B	T	B
298	-13.2 ± 0.5	440	- 3.0 ± 0.8
303	-11.5 ± 0.5	470	- 1.4 ± 0.9
332	- 7.9 ± 0.5	514	- 0.3 ± 1.0
362	- 5.9 ± 0.8	540	- 0.4 ± 1.0
400	- 4.8 ± 0.8		

METHYL FLUORIDE (FLUOROMETHANE) CH_3F

The results of Michels et al. (2) for the second virial coefficient of methyl fluoride are in good agreement with those of Hamann and co-workers (3,4). Values taken from a smooth curve through the data are given.

T	B	T	B
280	-244 ± 3	360	-129 ± 3
300	-206 ± 3	380	-112 ± 3
320	-174 ± 3	400	- 99 ± 2
340	-150 ± 3	420	- 87 ± 2

1. W. Cawood and H.S. Patterson, J. chem. Soc. 134 2180 (1932); J. chem. Soc. 619 (1933); (a) Phil. Trans. R. Soc. A236 77 (1937).

   Class II.

T	B	T	B
273.15	-251	294.15	-217.2(a)
273.15	-251.6	317.65	-178
294.15	-211	317.65	-182.8
294.15	-212.8		

2. A. Michels, A. Visser, R.J. Lunbeck, and G.J. Wolkers, Physica, 's Grav. 18 114 (1952) (*).

   Class I.

   (a) 4-term fit of PV data (P → 50 atm).

T	B	C
273.15	-259.99	29 090
297.75	-209.47	21 490
322.75	-171.22	15 950
347.91	-142.71	12 930

T	B	C
372.77	-120.97	11 300
397.64	-102.86	9 470
422.70	- 87.30	7 450

These values for B and C agree with those calculated by R.J. Lunbeck and C.A. ten Seldam, Physica, 's Grav. <u>17</u> 788 (1951) except for the 273.15 K value of C. They give 28 100.

(b) 5-term fit of PV data (P → 150 atm).

T	B	C
273.15	-259.99	29 090
298.15	-205.29	18 350
323.15	-166.27	12 920
348.15	-138.75	10 430
373.15	-116.91	8 510
398.15	- 99.49	7 150
423.15	- 85.28	6 150

3. S.D. Hamann and J.F. Pearse, Trans. Faraday Soc. <u>48</u> 101 (1952). Maximum pressure 0.5 atm. Accuracy ± 5.

T	B	T	B
293.15	-222.3	333.15	-159.9
294.15	-219.5	333.15	-159.6
313.15	-185.6	353.15	-136.6
313.15	-185.4		

4. H.G. David, S.D. Hamann, and J.F. Pearse, J. chem. Phys. <u>20</u> 969 (1952).

T	B	T	B
323.15	-168	348.15	-140

Constant volume method. P range 20 - 160 atm.

5. H. Sutter and R.H. Cole, J. chem. Phys. <u>52</u> 132 (1970).

T	B	T	B
323.15	-173 ± 3	369.45	-123 ± 3

METHYL IODIDE (IODOMETHANE) CH_3I

1. P.G.T. Fogg, P.A. Hanks, and J.D. Lambert, Proc. R. Soc. A219 490 (1953) (†).

    Class II.

T	B	T	B
322.4	-604	363.2	-471
337.6	-546	375.2	-443
351.0	-472	383.2	-414

2. Sh. D. Zaalishvili and L.E. Kolysko, Zh. fiz. Khim. 36 846 (1962).
    Maximum pressure 500 torr.

    Class II.

T	B	T	B
313.15	-706	343.15	-517
328.15	-612	358.15	-462

NITROMETHANE CH_3NO_2

1. J.P. McCullough, D.W. Scott, R.E. Pennington, I.A. Hossenlopp, and G. Waddington, J. Am. chem. Soc. 76 4791 (1954).

    Values of B calculated by the authors from measurements of vapour pressure and heats of vaporization.

T	B	T	B
318.30	-3 102	353.36	-1 931
334.80	-2 463	374.44	-1 499

2. I. Brown and F. Smith, Aust. J. Chem. 13 30 (1960).
    From vapour pressure measurements.

T	B
318.2	-3 110

3. G.A. Bottomley and I.H. Coopes, J. chem. Soc. 2247 (1961).
    Maximum pressure 210 torr at 340.2 K, 95 torr at 323.2 K. Errors due to absorption ± 3 at 340.2 K, ± 10 at 323.2 K.

T	B	T	B
323.2	-2 926	340.2	-2 317

METHANE CH_4

The reported second virial coefficients generally agree to within the estimated experimental uncertainties except for the two values below 200K given
by Mueller, Leland, and Kobayashi (13) which are significantly more negative. These have been neglected in obtained the smooth values given below.

T	B	T	B
110	-330 ± 10	225	$- 83 \pm 2$
120	-273 ± 5	250	$- 66 \pm 1$
130	-235 ± 5	275	$- 53 \pm 1$
140	-207 ± 3	300	$- 42 \pm 1$
150	-182 ± 3	350	$- 26 \pm 1$
160	-161 ± 3	400	$- 15 \pm 1$
180	-129 ± 2	500	$- 0.5 \pm 1$
200	-105 ± 2	600	$+ 8.5 \pm 1$

1. F.G. Keyes and H.G. Burks, J. Am. chem. Soc. 49 1403 (1927).
    PVT data given: P → 250 atm, T 273.15 - 473.15 K.

2. H.M. Kvalnes and V.L. Gaddy, J. Am. chem. Soc. 53 394 (1931).
    PVT data given: P → 1000 atm, T 203.15 - 473.15 K.

3. F.A. Freeth and T.T.H. Verschoyle, Proc. R. Soc. A130 453 (1931)(*).
    4-term fit of PV data (P series; terms in P^0, P^1, P^2, P^4) at 293.15 K,
    an additional term (P^6) added for 273.15 K data. Maximum pressure
    215 atm.
    Class I.

T	B	C
273.15	-53.91	-
293.15	-48.68	4 524

4. A. Michels and G.W. Nederbragt, Physica, 's Grav. 2 1000 (1935) (*).
    Class I.
    4-term fit of PV data (P → 80 atm).

T	B	C
273.15	-54.07	3 047
298.15	-43.38	2 624
323.15	-34.72	2 430
348.15	-27.87	2 410
373.15	-21.74	2 232
398.15	-16.09	1 750
423.15	-11.46	1 656

5. A. Michels and G.W. Nederbragt, Physica, 's Grav. $\underline{3}$ 569 (1936) (*).
   Class I.
   5-term fit of PV data (P → 350 atm).

T	B	C
273.15	-53.86	2 870
298.15	-43.34	2 620
323.15	-34.62	2 370
348.15	-27.73	2 335
373.15	-21.58	2 144
398.15	-16.36	1 999
423.15	-11.62	1 767

6. J.A. Beattie and W.H. Stockmayer, J. chem. Phys. $\underline{10}$ 473 (1942).
   2-term fit of PV data. Maximum pressure 350 atm.
   Class I.

T	B	T	B
423.15	-11.4	523.15	+ 1.9
448.15	- 7.5	548.15	4.5
473.15	- 4.0	573.15	6.8
498.15	- 0.9		

7. R.H. Olds, H.H. Reamer, B.H. Sage, and W.N. Lacey, Ind. Engng Chem.
   ind. Edn. $\underline{35}$ 922 (1943).
   PVT data given: P range 0 - 670 atm, T range 294 - 511 K.

8. Eizo Kanda, Sc. Rep. Res. Insts Tôhoku Univ. Ser. $\underline{A1}$ 157 (1949).
   Values of B taken from Chem. Abstr. $\underline{45}$ 5993b (1951). They were cal-
   culated by the author from other thermodynamic measurements.
   Class II.

T	B	T	B
150	-169.1	350	- 26.80
200	-100.1	400	- 15.33
250	- 63.14	450	- 3.91
300	- 43.32		

9. S.D. Hamann, J.A. Lambert, and R.B. Thomas, Aust. J. Chem. $\underline{8}$ 149 (1955).
   Class I.

T	B		T	B
303.15	-38.2		343.15	-28.5
323.15	-35.2		363.15	-22.7
333.15	-33.9		383.15	-19.7

10. H.W. Schamp, Jr., E.A. Mason, A.C.B. Richardson, and A. Altman, Physics Fluids 1 329 (1958).
    Class I.
    (a) 3-term fit of PV data (P → 80 atm).

T	B	C
273.15	-53.43	2 710
298.15	-43.03	2 510
323.15	-34.42	2 310
348.15	-27.29	2 170
373.15	-21.26	2 030
398.15	-15.99	1 880
423.15	-11.41	1 760

(b) 4-term fit of PV data (P → 80 atm).

T	B	C
273.15	-53.62	2 880
298.15	-43.26	2 720
323.15	-34.58	2 450
348.15	-27.45	2 300
373.15	-21.26	2 010
398.15	-15.93	1 800
423.15	-11.24	1 560

11. G. Thomaes and R. van Steenwinkel, Nature, Lond. 187 229 (1960).
    Class II.

T	B		T	B
108.45	-364.99		186.4	-126.10
108.45	-361.54		186.4	-126.20
125.2	-267.97		223.6	- 82.62
125.2	-268.92		223.6	- 82.69
149.1	-188.04		249.3	- 68.53
149.1	-187.64		249.3	- 68.38

12. R.D. Gunn, M.S. Thesis, University of California (Berkeley) (1958).

Values of B given by J.A. Huff and T.M. Reed, J. chem. Engng Data $\underline{8}$ 306 (1963).

Class I.

T	B	T	B
273.2	-54.1	477.6	- 3.6
444.3	- 8.1	510.9	- 0.0

13. W.H. Mueller, T.W. Leland, Jr., and R. Kobayashi, A. I. Ch. E. Jl $\underline{7}$ 267 (1961) (*).

T	B	T	B
144.28	-221.0 ± 2	227.60	- 81.7 ± 1
172.05	-153.5 ± 1	255.38	- 63.3 ± 1
199.83	-107.8 ± 2	283.16	- 49.1 ± 2

14. G. Thomaes and R. van Steenwinkel, Molec. Phys. $\underline{5}$ 307 (1962).

Values of B for $C^{13}H_4$ relative to $C^{12}H_4$ given at temperatures in the range 368–471 K.

15. D.R. Douslin, R.H. Harrison, R.T. Moore, and J.P. McCullough, J. chem. Engng Data $\underline{9}$ 358 (1964).

4-term fit of PV data. P range 16 - 400 atm. Errors in B less than 0.2.

T	B	C
273.15	-53.35	2 620
298.15	-42.82	2 370
303.15	-40.91	2 320
323.15	-34.23	2 150
348.15	-27.06	1 975
373.15	-21.00	1 834
398.15	-15.87	1 727
423.15	-11.40	1 640
448.15	- 7.56	1 585
473.15	- 4.16	1 514
498.15	- 1.16	1 465
523.15	+ 1.49	1 420
548.15	3.89	1 385
573.15	5.98	1 360
598.15	7.88	1 345
623.15	9.66	1 330

These values are quoted by D.R. Douslin, Progress in international research in thermodynamic and transport properties, A.S.M.E. (1962) p. 135, where a short discussion of errors is given.

16. A.E. Hoover, T.W. Leland, Jr., and R. Kobayashi, J. chem. Phys. <u>45</u> 399 (1966).

Maximum probable error in C is 90 per cent.

T	C
131.93	-13 600

17. A.E. Hoover, I. Nagata, T.W. Leland, Jr., and R. Kobayashi, J. chem. Phys. <u>48</u> 2633 (1968).

T	B	C
131.93	-224 ± 9%	-13 600 ± 90%
191.06	-116.31 ± 1%	+ 4 741 ± 10%
200.00	-106.12 ± 1%	4 351 ± 10%
215.00	- 92.59 ± 0.4%	4 169 ± 4%
240.00	- 72.72 ± 0.3%	3 508 ± 3%
273.15	- 53.28 ± 0.2%	2 669.6 ± 2%

18. M.A. Byrne, M.R. Jones, and L.A.K. Staveley, Trans. Faraday Soc. <u>64</u> 1747 (1968).

Class I.

T	B	T	B
110.83	-330.1	148.28	-187.7
112.43	-319.9	162.29	-158.4
114.45	-307.8	178.41	-132.2
116.79	-295.5	202.49	-103.4
121.25	-274.5	221.10	- 85.8
128.84	-244.3	243.80	- 70.3
136.75	-218.9	273.17	- 53.7

19. R.N. Lichtenthaler and K. Schäfer, Ber. (dtsch.) Bunsenges. phys. Chem. <u>73</u> 42 (1969).

Estimated absolute error in B ± 1.

T	B	T	B
288.2	-46.20	313.2	-37.00
296.0	-43.13	323.1	-33.80
303.2	-40.40		

20. S.L. Robertson and S.E. Babb, Jr., J. chem. Phys. 51 1357 (1969).

PVT data given: P range 1500-10000 atm., T range 308-473 K.

21. V. Jansoone, H. Gielen, J. De Boelpaep and O.B. Verbeke, Physica, 's Grav, 46 213 (1970).

PVT data given: P range 44.5 - 50.0 atm., T range 190.0 - 193.6 K.

22. R.C. Lee and W.C. Edmister, A. I. Ch. E. Jl. 16 1047 (1970).

(a) Slope-intercept calculations.

T	B	C
298.15	-42.88 ± 1.5	2390 ± 800
323.15	-33.22 ± 1.0	1780 ± 400
348.15	-26.54 ± 1.1	1960 ± 500

(b) Curve fit.

T	B	C
298.15	-42.70 ± 2.3	2450 ± 870
323.15	-33.46 ± 0.3	1900 ± 120
348.15	-25.69 ± 0.9	1400 ± 420

23. A.J. Vennix, T.W. Leland, Jr., and R. Kobayashi, J. chem. Engng. Data 15 238 (1970).

PVT data given: P range 1 - 680 atm.,
                T range 150 - 273 K.

24. K. Strein, R.N. Lichtenthaler, B. Schramm, and Kl. Schäfer, Ber. (dtsch.) Bunsenges. phys. Chem. 75 1308 (1971).

Estimated accuracy of B ± 1.

T	B	T	B
296.1	-44.5	413.8	-11.5
308.0	-39.5	434.0	- 8.0
333.5	-29.8	453.4	- 5.0
353.8	-23.8	473.5	- 3.0
374.0	-19.4	493.0	- 1.0
393.9	-15.4	511.1	+ 0.5

25. T.K. Bose, J.S. Sochanski and R.H. Cole, J. chem. Phys. 57 3592 (1972).

B derived from low pressure dielectric measurements.

T	B		T	B
279.8	-52.9 ± 1.4		373.4	-21.8 ± 1.0
322.5	-35.8 ± 2.3			

26. D.R. Roe and G. Saville (unpublished results). See also D.R. Roe, Ph. D. thesis, University of London (1972).

Estimated errors given.

T	B	C
155.89	-167.95 ± 0.60	3720 ± 500
167.67	-146.55 ± 0.40	4190 ± 300
181.86	-125.70 ± 0.40	4260 ± 300
192.64	-112.85 ± 0.20	4230 ± 125
204.61	-100.15 ± 0.20	3930 ± 125
218.87	- 87.15 ± 0.20	3550 ± 125
234.05	- 75.90 ± 0.20	3300 ± 125
248.54	- 66.50 ± 0.20	3010 ± 125
263.08	- 58.35 ± 0.20	2790 ± 125
291.41	- 45.50 ± 0.20	2490 ± 125

27. G.A. Pope, P.S. Chappelear and R. Kobayashi, J. chem. Phys. <u>59</u> 423 (1973).

Burnett method.

T	B	C
126.584	-242.27 ± 0.72	-20400 ± 2800
135.994	-215.20 ± 0.12	- 1497 ± 289
147.583	-185.00 ± 0.94	1950 ± 640
158.909	-161.51 ± 0.23	3580 ± 109
173.485	-137.63 ± 0.15	4381 ± 48
191.097	-114.29 ± 0.25	3990 ± 130

28. J. Bellm, W. Reineke, K. Schäfer, and B. Schramm, Ber. (dtsch.) Bunsenges. phys. Chem. <u>78</u> 282 (1974).

Estimated accuracy of B is ± 2.

T	B		T	B
300	-42.8		430	- 9.6
320	-35.2		460	- 5.3
340	-28.9		490	- 2.0
370	-21.3		520	+ 0.6
400	-15.0		550	2.5

29. R. Hahn, K. Schäfer, and B. Schramm, Ber. (dtsch.) Bunsenges. phys. Chem. <u>78</u> 287 (1974).

B determined assuming B (296 K) = -44.5.

Quoted accuracy of B is ± 2.

T	B	T	B
200.5	-106	251.5	- 65.2
231.2	- 79.6	273.0	- 54.7

METHANOL (METHYL ALCOHOL, CARBINOL) CH_3OH

1. J.D. Lambert, G.A.H. Roberts, J.S. Rowlinson, and V.J. Wilkinson, Proc. R. Soc. <u>A196</u> 113 (1949) (†).

Maximum pressure 600 torr. Accuracy ± 50.

T	B	T	B
319.3	-1 424.3	350.9	-1 045.8
329.8	-1 316.0	351.4	-1 038.6
335.0	-1 251.1	351.4	-1 022.8
340.0	-1 189.1	360.6	- 933.3
340.1	-1 172.6	378.9	- 730.2
345.5	-1 093.4	392.8	- 626.6
350.8	-1 064.7	404.8	- 525.8

2. C.B. Kretschmer and R. Wiebe, J. Am. chem. Soc. <u>76</u> 2579 (1954).

Low-pressure vapour density measurement. Data fit to 3-term P series (P^0, P^1, and P^3). Uncertainty in B is 5 at 373.15 K and above, but increases to about 100 at 313.15 K.

T	B	T	B
313.15	-1463	373.15	- 543
333.15	- 926	393.15	- 433
353.15	- 701		

3. O.R. Fox, J. Morcillo, and A. Mendez, An. R. Soc. esp. Fis. Quim. <u>17B</u> 23 (1954).

Values read from diagram.

Class II.

T	B	T	B
350.0	-900	424.2	-400
371.0	-660	448.7	-370
402.0	-470	474.1	-320

4. G.A. Bottomley and T.H. Spurling, Aust. J. Chem. <u>20</u> 1789 (1967).
3-term fit. Maximum pressure 155 torr.

T	B	C
323.2	-1 144	-4.29×10^7
333.2	-1 033	$+0.37 \times 10^7$
348.2	- 886	1.40×10^7
373.2	- 691	1.47×10^7
398.2	- 546	1.42×10^7
423.2	- 412	1.10×10^7

5. D.H. Knoebel and W.C. Edmister, J. chem. Engng Data <u>13</u> 312 (1968).
Low pressure PVT measurements.

T	B	T	B
313.2	-2079	353.2	- 752
333.2	-1079	373.2	- 542

6. G.S. Kell and G.E. McLaurin, J. chem. Phys. <u>51</u> 4345 (1969).

T	B	C
423.16	-323 ± 2	-174000 ± 5000
423.16	-318 ± 2	-188000 ± 4000
448.16	-258.8 ± 0.7	- 64100 ± 1600
448.16	-258.2 ± 0.4	- 65600 ± 900
473.16	-220.5 ± 0.5	- 15300 ± 700
473.16	-220.0 ± 1.3	- 15700 ± 1900
473.16	-216.9 ± 0.2	- 19900 ± 300
473.16	-218.1 ± 0.3	- 18300 ± 500
498.16	-181.2 ± 0.4	- 7050 ± 530
498.16	-180.7 ± 0.5	- 7590 ± 640
523.16	-156.6 ± 0.4	
523.16	-155.9 ± 0.1	
573.16	-113.1 ± 0.4	
573.16	-117.2	

7. E.E. Tucker, S.B. Farnham, and S.D. Christian, J. phys. Chem., Ithaca
<u>73</u> 3820 (1969).
PVT measurements at low pressures, at 288, 298 and 308K.

8. A.P. Kudchadker and P.T. Eubank, J. chem. Engng Data <u>15</u> 7 (1970).

T	B	T	B
298.15	-2075 ± 104	398.15	- 413 ± 15
323.15	-1185 ± 53	423.15	- 321 ± 10
348.15	- 737 ± 30	448.15	- 251 ± 8
373.15	- 535 ± 19	473.15	- 185 ± 6

METHYLAMINE (AMINOMETHANE) CH_3NH_2

1. J.D. Lambert and E.D.T. Strong, Proc. R. Soc. <u>A200</u> 566 (1950) (†).
    Maximum pressure 600 torr. Accuracy ± 50.

T	B	T	B
293.4	-535	360.8	-311
312.2	-451	372.4	-279
320.4	-422.5	381.8	-266
331.4	-382	393.3	-237
342.6	-351	405.2	-220
351.6	-325		

2. G. Adam, Kl. Schäfer, and B. Schramm, Ber. (dtsch.) Bunsenges. phys. Chem. <u>80</u> 912 (1976).
    Repeatability of B ± 1. Estimated accuracy ± 10.

T	B	T	B
296.3	-470.9	432.3	-188.1
302.8	-441.0	462.8	-164.5
312.5	-404.3	488.4	-148.0
332.3	-341.6	501.9	-140.0
362.2	-276.7	520.2	-131.4
400.7	-220.6	549.6	-118.0

CARBON MONOXIDE CO

1. G.A. Scott, Proc. R. Soc. <u>A125</u> 330 (1929) (*).
    3-term fit of PV data. Maximum pressure 170 atm.

T	B	C
298.15	-9.84	2 043

2. E.P. Bartlett, H.C. Hetherington, H.M. Kvalnes, and T.H. Tremearne, J. Am. chem. Soc. <u>52</u> 1374 (1930).
    PVT data given: P range 1 - 1000 atm, T range 203 - 473 K.

3. D.T.A. Townend and L.A. Bhatt, Proc. R. Soc. A134 502 (1932)(*).
   4-term fit of PV data. Maximum pressure 600 atm.

T	B	C
273.15	-16.44	2 641
298.15	- 9.98	2 280

4. G.A. Bottomley, D.S. Massie, and R. Whytlaw-Gray, Proc. R. Soc. A200
   201 (1950) (*).

T	B
295.21	- 8.43

5. A. Michels, J.M. Lupton, T. Wassenaar, and W. de Graaff, Physica, 's
   Grav. 18 121 (1952) (*).
   Class I.
   (a) 3-term fit to PV data (P→80 atm).

T	B	C
273.15	-14.19	1 781
298.15	- 8.28	1 674
323.15	- 3.40	1 591
348.15	+ 0.90	1 444
373.15	4.49	1 339
398.15	7.52	1 264
423.15	10.04	1 259

   (b) 8-term fit to PV data (P→3000 atm).

T	B	C
273.15	-13.65	1 378
298.15	- 7.95	1 439
323.15	- 3.29	1 526
348.15	+ 1.06	1 283
373.15	4.57	1 197
398.15	7.67	1 031
423.15	10.16	1 073

6. V. Mathot, L.A.K. Staveley, J.A. Young, and N.G. Parsonage, Trans.
   Faraday Soc. 52 1488 (1956).

T	B
90.67	-233 ± 5

7. J.F. Connolly, Physics Fluids $\underline{7}$ 1023 (1964).
    Class I.

T	B	T	B
323.2	- 3.7	513.2	17.3
423.2	+ 9.6	573.2	20.5
473.2	14.5		

8. S.L. Robertson and S.E. Babb, Jr., J. chem. Phys. $\underline{53}$ 1094 (1970).
    PVT data given: P range 1500-10000 atm, T range 308-573 K.

9. B. Schramm and H. Schmiedel (1979)(†).
    Estimated error in B is ± 3.

T	B	T	B
295	- 9.0	425	7.5
330	- 4.3	450	9.6
365	+ 0.3	475	11.0
400	5.0		

10. B. Schramm and R. Gehrmann (1979)(†).
    Estimated error in B is ± 6.

T	B	T	B
213	-35.0	262	-16.5
223	-30.5	272	-13.0
242	-22.8	295	- 9.0

CARBON DIOXIDE CO_2

At temperatures above 260 K, the second virial coefficients reported by
Dadson et al. (15, 19) are significantly less negative than those obtained
by most other workers. Below 260 K there is a marked discrepancy between
the results of Schäfer (5) and those of Cook (11).

We recommend the set of values given by Angus et al. (23) and
estimate that these are accurate to ± 2 above 260 K, though the uncertainty
increases at lower temperatures.

1. A. Luduc and P. Sacerdote, C. r. hebd. Seanc. Acad. Sci., Paris $\underline{125}$
    297 (1897).

T	B
289.2	-124.8

2. Lord Rayleigh, Phil. Trans. R. Soc. $\underline{A204}$ 351 (1905).

T	B
288.2	-131.9

3. A. Michels and C. Michels, Proc. R. Soc. A153 201 (1935)(*).
    Class I.
    4-term fit to PV data (P→240 atm).

T	B	C
273.15	-151.18	5 608
298.20	-123.56	4 931
303.05	-119.45	5 160
304.19	-118.37	5 112
305.23	-117.29	4 902
313.25	-110.83	4 987
322.86	-103.52	4 928
348.41	- 86.68	4 429
372.92	- 73.68	4 154
398.16	- 62.20	3 623
412.98	- 55.76	3 044
418.20	- 54.02	3 084
423.29	- 52.23	3 046

4. W. Cawood and H.S. Patterson, J. chem. Soc. 619 (1933); (a) Phil.
    Trans. R. Soc. A236 77 (1937) (*).

T	B	T	B
273.15	-148.6	294.15	-126.0(a)
294.15	-127.0	304.15	-120.8

5. K. Schafer, Z. phys. Chem. B36 85 (1937).
    2-term fit of PV data. Maximum pressure less than 1 atm. Uncertainty
    in B ± 2 per cent.

T	B	T	B
203.83	-330	226.47	-225
206.63	-316	229.96	-212
207.72	-313	230.93	-216
209.03	-302	231.79	-213
210.12	-300	233.34	-210
211.60	-286	235.05	-198
223.75	-226	244.93	-175
225.63	-229	273.15	-142

6. H.H. Reamer, R.H. Olds, B.H. Sage, and W.N. Lacey, Ind. Engng Chem.
    ind. Edn 36 88 (1944).
    PVT data given: P range 1 - 680 atm. T range 311 - 511 K.

7. G.A. Bottomley, D.S. Massie, and R. Whytlaw-Gray, Proc. R. Soc. <u>A200</u> 201 (1950) (*).

T	B
295.21	-125.2

8. K.E. MacCormack and W.G. Schneider, J. chem. Phys. <u>18</u> 1269 (1950) (*); J. chem. Phys. <u>19</u> 849 (1951).

4-term fit of PV data (P series). Maximum pressure 50 atm. Accuracy of B ± 0.5 - 1 per cent at 273.15, ± 1.5 per cent at 473.15 K.

T	B	C
273.15	-156.36	12 040
323.15	-102.63	4 390
373.15	- 71.85	3 165
423.15	- 50.59	
473.15	- 34.08	
573.15	- 13.58	
673.15	- 1.58	
773.15	+ 6.05	
873.15	12.11	

9. W.C. Pfefferle, Jr., J.A. Goff, and J.G. Miller, J. chem. Phys. <u>23</u> 509 (1955). Burnett method. Maximum pressure 55 atm.

T	B	C
303.15	-117.7	4 250
303.15	-117.9	4 350

10. T.L. Cottrell, R.A. Hamilton, and R.P. Taubinger, Trans. Faraday Soc. <u>52</u> 1310 (1956).

T	B	T	B
303.15	-116.4 ± 4.6	363.15	- 75.9 ± 4.0
333.15	- 96.7 ± 4.6		

11. D. Cook, Can. J. Chem. <u>35</u> 268 (1957).

2-term fit of PV data. Maximum pressure 2.5 atm.

T	B	T	B
213.2	-310 ± 7	248.2	-204 ± 12
223.2	-302 ± 15	273.2	-168 ± 8
233.2	-266 ± 11	303.2	-127 ± 13

12. A. Perez Masia and M. Diaz Pena, An. R. Soc. esp. Fis. Quim. <u>54B</u> 661 (1958).

Class II.

T	B	T	B
298.15	-125.7	373.15	- 73.9
303.15	-119.3	398.15	- 59.4
323.15	-104.3	423.15	- 52.6
348.15	- 85.1		

13. R.D. Gunn, M.S. Thesis, University of California, Berkeley (1958). Values of B given by J.A. Huff and T.M. Reed, J. chem. Engng Data <u>8</u> 306 (1963).

Class I.

T	B	T	B
298.2	-124.6	398.2	- 61.2
310.9	-112.7	410.9	- 56.5
323.2	-103.0	444.3	- 44.6
344.3	- 88.8	477.5	- 34.9
377.6	- 70.7	510.9	- 26.4

14. J.H. Dymond and E.B. Smith (1962) (†).

T	B
298.15	-127 ± 5

15. E.G. Butcher and R.S. Dadson, Proc. R. Soc. <u>A277</u> 448 (1964). 4-term fit of PV data.

Class I.

T	B	C
262.65	-159.9	+4 300
273.15	-147.4	4 300
283.15	-136.7	4 400
299.65	-120.5	4 350
309.65	-111.3	4 100
323.15	-100.7	4 000
333.15	- 93.9	3 950
343.15	- 87.1	3 600
353.15	- 80.9	3 450
363.15	- 75.3	2 950

373.15	- 69.5	2 650
423.15	- 46.3	600
473.15	- 29.1	-2 150

16. M.P. Vukalovich and Ya. F. Masalov, Teploenergetika 13 (5), 58 (1966);
     Heat Pwr Engng, Wash. 13 (5), 73 (1966) (*).

     3-term fit of PV data. P range 5 - 200 atm.

     Class I.

T	B	C
423.2	-52.4	4 265
473.2	-36.7	2 585
523.2	-25.1	1 675
573.2	-16.3	1 230
623.2	- 9.4	1 060
673.2	- 3.7	1 090
723.2	+ 0.9	1 240
773.2	4.8	1 485

17. P.S. Ku and B.F. Dodge, J. chem. Engng Data 12 158 (1967) (*).

     3-term fit of PV data.

T	B	C
373.15	-77.5	5 850

18. A. Sass, B.F. Dodge, and R.H. Bretton, J. chem. Engng Data 12 168
     (1967) (*).

     7-term fit of PV data (P series). P range 8 - 500 atm.

     Class I.

T	B	T	B
348.15	-81.8	398.15	-61.4
373.15	-69.8		

19. R.S. Dadson, E.J. Evans, and J.H. King, Proc. phys. Soc. 92 1115
     (1967).

     Determined using Piezometer method and used to confirm earlier
     results obtained by the series-expansion method.

     Class I.

T	B	T	B
263.2	-159.8	293.2	-125.6
273.2	-147.4	298.2	-121.8

313.2	-108.4	373.2	- 70.2
333.2	- 93.2	398.2	- 58.4
353.2	- 80.9		

20. R.N. Lichtenthaler and K. Schäfer, Ber. (dtsch.) Bunsenges. phys. Chem. 73 42 (1969).

Estimated absolute error in B ± 1.

T	B	T	B
288.2	-137.1	313.2	-112.0
296.0	-129.1	323.1	-103.1
303.2	-122.2		

21. T.K. Bose and R.H. Cole, J. chem. Phys. 52 140 (1970).

T	B
322.85	-109.3 ± 4.4

22. M. Waxman, H.A. Davis and J.R. Hastings, Proc. Sixth Symp. Thermophys. Props., A.S.M.E., New York 245 (1973).

Burnett Method. Reliability of B values given as ± .3.

T	B	T	B
273.15	-150.07	373.15	- 71.85
323.15	-102.25	398.15	- 60.69
348.15	- 85.42	423.15	- 51.41

23. S. Angus, B. Armstrong, and K.M. de Reuck, International Thermodynamic Tables of the Fluid State Carbon Dioxide, Pergamon Press, Oxford, 1976.

T	B	T	B
220	-248.2	330	- 98.5
230	-223.3	340	- 91.7
240	-202.4	350	- 85.5
250	-184.5	360	- 79.7
260	-169.0	370	- 74.4
270	-155.4	380	- 69.5
273.15	-151.4	390	- 64.8
280	-143.3	400	- 60.5
290	-132.5	410	- 56.5
298.15	-124.5	420	- 52.8
300	-122.7	430	- 49.3
310	-113.9	440	- 45.9
320	-105.8	450	- 42.8

T	B	T	B
460	− 39.9	790	5.7
470	− 37.2	800	6.3
480	− 34.6	810	6.9
490	− 32.1	820	7.5
500	− 29.8	830	8.0
510	− 27.6	840	8.6
520	− 25.5	850	9.1
530	− 23.5	860	9.6
540	− 21.6	870	10.1
550	− 19.9	880	10.6
560	− 18.2	890	11.0
570	− 16.5	900	11.5
580	− 15.0	910	11.9
590	− 13.5	920	12.3
600	− 12.1	930	12.7
610	− 10.8	940	13.1
620	− 9.5	950	13.5
630	− 8.3	960	13.9
640	− 7.1	970	14.2
650	− 6.0	980	14.6
660	− 5.0	990	14.9
670	− 3.9	1000	15.3
680	− 2.9	1010	15.6
690	− 2.0	1020	15.9
700	− 1.1	1030	16.2
710	− 0.2	1040	16.5
720	0.6	1050	16.8
730	1.4	1060	17.1
740	2.2	1070	17.3
750	2.9	1080	17.6
760	3.7	1090	17.9
770	4.4	1100	18.1
780	5.0		

24. B. Schramm and R. Gehrmann (1979)(†).

Estimated error in B is ± 6.

T	B	T	B
213	-245.9	262	-155.9
223	-221.3	276	-141.0
242	-183.5	295	-126.0

25. B. Schramm and H. Schmiedel (1979)(†).
Estimated error in B is ± 4.

T	B	T	B
295	-126.0	425	-47.5
330	- 98.0	450	-40.7
365	- 76.8	475	-34.6
400	- 58.0		

CARBON DISULPHIDE CS_2

The second virial coefficients of carbon disulphide are subject to large uncertainty especially since the latest set of measurements by Hajjar, Kay and Leverett (7) are up to 100 cm^3mol^{-1} less negative than previous results at the highest temperatures. The smoothed values given below are based on the values of Bottomley and Reeves (2), Waddington, Smith, Williamson and Scott (4), and Bottomley and Spurling (5) and (6).

T	B	T	B
280	-930 ± 30	350	-567 ± 15
290	-862 ± 30	375	-489 ± 15
300	-796 ± 25	400	-429 ± 15
310	-740 ± 20	430	-377 ± 15
325	-666 ± 15		

1. F.L. Cassado, D.S. Massie, and R. Whytlaw-Gray, Proc. R. Soc. A214 466 (1952) (*).

T	B
295.2	-659

2. G.A. Bottomley and C.G. Reeves, J. chem. Soc. 3794 (1958).

T	B	T	B
295.2	-849 ± 3	323.2	-661 ± 3
308.2	-748 ± 3		

3. G.A. Bottomley and T.A. Remmington, J. chem. Soc. 3800, (1958).
Obtained using a density balance.

T	B	T	B
295.2	-646	308.2	-582

4. G. Waddington, J.C. Smith, K.D. Williamson, and D.W. Scott, J. phys. Chem., Ithaca 66 1074 (1962).
Values of B calculated by the authors from measurements of vapour pressure and heats of vaporization.

T	B		T	B
281.94	-923		319.37	-696
298.15	-802			

5. G.A. Bottomley and T.H. Spurling, J. phys. Chem., Ithaca <u>68</u> 2029 (1964).
    Class II.

T	B		T	B
297.69	-793		346.05	-583
323.16	-671		374.16	-492
325.29	-661		426.16	-381
330.23	-638			

6. G.A. Bottomley and T.H. Spurling, Aust. J. Chem. <u>20</u> 1789 (1967).
    Errors in B are ± 12.

T	B		T	B
324.85	-664		407.39	-415
349.86	-568		432.10	-373
378.85	-479			

7. R.F. Hajjar, W.B. Kay, and G.F. Leverett, J. chem. Engng Data <u>14</u> 377 (1969).
    Low pressure PVT measurements.

T	B		T	B
313.15	-810		398.15	-335
323.15	-700		413.15	-310
337.15	-605		427.15	-270
353.15	-480		453.15	-230
368.15	-430		473.15	-195
382.15	-380			

CHLOROPENTAFLUOROETHANE $CClF_2.CF_3$

1. W.H. Mears, E. Rosenthal, and J.V. Sinka, J. chem. Engng Data <u>11</u> 338 (1966).
    PVT data given: P → 69 atm, T 313 - 450 K.

1,1-DICHLORO 1,2,2,2-TETRAFLUOROETHANE CF_3CFCl_2

1. W.H. Mears, R.F. Stahl, S.R. Orfeo, R.C. Shair, L.F. Kells, W. Thompson, and H. McCann, Ind. Engng Chem. ind. Edn <u>47</u> 1449 (1955) (*).
    Values of B calculated from the equation given by the authors to represent their PVT data.

T	B	T	B
363.15	-502	423.15	-317
383.15	-432	443.15	-266
403.15	-373		

1,2-DICHLORO-1,1,2,2 -TETRAFLUOROETHANE $CClF_2CClF_2$

1. J.J. Martin, J. chem. Engng Data 5 334 (1960).

   Values of B given by G.A. Bottomley and D.B. Nairn, Aust. J. Chem. 30 1645 (1977).

T	B	T	B
296.2	-721.7	400.5	-415.4
320.5	-662.9	432.2	-337.0
344.5	-586.5	467.3	-266.3
371.4	-499.9	503.9	-208.0

2. G.A. Bottomley and D.B. Nairn, Aust. J. Chem. 30 1645 (1977).

   B determined relative to B(296.23K) = - 825. Precision in B ranges from ± 4 to ± 8.

T	B	T	B
296.23	-825.0	371.34	-501.0
296.22	-824.4	400.49	-422.5
296.26	-823.6	400.54	-428.5
320.53	-689.2	400.58	-421.6
320.45	-697.3	432.17	-353.6
344.46	-586.3	432.33	-356.0
344.42	-585.0	432.18	-351.6
371.44	-496.5	467.25	-300.0
371.40	-499.3	467.32	-300.4
371.35	-504.4	503.85	-244.6
371.37	-499.3	503.88	-246.6

1,1,2-TRICHLORO-1,2,2 TRIFLUOROETHANE $CCl_2F.CClF_2$

1. J.O. Hirschfelder, F.T. McClure, and I.F. Weeks, J. chem. Phys. 10 201 (1942).

   Calculated from available data.

T	B	T	B
283.15	-1 080	394.26	- 654
310.93	- 944	422.04	- 561
338.70	- 795	449.81	- 495
366.48	- 737		

2. M.J. Mastroianni, R.F. Stahl, and P.N. Sheldon, J. chem. Engng Data 23 113 (1978).

PVT data given; T 404-523 K, P → 60 atm.

TETRADEUTEROETHYLENE (TETRADEUTEROETHENE) $CD_2:CD_2$

1. W. Göpel and T. Dorfmüller, Z. phys. Chem. Frankf. Ausg. 82 58 (1972). Standard deviation given for B.

T	B	T	B
199.7	-314.46 ± 0.32	233.2	-229.83 ± 0.37
209.2	-285.39 ± 0.34	273.2	-171.36 ± 0.42
224.0	-247.26 ± 0.36	296.7	-146.50 ± 0.46

2. I. Gainar, Kl. Schäfer, B. Schmeiser, B. Schramm, and K. Strein, Ber. (dtsch.) Bunsenges. phys. Chem. 77 372 (1973).

Values of ΔB ($B_{C_2D_4} - B_{C_2H_4}$) determined on the basis that ΔB (511.0K) = 0. Repeatability of ΔB ± 0.3.

T	B	T	ΔB
210.3	4.1	373.3	1.6
243.1	3.3	422.4	0.6
273.2	3.0	475.5	0.2
296.8	2.4		

DEUTEROETHANE $CD_3.CD_3$

1. I. Gainar, Kl. Schäfer, B. Schmeiser, B. Schramm, and K. Strein, Ber. (dtsch.) Bunsenges. phys. Chem. 77 372 (1973).

Values of ΔB ($B_{C_2D_6} - B_{C_2H_6}$) determined on the basis that ΔB (521.8K) = 0. Repeatability of ΔB ± 0.3.

T	ΔB	T	ΔB
205.6	5.7	377.8	1.7
242.9	4.8	425.3	1.3
273.2	3.5	475.3	0.3
313.1	3.0		

TETRAFLUOROETHYLENE $CF_2:CF_2$

1. P.G.T. Fogg and J.D. Lambert, Proc. R. Soc. A232 537 (1955) (†).
    Maximum pressure 500 torr.
    Class II.

T	B	T	B
292.6	-299	351.2	-123
325.7	-158	373.4	- 82
340.2	-149		

PERFLUOROETHANE CF_3CF_3

1. E.L. Pace and J.G. Aston, J. Am. chem. Soc. 70 566 (1948).
    B calculated from measured vapour pressures and heats of vaporization.

T	B	T	B
179.96	-837	194.87	-641
180.22	-815	194.90	-653
188.30	-758	195.10	-611
190.05	-766	195.21	-624

B calculated from vapour density measurements.

T	B
298.15	-199 to -224

HALOTHANE (1,1,1-TRIFLUOROCHLOROBROMOETHANE, 2-BROMO-2-CHLORO 1,1,1-TRIFLUOROETHANE) $CF_3CHClBr$

1. G.A. Bottomley and G.H.F. Seiflow, J. appl. Chem.,Lond. 13 399 (1963).
    Class II.

T	B	T	B
298.15	-1 371	323.15	-1 090

2-CHLORO 1,1-DIFLUOROETHYLENE $CHCl:CF_2$

1. W.H. Mears, R.F. Stahl, S.R. Orfeo, R.C. Shair, L.F. Kells, W. Thompson, and H. McCann, Ind. Engng Chem. ind. Edn 47 1449 (1955) (*).
    Values of B calculated from the equation given by the authors to represent their PVT data.

T	B		T	B
343.15	-345		403.15	-251
363.15	-310		423.15	-225
383.15	-279			

TRICHLOROETHYLENE CHCl:CCl$_2$

1. P.G.T. Fogg and J.D. Lambert, Proc. R. Soc. A232 537 (1955) (†).
   Maximum pressure 500 torr.
   Class III.

T	B		T	B
323.6	-1 907		351.6	-1 194
329.6	-1 786		361.2	-1 115
336.2	-1 261		374.0	-1 049
343.2	-1 263			

ACETYLENE (ETHYNE) CH:CH

The two sets of data for the second virial coefficient of acetylene overlap only at one point, 273 K, and differ by 60 cm^{3}mol.$^{-1}$, some ten times the sum of the estimated limits!

1. K. Schafer, Z. phys. Chem. B36 85 (1937).
   Estimated error ± 2 per cent.

T	B		T	B
199.63	-572		225.98	-406
201.66	-566		230.57	-390
203.09	-550		232.96	-381
205.04	-532		235.68	-369
206.98	-518		237.07	-361
209.56	-503		237.64	-358
211.71	-479		238.92	-352
218.15	-454		240.24	-343
219.42	-446		242.71	-334
221.45	-436		245.27	-328
222.90	-426		248.96	-320
224.21	-414		273.15	-258

2. G.A. Bottomley, C.G. Reeves, and G.H.F. Seiflow, Nature, Lond. 182 596 (1958).

2-term fit. Maximum pressure 230 torr. Accuracy about ± 1.

T	B	T	B
273.2	-191	313.2	-133
293.2	-158		

1,2-DICHLOROETHENE CHC1:CHC1

1. P.G.T. Fogg and J.D. Lambert, Proc. R. Soc. A232 537 (1955) (+).
   Maximum pressure 500 torr.
   Class III.
   (a) cis-form

T	B	T	B
319.2	- 987	351.6	- 822
326.1	-1 054	362.6	- 868
333.0	- 908	373.4	- 721
342.4	- 810		

   (b) trans-form

T	B	T	B
319.8	-953	340.9	-833
327.1	-836	351.3	-746
333.2	-868		

1,1 DIDEUTEROETHYLENE (1,1-DIDEUTEROETHENE) $CH_2:CD_2$

1. W. Göpel and T. Dorfmüller, Z. phys. Chem. Frankf. Ausg. 82 58 (1972).
   Standard deviation given for B.

T	B	T	B
199.7	-315.83 ± 0.32	209.2	-286.73 ± 0.34

1,2-TRANS DIDEUTEROETHYLENE (1,2-TRANS DIDEUTEROETHENE) CHD:CHD

1. W. Göpel and T. Dorfmüller, Z. phys. Chem. Frankf. Ausg. 82 58 (1972).
   Standard deviation given for B.

T	B	T	B
199.7	-315.58 ± 0.32	233.2	-231.37 ± 0.37
209.2	-286.60 ± 0.34	248.2	-205.41 ± 0.40
224.0	-248.64 ± 0.36	273.2	-172.42 ± 0.42

VINYLIDENE FLUORIDE (1,1-DIFLUOROETHYLENE) $CH_2:CF_2$

1. P.G.T. Fogg and J.D. Lambert, Proc. R. Soc. A232 537 (1955) (†).
   Maximum pressure 500 torr.
   Class III.

T	B	T	B
293.2	-175	349.2	-110
327.7	-165	362.9	-104
336.2	-191	374.0	- 96

2. W.H. Mears, R.F. Stahl, S.R. Orfeo, R.C. Shair, L.F. Kells, W. Thompson, and H. McCann, Ind. Engng Chem. ind. Edn 47 1449 (1955) (*).
   Values of B calculated from the equation given by the authors to represent their PVT data.

T	B	T	B
303.15	-190	343.15	-145
323.15	-167		

3. W.S. Haworth and L.E. Sutton, Trans. Faraday Soc. 67 2907 (1971).
   Estimated precision of B ± 10.

T	B	T	B
298.2	-190	328.2	-154
313.2	-174		

VINYL BROMIDE (BROMOETHYLENE) $CH_2:CHBr$

1. P.G.T. Fogg and J.D. Lambert, Proc. R. Soc. A232 537 (1955) (†).
   Maximum pressure 500 torr.
   Class III.

T	B	T	B
294.1	-672	349.3	-508
326.0	-596	374.2	-410
337.4	-499		

VINYL CHLORIDE (CHLOROETHYLENE) $CH_2:CHCl$

1. P.G.T. Fogg and J.D. Lambert, Proc. R. Soc. A232 537 (1955) (†).
    Maximum pressure 500 torr.
    Class III.

T	B	T	B
297.3	-416	350.7	-380
323.0	-355	362.7	-313
338.1	-331	374.0	-309

2. W. Hayduk and H. Laudie, J. chem. Engng Data 19 253 (1974).
    Compressibility factor values given
    $\quad\quad\quad\quad$ T 273.15 - 348.15 K, P → 10 atm.

1-CHLORO 1,1-DIFLUOROETHANE CH_3CF_2Cl

1. W.H. Mears, R.F. Stahl, S.R. Orfeo, R.C. Shair, L.F. Kells, W. Thompson,
    and H. McCann, Ind. Engng Chem. ind. Edn 47 1449 (1955) (*).

    Values of B calculated from the equation given by the authors to re-
    present their PVT data.

T	B	T	B
353.15	-452	393.15	-320
373.15	-382	413.15	-263

1,1,1-TRICHLOROETHANE CH_3CCl_3

1. A. Perez Masia, M. Diaz Pena, and J.A. Burriel Lluna, An. R. Soc. esp.
    Fis. Quim. 60B 229 (1964).

T	B	C
354.85	-1 067 ± 20	30 000
375.13	- 972 ± 18	43 000
398.98	- 847 ± 21	30 000

VINYL FLUORIDE (FLUOROETHYLENE) $CH_2:CHF$

1. P.G.T. Fogg and J.D. Lambert, Proc. R. Soc. A232 537 (1955) (†).
    Maximum pressure 500 torr.
    Class III.

T	B	T	B
292.9	-285	343.1	-217
320.0	-201	357.8	-187
327.9	-182	373.6	-130

1,1,1-TRIFLUOROETHANE CH_3CF_3

1. W.H. Mears, R.F. Stahl, S.R. Orfeo, R.C. Shair, L.F. Kells, W. Thompson, and H. McCann, Ind. Engng Chem. ind. Edn 47 1449 (1955) (*).

   Values of B calculated from the equation given by the authors to represent their PVT data.

T	B	T	B
323.15	-299	363.15	-212
343.15	-253		

VINYL IODIDE (IODOETHYLENE) $CH_2{:}CHI$

1. P.G.T. Fogg and J.D. Lambert, Proc. R. Soc. A232 537 (1955) (†).
   Maximum pressure 500 torr.
   Class II.

T	B	T	B
326.6	-917	360.8	-719
338.8	-826	374.3	-728
348.7	-815		

ACETONITRILE (METHYL CYANIDE) CH_3CN

1. J.D. Lambert, G.A.H. Roberts, J.S. Rowlinson, and V.J. Wilkinson, Proc. R. Soc. A196 113 (1949) (*).
   Maximum pressure 600 torr. Accuracy ± 50.

T	B	T	B
324.4	-4 010	340.4	-3 000
325.6	-3 810	350.9	-2 660
332.8	-3 130	351.0	-2 820
333.6	-3 120	351.2	-2 540
333.8	-3 210	351.5	-2 470
334.9	-3 090	351.7	-2 280
340.1	-2 990	351.7	-2 460

351.7	-2 550	393.2	-1 520
364.6	-2 240	395.1	-1 550
364.7	-2 080	404.2	-1 480
368.5	-1 940	405.4	-1 480
373.5	-1 880	405.6	-1 410
382.8	-1 740		

2. I. Brown and F. Smith, Aust. J. Chem. 13 30 (1960).
From vapour pressure measurements.

T	B
318.2	-4 500

3. J.M. Prausnitz and W.B. Carter, A. I. Ch. E. J1 6 611 (1960).

T	B	T	B
313.15	-5 250 ± 100	353.25	-2 610 ± 40
333.15	-3 620 ± 40	373.65	-1 930 ± 40

4. Sh. D. Zaalishvili, L.E. Kolysko, and E. Ya. Gorodinskaya, Russ. J. phys. Chem. 45 1500 (1971); Zh. fiz. Khim. 45 2648 (1971).

T	B	T	B
308.2	-5350	338.2	-3150
323.2	-4280	353.2	-2400

5. L.E. Kolysko, L.V. Mozhginskaya, and E. Ya. Gorodinskaya, Russ. J. phys. Chem. 46 614 (1972); Zh. fiz. Khim. 46 1046 (1972).

T	B	T	B
323.2	-4148 ± 3	403.2	-1290 ± 2
353.2	-2521 ± 3	423.2	-1230 ± 1
373.2	-1968 ± 2		

ETHYLENE (ETHENE) $CH_2:CH_2$

There is considerable variation in the reported second virial coefficients of ethylene, although the more recent measurements of Trappeniers, Wassenaar, and Wolkers (18), Douslin and Harrison (19), and Lee and Saville (20) are in close agreement. The recommended values given below are based on a smooth curve through these results.

T	B	T	B
240	-218.5 ± 2	350	- 99 ± 1
250	-201 ± 2	375	- 84 ± 1
275	-166 ± 1	400	- 71.5 ± 1
300	-138 ± 1	450	- 51.7 ± 1
325	-117 ± 1		

1. C.A. Crommelin and H.G. Watts, Communs phys. Lab. Univ. Leiden 189c (1927) (*).

   3-term fit. Maximum pressure 40 atm.

   Class II.

T	B	C
271.80	-176.34	10 337
273.06	-170.09	8 363
283.33	-165.40	11 023
293.34	-152.54	9 439

2. A. Eucken and A. Parts, Z. phys. Chem. B20 184 (1933). Pressure less than 100 torr.

   Class II.

T	B	T	B
181.13	-428	230.30	-255
191.32	-381	232.38	-250
193.01	-374	240.87	-232
201.88	-339	242.86	-229
203.46	-333	250.57	-212
209.92	-312	252.99	-207
212.22	-305	262.08	-192
220.13	-280	273.20	-176
223.12	-275		

3. W. Cawood and H.S. Patterson, J. chem. Soc. 619 (1933) (*); (a) Phil. Trans. R. Soc. A236 77 (1937).

   Maximum pressure: 4 atm.

   Class II.

T	B	T	B
273.15	-162.9	294.15	-153.5
294.15	-147.7(a)		

4. A. Michels, J. Gruyter, and F. Niesen, Physica, 's Grav. <u>3</u> 346 (1936) (*).

Class I.

5-term fit to PV data (P → 80 atm).

T	B	C
273.15	-162.39	1 906
298.15	-135.79	2 535
323.15	-113.65	1 832
348.15	- 95.48	739
373.15	- 80.57	4
398.15	- 69.20	1 096
423.15	- 59.58	1 831

5. E.E. Roper, J. phys. Chem.,Ithaca <u>44</u> 835 (1940).

Maximum pressure 1072 torr.

Class II.

T	B	T	B
198.77	-314.6	273.15	-181.1
223.22	-250.9	273.15	-170.7
223.23	-250.5	343.10	-109.5

6. A. Michels and M. Geldermans, Physica, 's Grav. <u>9</u> 967 (1942) (*).

Values of B and C calculated by R.B. Bird, E.L. Spotz, and J.O. Hirschfelder, J. chem. Phys. <u>18</u> 1395 (1950).

Class I.

(a) 4-term fit of PV data (P → 80 atm).

T	B	C
273.15	-167.84	8 091
298.15	-140.33	7 453
323.15	-117.97	6 658
348.15	- 99.74	5 883
373.15	- 84.92	5 302
398.15	- 72.34	4 703
423.15	- 62.29	4 827

(b) 4-term fit of PV data (P → 170 atm).

T	B	C
298.15	-140.55	7 644
323.15	-118.07	6 743
348.15	- 99.90	5 999
373.15	- 85.11	5 464
398.15	- 72.61	4 990
423.15	- 62.07	4 670

7. G.A. Bottomley, Miss D.S. Massie, and R. Whytlaw-Gray, Proc. R. Soc. A200 210 (1950) (*).

T	B
295.51	-147.0

8. H.M. Ashton and E.S. Halberstadt, Proc. R. Soc. A245 373 (1958).

Values obtained from refractive index measurements. Errors not less than ± 4.

T	B (average)	T	B (average)
299.8	-145	337.0	-108
323.4	-124		

9. D. McA. Mason and B.E. Eakin, J. chem. Engng Data 6 499 (1961) (*).

T	B
288.70	-142.1 ± 3

10. E.G. Butcher and R.S. Dadson, Proc. R. Soc. A277 448 (1964).

4-term fit. Errors ± 0.3 in B, ± 300 in C.

T	B	C
273.15	-172.8	7 500
273.15	-161.4	7 400
283.15	-150.9	7 350
299.65	-134.8	7 150
313.15	-123.3	6 950
323.15	-115.1	6 750
333.15	-108.0	6 450
343.15	-101.2	6 150
353.15	- 95.0	5 800
363.15	- 88.6	5 400

373.15	- 83.2	5 150
423.15	- 59.8	2 650
473.15	- 42.9	- 200

11. W. Thomas and M. Zander, Z. angew. Phys. 20 417 (1966) (*).
3-term fit. Maximum pressure 21 atm.
Class I.

T	B	C
273.15	-156.3	-7 250
283.15	-144.9	-4 780
293.15	-136.7	- 860
303.15	-128.5	+1 680
313.15	-120.7	3 320
323.15	-113.6	4 460

12. A. Sass, B.F. Dodge, and R.H. Bretton, J. chem. Engng Data 12 168 (1967) (*).
P range 8-500 atm.
9-term fit to PV data (ρ series).

T	B
313.15	-116.1

7-term fit of PV data (ρ series).

T	B
373.15	- 84.9

13. P.S. Ku and B.F. Dodge, J. chem. Engng Data 12 158 (1967) (*).
3-term fit of PV data: P values up to 100 atm.

T	B	C
373.15	-87.5	4 490

14. S. Angus, B. Armstrong, and K.M. de Reuck, International Thermodynamic Tables of the Fluid State, Ethylene, 1972, Butterworths, London, 1974.

T	B	T	B
273.15	-168.7	290	-148.9
275	-166.4	295	-143.5
280	-160.4	298.15	-140.2
285	-154.5	300	-138.3

305	-133.4	370	- 86.6
310	-128.7	375	. 84.0
315	-124.2	380	- 81.5
320	-119.9	385	- 79.0
325	-115.8	390	- 76.6
330	-112.0	395	- 74.3
335	-108.3	400	- 72.0
340	-104.8	405	- 69.8
345	-101.4	410	- 67.6
350	- 98.2	415	- 65.5
355	- 95.1	420	- 63.3
360	- 92.2	425	- 61.3
365	- 89.4		

15. S.E. Babb, Jr., and S.L. Robertson, J. chem. Phys. <u>53</u> 1097 (1970). PVT data given: P range 1500-8000 atm. T is 308 K.

16. R.C. Lee and W.C. Edmister, A. I. Ch. E. Jl <u>16</u> 1047 (1970).
    (a) Slope-intercept calculations.

T	B	C
298.15	-145.60 ± 4.8	9790 ± 2600
323.15	-120.40 ± 1.3	7050 ± 400
348.15	-100.80 ± 1.1	5980 ± 800

   (b) Curve fit.

T	B	C
298.15	-152.50 ± 5.0	14300 ± 4900
323.15	-122.10 ± 2.4	7500 ± 1500
348.15	-101.50 ± 0.8	6320 ± 470

17. W. Göpel and T. Dorfmüller, Z. phys. Chem. Frankf. Ausg. <u>82</u> 58 (1972).
    Standard deviations given for B.

T	B	T	B
199.7	-317.43 ± 0.32	248.2	-206.54 ± 0.40
209.2	-288.51 ± 0.34	273.2	-173.06 ± 0.42
224.0	-250.52 ± 0.36	296.7	-147.31 ± 0.46
233.2	-233.01 ± 0.37	343.1	-109.73

18. N.J. Trappeniers, T. Wassenaar, and G.J. Wolkers, Physica, 's Grav. 82A 305 (1975) (*).

Class I.

(a) 5-term fit to PVT data.

T	B	C
273.15	-168.3	8155
279.15	-160.8	7952

(b) 9-term fit to PVT data.

T	B	C
283.65	-154.6	7695
285.65	-152.3	7423
289.15	-148.4	7024
293.15	-144.2	6747
298.15	-139.3	6664
303.15	-134.5	6533
323.15	-117.0	5584
348.15	-100.4	6589

(c) 6-term fit to PVT data.

T	B	C
373.15	- 84.9	5211
398.15	- 72.1	4631
423.15	- 61.4	4340

19. D.R. Douslin and R.H. Harrison, J. chem. Thermodyn. 8 301 (1976).

Class I.

T	B	C
238.15	-220.9	7350
243.15	-212.0	7800
248.15	-203.5	7960
253.15	-195.5	8090
258.15	-188.1	8220
263.15	-180.9	8180
268.15	-174.1	8020
273.15	-167.6	7900
278.15	-161.6	7760
282.35	-156.7	7640
283.15	-155.7	7600

288.15	-150.3	7580
293.15	-144.9	7390
298.15	-139.8	7200
303.15	-135.0	7040
323.15	-117.7	6540
348.15	- 99.7	5870
373.15	- 84.8	5320
398.15	- 72.3	4820
423.15	- 61.6	4460
448.15	- 52.4	4120

20. J.W. Lee and G. Saville (unpublished results). See also J.W. Lee, Ph.D. thesis, University of London (1976).

Estimated uncertainty is ± 1.5 in B and ± 500 in C.

T	B	C
243.6	-212.7	8400
247.1	-206.5	8700
251.1	-199.9	8900
254.3	-194.9	9100
258.9	-188.1	9300
266.6	-177.5	9400
270.4	-172.6	9400
273.8	-168.3	9300
278.8	-162.2	9200
283.1	-157.2	9100
288.8	-150.7	8800
292.6	-146.5	8600

21. W. Thomas, M. Zander, G. Quietzsch and H. Hartmann, Dichte de Äthylens in Einpkasengebiet für Temperaturen von -30°C bis 75°C, Farbewerke Hoechst A.G., Frankfurt, 2nd edn. (1976).

PVT data given.

1,1-DICHLOROETHANE CH_3CHCl_2

1. A.R. Paniego, J.A.B. Lluna, and J.E.H. Garcia, An. Quim. 71 349 (1975). Relative compressibility method, using nitrogen.

Class II.

T	B		T	B
334.6	-901		392.6	-617
353.0	-788		412.5	-563
373.6	-707			

1,2 DICHLOROETHANE CH_2ClCH_2Cl

1. K. Bohmhammel and W. Mannchen, Z. phys. Chem. <u>248</u> 230 (1971).

T	B	T	B
364.8	-843.5	484.1	-425.8
398.5	-678.4	508.9	-384.6
407.5	-639.5	511.7	-381.7
420.9	-595.1	531.3	-346.2
451.9	-503.7	563.2	-304.2
471.8	-452.1	578.2	-278.9

2. A.R. Paniego, J.A.B. Lluna, and J.E.H. Garcia, An. Quim. <u>71</u> 349 (1975).
   Relative compressibility method, using nitrogen.
   Class II.

T	B	T	B
365.5	-1053	397.3	- 888
384.1	- 936	413.3	- 806

1,1-DIFLUOROETHANE CH_3CHF_2

1. W.H. Mears, R.F. Stahl, S.R. Orfeo, R.C. Shair, L.F. Kells, W. Thompson, and H. McCann, Ind. Engng Chem. ind. Edn <u>47</u> 1449 (1955) (*).
   Values of B calculated from the equation given by the authors to represent their PVT data.

T	B	T	B
343.15	-337	383.15	-245
363.15	-289	403.15	-207

ACETALDEHYDE (ETHANAL) $CH_3.CHO$

1. E.A. Alexander and J.D. Lambert, Trans. Faraday Soc. <u>37</u> 421 (1941) (†).
   Class II.
   Measurements made with a constant-pressure gas thermometer.

T	B	T	B
288.2	-1 461	303.2	-1 125
293.2	-1 295	308.2	-1 050
297.9	-1 234	313.2	- 959
298.2	-1 206	345.2	- 770

372.9	- 530	438.2	- 320
381.2	- 550	476.2	- 260

2. J.M. Prausnitz and W.B. Carter, A. I. Ch. E. Jl <u>6</u> 611 (1960).

T	B	T	B
313.15	-944 ± 20	353.25	-633 ± 20
333.15	-765 ± 20	373.65	-533 ± 20

METHYL FORMATE HCOO.CH$_3$

1. J.D. Lambert, J.S. Clarke, J.F. Duke, C.L. Hicks, S.D. Lawrence, D.M. Morris, and M.G.T. Shone, Proc. R. Soc. <u>A249</u> 414 (1959).
    Accuracy ± 50.

T	B	T	B
319.2	-840	351.3	-610
326.7	-760	367.2	-550
333.2	-710	378.0	-500
335.1	-705	386.3	-470
342.3	-660	396.7	-440
348.2	-640		

ACETIC ACID CH$_3$.COOH

1. E.W. Johnson and L.K. Nash, J. Am. chem. Soc. <u>72</u> 547 (1950).
    PVT data given: P range up to 1.5 atm, T range 353-473 K.
    Vapour density method.

2. J.R. Barton and C.C. Hsu, J. chem. Engng Data <u>14</u> 184 (1969).
    PVT data given; T 323-398 K, P below 1 atm.

ETHYL BROMIDE (BROMOETHANE) CH$_3$.CH$_2$Br

1. M. Ratzsch and H.-J. Bittrich, Z. phys. Chem. <u>228</u> 81 (1965).
    Class II.

T	B	T	B
293.1	-795	313.2	-669

ETHYL CHLORIDE (CHLOROETHANE) $CH_3.CH_2Cl$

1. J.D. Lambert, G.A.H. Roberts, J.S. Rowlinson, and V.J. Wilkinson, Proc.
    R. Soc. <u>A196</u> 113 (1949) (†).

    Maximum pressure 600 torr. Accuracy ± 50.

T	B	T	B
324.4	-589	351.5	-501
334.0	-556	368.2	-452
344.0	-474	381.4	-388
351.0	-473	392.2	-373
351.4	-511	405.4	-346
351.4	-504		

2. K. Schafer and O.R. Foz Gazulla, Z. phys. Chem. <u>B52</u> 299 (1942).

T	B	T	B
273.15	-1 092	367.7	- 597
293.40	- 898	373.1	- 552
348.8	- 648		

3. M. Ratzsch, Z. phys. Chem. <u>238</u> 321 (1968).

T	B	T	B
303.2	-724	333.2	-535
313.2	-674		

4. K. Bohmhammel and W. Mannchen, Z. phys. Chem. <u>248</u> 230 (1971).

T	B	T	B
316.7	-659.0	447.1	-242.9
324.2	-613.3	468.2	-213.2
331.8	-569.5	493.4	-185.3
344.2	-509.3	517.7	-163.0
378.9	-387.0	518.9	-162.8
386.5	-358.0	561.8	-132.4
399.8	-324.3	603.4	-108.0
420.6	-280.7	604.8	-106.2

5. W.S. Haworth and L.E. Sutton, Trans. Faraday Soc. <u>67</u> 2907 (1971).

    Estimated precision of B ± 10.

T	B		T	B
298.2	-737		313.2	-642

ETHANE C_2H_6

There are many discrepancies in the second virial coefficient values reported for ethane prior to 1954 but more recent measurements confirm that the
results of Eucken and Parts (1) and of Beattie, Hadlock and Poffenberger
(2) are too negative, and those of Hamann and McManamey are not sufficient-
ly negative. These results were all neglected in obtaining the smoothed
values given below.

T	B		T	B
200	-410 ± 10		350	-130.5 ± 1
210	-370 ± 5		375	-111.0 ± 1
220	-336 ± 5		400	- 96.0 ± 1
240	-282 ± 3		450	- 71.0 ± 1
260	-243 ± 2		500	- 52.0 ± 0.5
280	-211 ± 2		550	- 36.5 ± 0.5
300	-182 ± 2		600	- 24.5 ± 0.5
325	-154 ± 1			

1. A. Eucken and A. Parts, Z. phys. Chem. <u>B20</u> 184 (1933).

   Class II.

T	B		T	B
191.86	-498		233.41	-325
193.65	-487		236.67	-316
201.61	-446		244.70	-293
202.17	-443		247.00	-287
211.33	-404		257.09	-262
213.34	-395		259.03	-258
222.58	-360		273.20	-227.5
224.50	-354			

2. J.A. Beattie, C. Hadlock, and N. Poffenberger, J. chem. Phys. <u>3</u> 93
   (1935).

   PVT data given. Maximum pressure 200 atm.

   Values of B given by J.O. Hirschfelder, F.T. McClure, and I.F. Weeks,
   J. chem. Phys. <u>10</u> 201 (1942).

   Class I.

T	B	T	B
298.15	-191	423.15	- 94
323.15	-160	448.15	- 86
348.15	-139	473.15	- 77
373.15	-122	498.15	- 71
398.15	-108	523.15	- 60

3. A. Michels and G.W. Nederbragt, Physica, 's Grav. 6 656 (1939).

   PVT data given. Maximum pressure 100 atm.

   Values of B given by S.D. Hamann and W.J. Mcmanamey, Trans. Faraday Soc. 49 149 (1953).

   Values of C calculated by H.G. David, S.D. Hamann, and R.G.H. Prince, J. chem. Phys. 20 1973 (1952).

   Class I.

T	B	C
273.15	-223	11 710
298.15	-187	11 070
323.15	-157	9 660

4. H.H. Reamer, R.H. Olds, B.H. Sage, and W.N. Lacey, Ind. Engng Chem. ind. Edn 36 956 (1944).

   PVT data given (P → 700 atm).

   Values of B calculated by S.D. Hamann and W.J. McManamey, Trans. Faraday Soc. 49 149 (1953).

   Values of C calculated by H.G. David, S.D. Hamann, and R.G.H. Prince, J. chem. Phys. 20 1973 (1952).

   Class I.

T	B	C
310.94	-164.9	9 100
344.27	-132.5	8 300
377.60	-110.0	7 400
410.94	- 90.4	6 600
444.27	- 74.2	6 000
477.60	- 59.9	5 400
510.94	- 47.4	4 700

5. J.D. Lambert, G.A.H. Roberts, J.S. Rowlinson, and V.J. Wilkinson, Proc. R. Soc. A196 113 (1949) (†).

   Accuracy B ± 50.

76

T	B		T	B
291.95	-220		341.35	-170
319.25	-190		350.75	-140
329.45	-140			

6. S.D. Hamann and W.J. McManamey, Trans. Faraday Soc. 49 149 (1953). Class II.

T	B		T	B
303.15	-175.8		363.15	-115.8
323.15	-146.4		373.15	-111.0
323.15	-144.8		383.15	-105.3
333.15	-139.9		393.15	- 98.3
343.15	-135.3		398.15	-103.3
348.15	-125.0		413.15	- 89.1
358.15	-121.5		423.15	- 83.5

7. A. Michels, W. van Straaten, and J. Dawson, Physica, 's Grav. 20 17 (1954) (*).
Class I.

(a) 3-term fit of PV data (P → 80 atm).

T	B	C
273.15	-221.46	10 607
298.14	-185.61	10 738
322.75	-156.91	9 671
347.65	-133.29	8 576
372.52	-114.06	7 709
397.84	- 97.72	7 001
422.70	- 83.91	6 376

(b) 5-term fit of PV data (P → 200 atm).

T	B	C
273.2	-221.46	10 607
298.2	-184.65	9 852
323.2	-157.67	10 123
348.2	-134.92	9 832
373.2	-115.45	8 985
398.2	- 99.28	8 248
423.2	- 84.92	7 279

8. R.D. Gunn, M.S. Thesis, University of California, Berkeley (1958).
    Values of B given by J.A. Huff and T.M. Reed, J. chem. Engng Data
    $\underline{8}$ 306 (1963).

    Class I

T	B	T	B
273.2	-222.2	410.9	- 89.6
298.2	-186.9	444.3	- 74.0
323.2	-157.5	477.6	- 61.6
377.6	-109.4	510.9	- 51.0

9. M. Rigby, J.H. Dymond, and E.B. Smith (1963) (†).

T	B
273.15	-221 ± 4
293.15	-191 ± 4
308.15	-170 ± 4
323.15	-152 ± 4

10. A.E. Hoover, T.W. Leland, Jr., and R. Kobayashi, J. chem. Phys. $\underline{45}$
    399 (1966).

    3-term fit of PV data. Maximum probable error is 10 per cent
    at 215.00 K and 4 per cent at 240.00 K.

T	C
215.00	-71 000
240.00	- 2 570

11. A.E. Hoover, I. Nagata, T.W. Leland, Jr., and R. Kobayashi,
    J. chem. Phys. $\underline{48}$ 2633 (1968).

T	B	C
215.00	-340.63 ± 1%	-71 100 ± 10%
240.00	-276.5 ± 0.4%	- 2 570 ± 4%
273.15	-223.41 ± 0.1%	11 373 ± 1%

12. R.N. Lichtenthaler and K. Schäfer, Ber. (dtsch.) Bunsenges. phys. Chem. 73 42 (1969).

Estimated absolute error in B ± 1.

T	B
288.2	-203.3
296.0	-191.5
303.2	-181.5
313.2	-168.4
323.1	-156.1

13. K. Strein, R.N. Lichtenthaler, B. Schramm, and Kl. Schäfer, Ber. (dtsch.) Bunsenges. phys. Chem. 75 1308 (1971).

Estimated accuracy of B is ± 1%.

T	B	T	B
296.1	-188.0	413.6	- 88.1
307.6	-172.0	433.8	- 78.4
333.6	-144.3	453.6	- 69.1
353.4	-126.2	473.8	- 62.6
373.7	-111.9	493.3	- 54.1
394.2	- 98.7		

14. D.S. Tsiklis, A.I. Semenova, S.S. Tsimmerman and E.A. Emel'yanova, Russ. J. phys. Chem 46 1677 (1972); Zh. fiz. Khim. 46 2940 (1972).

PVT data given: P range 2000-9000 atm, T range 323-673 K.

15. D.R. Douslin and R.H. Harrison, J. chem. Thermodyn. 5 491 (1973). Class 1.

T	B	C
273.15	-222.2	10360
298.15	-185.8	10600
303.15	-179.4	10400
323.15	-156.7	9650
348.15	-133.0	8660
373.15	-113.6	7720
398.15	- 97.3	6960
423.15	- 83.6	6260
448.15	- 71.7	5680

473.15	- 61.5	5290
498.15	- 52.4	4840
523.15	- 44.5	4500
548.15	- 37.3	4130
573.15	- 30.9	3860
598.15	- 25.0	3540
623.15	- 19.6	3270

16. G.A. Pope, P.S. Chappelear and R. Kobayashi, J. chem. Phys. 59 423 (1973).

Burnett method.

T	B	C
209.534	-368.66 ± 4.59	-58700 ± 20800
238.769	-287.05 ± 0.26	3704 ± 480
254.807	-252.27 ± 0.99	8504 ± 786
273.150	-219.38 ± 0.09	10360 ± 55
306.062	-175.27 ± 0.56	10030 ± 559

17. R. Hahn, K. Schäfer, and B. Schramm, Ber. (dtsch.) Bunsenges. phys. Chem. 78 287 (1974).

B determined assuming B (296 K) = -188.

Quoted accuracy of B is ± 2.

T	B
199.4	-422
210.8	-374
231.4	-313
251.2	-262

18. K. Schäfer, B. Schramm, and J.S.U. Navarro, Z. phys. Chem. Frankf. Ausg. 93 203 (1974).

Values of B obtained relative to the value at 296K.

T	B	T	B
296[†]	-192	432.1	- 85.0
353.2	-132.6	472.3	- 68.0
392.8	-104.3	510.6	- 54.5

[†] reference value

19. K. Bier, J. Kunze, G. Maurer, and H. Sand, J. chem. Engng Data
     21 5 (1976).

     Heat capacity and differential Joule-Thomson coefficient
     measurement: T 298 - 473 K, P → 10 atm.

20. G.C. Straty and R. Tsumura, J. Res. Natn. Bur. Stand. 80A 35
     (1976).

     PVT data given; P → 350 atm, T → 320 K.

21. H. Mansoorian, K.R. Hall, and P.T. Eubank, Proc. Seventh Symp.
     Thermophys. Props., Am. Soc. Mech. Engrs., New York, 456 (1977).

     Burnett method; P → 135 atm.

T	B	C
323.15	-156.1 ± 1.5	9430 ± 1800
348.15	-132.4 ± 1.5	7920 ± 1800
373.15	-111.4 ± 1.5	6240 ± 1800
398.15	- 98.0 ± 1.4	8280 ± 1600
423.15	- 85.7 ± 1.3	8390 ± 1400
448.15	- 72.4 ± 1.2	6530 ± 1200
473.15	- 62.4 ± 1.1	6510 ± 1000

ETHANOL (ETHYL ALCOHOL) $CH_3.CH_2OH$

1. P.A. Hanks and J.D. Lambert, (1951) (†).

    Class II.

T	B	T	B
321.7	-2 731	351.3	-1 325
331.9	-1 988	360.9	-1 154
336.9	-1 687	371.0	- 926
343.4	-1 697	381.7	- 675
344.1	-1 357	399.4	- 523

2. C.B. Kretschmer and R. Wiebe, J. Am. chem. Soc. 76 2579 (1954).

    Low-pressure vapour density method. Data fit to 3-term P series
    (P^0, P^1, and P^3). Uncertainty in B is ± 5 at 373.15 K and above,
    but increases to about ± 100 at 313.15 K.

T	B
313.15	-2134
333.15	-1285
353.15	- 938
373.15	- 723
393.15	- 578

3. D.H. Knoebel and W.C. Edmister, J. chem. Engng Data 13 312 (1968).

Low pressure PVT measurements.

T	B
333.2	-1522
353.2	- 941
373.2	- 687

4. H.Y. Lo and L.I. Stiel, Ind. Engng Chem., Fundamen. 8 713 (1969).
PVT data given; P → 600 atm. for T range 473 - 623 K.

DIMETHYL ETHER $(CH_3)_2O$

1. W. Cawood and H.S. Patterson, J. chem. Soc. 619 (1933).
Maximum pressure 3000 torr.

T	B
273.15	-613.9
294.15	-499.6
313.15	-421.2

2. R.M. Kennedy, M. Sagen-Kahn, and J.G. Aston, J. Am. chem. Soc. 63 2267 (1941).
Maximum pressure 0.9 atm.

T	B
298.15	-446.1 ± 1.5

3. T.B. Tripp and R.D. Dunlap, J. phys. Chem., Ithaca 66 635 (1962).
Values of B obtained from compressibility data by use of (1) linear and (2) quadratic equations in (a) pressure and (b) concentration.

Maximum pressure 700 torr.

T	1(a)	1(b)	B	2(a)	2(b)
283.25	-542 ± 4	-531 ± 5		-591 ± 14	589 ± 13
303.15	-466 ± 11	-457 ± 11		-515 ± 62	513 ± 61
323.15	-411 ± 9	-405 ± 6		-508 ± 21	505 ± 20

4. W.S. Haworth and L.E. Sutton, Trans. Faraday Soc. 67 2907 (1971).
    Estimated precision of B ± 10

T	B
298.2	-456
313.2	-405
328.2	-368

DIMETHYL SULPHIDE $(CH_3)_2S$

1. G.A. Bottomley and I.H. Coopes, Aust. J. Chem. 15 190 (1962).

T	B
299.15	-906 ± 8
339.15	-660 ± 7

2. J.P. McCullough, W.N. Hubbard, F.R. Frow, I.A. Hossenlopp, and
    G. Waddington, J. Am. chem. Soc. 79 561 (1957).

    Values of B were calculated by the authors from heats of vapori-
    zation and vapour pressure measurements.

T	B
275.85	-1 101
292.03	- 917
310.50	- 796

ETHANETHIOL (ETHYL MERCAPTAN, ETHYL HYDROSULPHIDE, ETHYL THIOALCOHOL) C_2H_5SH

1. J.P. McCullough, D.W. Scott, H.L. Finke, M.E. Gross, K.D. Williamson,
    R.E. Pennington, G. Waddington, and H.M. Huffman, J. Am. chem. Soc.
    74 2801 (1952).

    Values of B calculated by the authors from measurements of vapour
    pressure and heats of vaporization.

T	B
281.15	-1 066
298.15	- 897
308.15	- 839

2,3-DITHIABUTANE (DIMETHYLDISULPHIDE) $CH_3.S.S.CH_3$

1. W.N. Hubbard, D.R. Douslin, J.P. McCullough, D.W. Scott, S.S. Todd,
    J.F. Messerly, I.A. Hossenlopp, A. George, and G. Waddington,
    J. Am. chem. Soc. <u>80</u> 3547 (1958).

    Values of B calculated by the authors from measurements of vapour
    pressure and heats of vaporization.

T	B
340.80	-1 331
360.40	-1 208
382.90	-1 030

DIMETHYLAMINE $(CH_3)_2NH$

1. J.D. Lambert and E.D.T. Strong, Proc. R. Soc. <u>A200</u> 566 (1950)(†).

    Maximum pressure 600 torr. Accuracy ± 50.

T	B	T	B
312.4	-596	364.6	-404
320.8	-563	374.6	-389
332.2	-505	383.4	-359
342.4	-478	394.3	-337
350.8	-460	405.2	-309

ETHYLAMINE (AMINOETHANE) $C_2H_5NH_2$

1. J.D. Lambert and E.D.T. Strong, Proc. R. Soc. <u>A200</u> 566 (1950)(†).
    Maximum pressure 600 torr. Accuracy ± 50.

T	B	T	B
293.4	-821	363.4	-461
315.0	-681	373.8	-432
323.8	-623	383.6	-408
333.0	-588	393.0	-387
343.3	-547	405.2	-345
351.0	-529		

CYANOGEN (OXALIC ACID DINITRILE) C_2N_2

1. J.D. Hamann, W.J. McManamey, and J.F. Pearse, Trans.Faraday Soc.
    49 351 (1953).

    Class II.

T	B	T	B
308.15	-352.0	373.15	-210.8
323.15	-316.2	398.15	-174.9
348.15	-262.5	423.15	-147.2
373.15	-216.7		

PERFLUORO-n-PROPANE $CF_3.CF_2.CF_3$

1. E.L. Pace and A.C. Plausch, J. chem. Phys. 47 38 (1967)(*).

    B calculated from heats of vaporization and vapour pressure
    measurements.

T	B
236	-998

2. J.A. Brown, J. chem. Engng Data 8 106 (1963).

    PVT data given; P range 17-57 atm, T range 347-439K

3. E.M. Dantzler and C.M. Knobler, J. phys. Chem., Ithaca 73 1335
    (1969).

    B calculated from an equation of state given by J.A. Brown,
    J. chem. Engng Data 8 106 (1963). Values read from their diagram.

T	B	T	B
233.2	-1011	323.2	- 426
248.2	- 870	348.2	- 351
273.2	- 696	373.2	- 300
298.2	- 525		

1,HYDROPERFLUOROPROPANE(1,1,1,2,2,3,3,HEPTAFLUOROPROPANE) $CF_3.CF_2CHF_2$

1. T.B. Tripp and R.D. Dunlap, J. phys. Chem.,Ithaca 66 635 (1962).

    Values of B obtained from compressibility data by use of (1) linear
    and (2) quadratic equation in both (a) pressure and (b) concentration.
    Pressure range 100 - 700 torr.

T	1(a)	1(b)	B 2(a)	2(b)
283.25	-788 ± 11	-759 ± 10	-765 ± 68	-766 ± 64
303.15	-674 ± 14	-653 ± 14	-789 ± 61	-781 ± 75
323.15	-577 ± 8	-561 ± 8	-626 ± 35	-632 ± 33

PROPADIENE (ALLENE, DIMETHYLENEMETHANE) $CH_2:C:CH_2$

1. E.E Roper, J. phys. Chem., Ithaca 44 835 (1940)(*).

    Maximum pressure 1039 torr.

    Class II.

T	B
222.65	-710.5
237.15	-616.1
273.15	-443.5
273.15	-447.0
343.15	-260.9

2. S.D. Hamann, W.J.McManamey, and J.F. Pearse, Trans. Faraday Soc.
    49 351 (1953).

    Maximum pressure 2 atm. Reproducibility ± 4.

T	B	T	B
293.15	-396.8	333.15	-306.8
303.15	-371.8	343.15	-283.7
313.15	-351.4	353.15	-266.9
323.15	-324.8		

PROPYNE (PROPINE, METHYLACETYLENE) $CH_3.C\!:\!CH$

1. S.P. Vohra, T.L. Kang, K.A. Kobe, and J.J. McKetta, J. chem.
    Engng Data 7 150 (1962).

    PV data given (P→315 atm).

    Values of B and C calculated by J. Brewer, J. chem. Engng Data
    10 113 (1965).

    Accuracy of B ± 1.

T	B	C
348.15	-287.8	15 120
373.15	-244.0	15 700
398.15	-210.0	14 750

402.40	-203.6	14 230
408.15	-196.0	13 500
413.15	-191.1	13 420
418.15	-186.0	13 150
423.15	-182.8	13 900
448.15	-161.3	13 350
473.15	-144.0	13 040

CYCLOPROPANE (TRIMETHYLENE) $\underset{\llcorner}{CH_2}\underset{}{CH_2}\underset{\lrcorner}{CH_2}$

1. J.D. Lambert and J.S. Rowlinson (1950)(†).

T	B
352.0	-320

2. S.D. Hamann and J.F. Pearse, Trans. Faraday Soc. 48 101 (1952).
   Maximum pressure 0.5 atm. Reproducible to within 4.

T	B	T	B
303.15	-363.9	343.15	-281.8
303.15	-366.1	343.15	-283.0
323.15	-321.6	343.15	-281.9
323.15	-323.5	363.15	-253.7

3. H.G. David, S.D. Hamann, and R.B. Thomas, Aust. J. Chem. 12 309
   (1959).
   Class I.

T	B	T	B
303.15	-375.0	363.15	-251.4
313.15	-345.0	373.15	-238.2
323.15	-325.0	383.15	-223.2
333.15	-307.0	393.15	-210.4
343.15	-281.5	403.15	-200.5
353.15	-266.0		

PROPENE (PROPYLENE) $CH_2:CH.CH_3$

In the temperature range 300-500K the various sets of data for the second virial coefficient of propene are in fairly good accord. The following values are recommended.

T	B	T	B
280	-395 ± 5	380	-205 ± 5
300	-343 ± 5	420	-163 ± 5
320	-299 ± 5	460	-132 ± 5
340	-262 ± 5	500	-106 ± 3

1. E.E. Roper, J. phys. Chem., Ithaca 44 835 (1940).

    Maximum pressure 1104 torr.

    Class II.

T	B	T	B
223.20	-664.6	273.15	-423.2
226.98	-630.5	273.15	-411.9
273.15	-433.5	308.17	-324.9
273.15	-418.6	343.11	-257.1

2. W.E. Vaughan and N.R. Graves, Ind. Engng Chem. ind. Edn 32 1252 (1940).

    PVT data given: P range 2-80 atm, T range 273 - 573 K.

3. P.S. Farrington and B.H. Sage, Ind. Engng Chem. ind. Edn 41 1734 (1949)(†).

    Class I.

T	B	C
277.60	-401.1	- 1 000
294.27	-362.2	+33 000
310.94	-322.4	44 000
327.60	-283.3	26 000
344.27	-255.5	24 000
355.38	-238.8	25 000
360.94	-229.6	17 000
377.60	-207.2	6 000
410.94	-170.8	10 000
444.27	-140.6	9 000
477.60	-117.6	11 000
510.94	- 98.8	11 000

88

4. H. Marcham, H.W. Prengle, and R.L. Motard, Ind. Engng Chem. ind.Edn
   41 2658 (1949)(†).

   Class I.

T	B	C
323.15	-308.3	49 000
348.15	-260.3	31 000
373.15	-218.7	19 000
398.15	-189.6	18 000
423.15	-164.5	16 000
448.15	-141.5	12 000
473.15	-120.4	8 000
498.15	-109.5	12 000
523.15	- 78.0	12 000

5. A. Michels, T. Wassenaar, P. Louwerse, R.J. Lunbeck, and G.J. Wolkers,
   Physica,'s Grav. 19 287 (1953)(*).

   4-term fit of PV data (P→30 atm).

   Class I.

T	B	C
298.15	-346.78	14 070
318.15	-301.92	15 800
323.15	-292.13	16 130
348.15	-247.73	13 750
373.15	-212.71	12 510
398.15	-184.05	11 270
423.15	-160.36	10 850

6. R.D. Gunn, M.S. Thesis, University of California, Berkeley (1958).
   Values of B given by J.A. Huff and T.M. Reed, J. chem. Engng Data
   8 306 (1963).

   Class I.

T	B
377.6	-207.9
410.9	-170.9
444.3	-143.9
477.6	-121.6

7. D.McA. Mason and B.E. Eakin, J. chem. Engng Data 6 499 (1961)(*).

T	B
288.70	-385.4 ± 3

8. M.L. McGlashan and C.J. Wormald, Trans. Faraday Soc. 60 646 (1964).
    Class I.

T	B	T	B
303.7	-329	353.5	-243
313.1	-308	363.6	-232
323.5	-282	373.4	-213
328.4	-280	384.1	-195
333.6	-257	393.5	-186
338.2	-266	403.2	-174
343.3	-262	413.6	-166
349.1	-244		

9. S.L. Robertson and S.E.Babb,Jr., J. chem. Phys. 51 1357 (1969).
    PVT data given: P range 1000-10000 atm., T range 308-473 K.

10. K. Bier, G. Ernst, J. Kunze and G. Maurer, J. chem. Thermodyn.
    6 1039 (1974).

    B determined from measured heat capacities and Joule-Thomson
    coefficient using B (298.15 K) = -347.0.

T	B	T	B
323.15	-293	373.15	-216
348.15	-250	398.15	-188
365.15	-223	423.15[+]	-163

    + extrapolated

11. W. Warowny and J. Stecki, J. chem. Engng Data 23 212 (1978).
    Burnett method, P → 70 atm.

T	B	
393.20	-189.15,	-190.98
407.50	-174.73,	-176.25
423.02	-161.05,	-162.22

2,2-DICHLOROPROPANE (ACETONE DICHLORIDE, ISOPROPYLIDENE CHLORIDE)
$(CH_3)_2CCl_2$

1. A. Perez Masia, M. Diaz Pena, and J.A Burriel Lluna, An. R. Soc. esp. Fis. Quim. 60B 229 (1964).

T	B	C
353.52	- 1115 ± 15	16 000
372.40	- 941 ± 15	20 000
401.11	- 798 ± 12	16 000
422.88	- 709 ± 16	2 000

PROPANAL $CH_3.CH_2.CHO$

1. D. Ambrose and C.H.S. Sprake, J. chem. Thermodyn. 6 453 (1974).

    B calculated from measured vapour pressures and enthalpies of vaporization.

    Class III.

T	B
286.25	-2210
298.15	-1800
302.69	-1680
321.21	-1340

ACETONE (2-PROPANONE) $(CH_3)_2CO$

There is considerable uncertainty in the second virial coefficient of acetone. Below 360 K there is reasonable agreement between Bottomley et al. (5, 8) Hajjar et al. (11) Knoebel et al. (10). Above 360 K the discrepancies are very large. We favour the less negative values of Bottomley et al. and Hajjar et al. The values obtained by Zaalishvili (7) and Anderson (9) are as much as 200 cm^3 more negative. We have neglected these in producing our recommended values.

T	B	T	B
300	-2000 ± 200	350	-1808 ± 50
310	-1730 ± 100	360	- 960 ± 50
320	-1520 ± 50	400	- 700
330	-1350 ± 50	440	- 490
340	-1200 ± 50	480	- 380

1. J.D. Lambert, G.A.H. Roberts, J.S. Rowlinson, and V.J. Wilkinson,
    Proc. R. Soc. A196 113 (1949)(†).

    Maximum pressure 600 torr. Accuracy ± 50.

T	B	T	B
291.7	-2 075.9	351.0	-1 258.1
319.4	-1 644.1	351.2	-1 257.5
329.5	-1 527.3	361.2	-1 133.6
335.0	-1 406.3	368.4	- 965.4
341.0	-1 362.0	385.0	- 853.9
341.4	-1 342.2	405.6	- 650.6

2. R.E. Pennington and K.A. Kobe, J. Am. chem. Soc. 79 300 (1957).
    Calculated from heat of vaporization and vapour pressure data.

T	B
300.4	-2 030
317.9	-1 580
329.3	-1 370
337.3	-1 230
345.0	-1 130

3. I. Brown and F. Smith, Aust. J. Chem. 13 30 (1960).
    From vapour pressure measurements.

T	B
318.2	-1 660

4. Sh.D. Zaalishvili and L.E. Kolysko, Zh. fiz. Khim. 34 2596 (1960).
    Maximum pressure 1200 torr. Average deviation of B 1 - 3 per cent.

T	B
323.15	-1 535
333.15	-1 370
343.15	-1 200

5. G.A. Bottomley and T.H. Spurling, Nature, London. 195 900 (1962).
    Maximum pressure: 323 K, 200 torr.; 295 K, 100 torr.

T	B
295.2	-2 111
323.2	-1 439

6. W. Kapallo, N. Lund, and K. Schafer, Z. phys. Chem. Frankf. Ausg. 37 196 (1963).
Class II.

T	B
282.3	-2 733
297.0	-2 268
312.0	-1 876
321.0	-1 680

7. Sh.D. Zaalishvili and Z.S. Belousova, Russ. J. phys. Chem. 38 296 (1964); Zh. fiz. Khim. 38 503 (1964).
Class II.

T	B	T	B
333.2	-1 360	363.2	- 874
343.2	-1 206	373.2	- 740
353.2	-1 024	383.2	- 620

8. G.A. Bottomley and T.H. Spurling, Aust. J. Chem. 20 1789 (1967).
Maximum pressure: (a) 97.8 torr, (b) 307.6 torr. Accuracy ± 16.

T	B	T	B
304.64	-1 796 a	377.05	- 808 a
323.16	-1 439 a/b	402.38	- 664 a
326.02	-1 378 b	405.12	- 656 a
326.66	-1 368 a	405.20	- 641 b
350.12	-1 057 b	430.35	- 495 a
351.57	- 981 a	430.88	- 527 b
370.06	- 872 b		

9. L.N. Anderson, A.P. Kudchadker and P.T. Eubank, J. chem. Engng Data 13 321 (1968)(*).

T	B	C
298.15	-2132	6.40×10^7
323.15	-1662	3.82×10^7
348.15	-1207	1.97×10^7
373.15	- 790	0.91×10^7
398.15	- 514	0.51×10^7
423.15	- 357	0.30×10^7

10. D.H. Knoebel and W.C. Edmister, J. chem. Engng Data 13 312 (1968).
    Low pressure PVT measurements.

T	B
313.2	-1690
333.2	-1263
353.2	-1005
373.2	- 834

11. R.F. Hajjar, W.B. Kay, and G.F. Leverett, J. chem. Engng Data
    14 377 (1969).
    Low pressure PVT measurements.

T	B	T	B
313.15	-1575	398.15	- 700
323.15	-1375	413.15	- 630
337.15	-1200	428.15	- 555
353.15	-1065	453.15	- 455
368.15	- 920	473.15	- 400
382.15	- 800		

12. R.M. Keller, Jr., and L.I. Stiel, J. chem. Engng Data 22 241 (1977).
    PVT data given: T 506.9 - 538.5 K, P 92 - 364 atm.

ETHYL FORMATE $HCOO.C_2H_5$

1. J.D. Lambert, J.S. Clarke, J.F. Duke, C.L. Hicks, S.D. Lawrence,
   D.M. Morris, and M.G.T. Shone, Proc. R. Soc. A249 414 (1959).
   Accuracy ± 50.

T	B	T	B
323.0	-1 090	360.0	- 760
329.8	-1 000	368.2	- 730
333.2	- 960	370.0	- 700
337.7	- 930	382.0	- 650
344.2	- 875	388.2	- 625
351.8	- 830	394.8	- 590
353.2	- 850		

METHYL ACETATE $CH_3.COO.CH_3$

1.   J.D. Lambert, J.S. Clarke, J.F. Duke, C.L. Hicks, S.D. Lawrence,
     D.M. Morris, and M.G.T. Shone, Proc. R. Soc. A249 414 (1959).
     Accuracy ± 50.

T	B	T	B
323.2	-1 240	351.8	- 980
328.4	-1 210	353.2	- 960
330.9	-1 190	357.8	- 900
335.6	-1 150	359.8	- 890
335.7	-1 175	366.2	- 830
336.3	-1 115	368.2	- 850
338.2	-1 080	374.3	- 820
338.6	-1 100	376.9	- 795
340.3	-1 045	378.8	- 810
345.5	-1 000	383.6	- 785
347.4	- 990	391.0	- 740

n-PROPYL BROMIDE (1-BROMOPROPANE) $CH_3.CH_2.CH_2Br$

1.   A.R. Paniego, J.A.B. Lluna, and J.E.H. Garcia, An. Quim. 71 349
     (1975).
     Relative compressibility method using nitrogen.
     Class II.

T	B
353.6	-1070
374.1	- 958
392.6	- 847
412.2	- 776

ISOPROPYL BROMIDE (2-BROMOPROPANE) $CH_3.CHBr.CH_3$

1.   A.R. Paniego, J.A.B. Lluna, and J.E.H. Garcia, An. Quim. 71 349
     (1975).
     Relative compressibility method using nitrogen.
     Class II.

T	B
338.5	-1004
353.7	- 901
374.1	- 814
392.6	- 724

n-PROPYL CHLORIDE (1-CHLOROPROPANE) $CH_3.CH_2.CH_2Cl$

1.   A. Perez Masia and M. Diaz Pena, An. R. Soc. esp. Fis. Quim.
     54B 661 (1958).

     Class II.

T	B
323.15	-873.4
348.15	-706.7
373.15	-611.7
398.15	-540.6

2.   M. Ratzsch, Z. phys. Chem. 238 321 (1968).

T	B
303.2	-1031
313.2	- 931
333.2	- 768

3.   K. Bohmhammel and W. Mannchen, Z. phys. Chem. 248 230 (1971).

T	B	T	B
310.2	-999.7	443.2	-387.8
318.6	-926.1	478.7	-321.3
344.8	-745.0	483.2	-308.1
348.5	-713.1	503.2	-284.5
371.4	-610.8	510.3	-272.4
389.4	-542.5	538.0	-241.4
397.7	-508.6	548.6	-226.8
412.4	-457.2	578.0	-200.3
432.2	-410.3		

ISOPROPYL CHLORIDE (2-CHLOROPROPANE) $CH_3CHClCH_3$

1. A. Perez Masia and M. Diaz Pena, An. R. Soc. esp. Fis. Quim. 54B 661 (1958).

    Class II.

T	B
313.15	-869.7
333.15	-752.5
353.15	-680.6
373.15	-603.8
398.15	-509.9

PROPANE $CH_3CH_2CH_3$

The different sets of values for the second virial coefficient of propane are in reasonable agreement, except for the results of Deschner and Brown (5) below 500 K which are significantly more negative than the values obtained by other workers. These results have been neglected in obtaining the following smoothed values.

T	B	T	B
240	-640 ± 20	350	-276 ± 10
250	-584 ± 20	375	-238 ± 10
260	-526 ± 20	400	-208 ± 10
270	-478 ± 20	430	-177 ± 5
285	-424 ± 20	470	-143 ± 5
300	-382 ± 15	500	-124 ± 5
315	-344 ± 10	550	- 97 ± 5
330	-313 ± 10		

1. G. Glocker, D.L. Fuller, and C.P. Roe, J. chem. Phys. 1 709 (1933) (*).

    PVT data given: P → 40 atm.

    Class II.

T	B
368.7	-242
389.5	-220

2. B.H. Sage, J.G. Schaafsma, and W.N. Lacey, Ind. Engng. Chem. ind. Edn. 26 1218 (1934).

    PVT data given: P range 1 - 200 atm, T range 294 - 377 K.

3. J.A. Beattie, W.C. Kay, and J. Kaminsky, J. Am. chem. Soc.
   59 1589 (1937).

   PVT data given. Maximum pressure 310 atm.

   Values of B calculated by J.O. Hirschfelder, F.T. McClure, and
   I.F. Weeks, J. chem. Phys. 10 201 (1942).

T	B	T	B
369.96	-260	473.15	-139
373.15	-247	498.15	-121
398.15	-211	523.15	-109
423.15	-183	548.15	- 96
448.15	-160		

4. F.W. Jessen and J.H. Lightfoot, Ind. Engng Chem. ind. Edn 30 312
   (1938)(*).

T	B
273.15	-470
323.15	-325 ± 10

5. W.W. Deschner and G.G. Brown, Ind. Engng Chem. ind. Edn 32 836
   (1940).

   PVT data given. Maximum pressure 140 atm.

   Values of B calculated by J.O. Hirschfelder, F.T. McClure, and
   I.F. Weeks, J. chem. Phys. 10 201 (1942).

T	B	T	B
303.15	-384	423.15	-197
348.15	-293	473.15	-155
373.15	-256	526.37	-108
398.15	-224	570.45	- 89

6. B.J. Cherney, H. Marchman, and R. York, Jr., Ind. Engng Chem. ind.
   Edn 41 2653 (1949).

   PVT data given: P range 10 - 50 atm, T range 323 - 398 K.

7. H.H. Reamer, B.H. Sage, and W.N. Lacey, Ind. Engng. Chem. ind. Edn
   41 482 (1949)(†).

   PVT data given. Maximum pressure 700 atm.

   Class I.

98

T	B	C
310.94	-333.0	-65 000
327.60	-306.0	-14 000
344.27	-280.6	+ 2 000
360.94	-256.8	13 000
377.60	-234.6	9 000
410.94	-199.6	25 000
444.27	-168.4	23 000
477.60	-142.0	21 000
510.94	-119.0	23 000

8. G.A. Bottomley, D.S. Massie, and R. Whytlaw-Gray, Proc. R. Soc. A200 201 (1950)(*).

T	B
295.21	-407.9

9. C.B. Kretschmer and R. Wiebe, J. Am. chem. Soc. 73 3778 (1951)(*).

T	B
303.16	-395 ± 11

10. R.D. Gunn, M.S. Thesis, University of California, Berkeley (1958). Values of B given by J.A. Huff and T.M. Reed, J. chem. Engng Data 8 306 (1963).
Class I.

T	B
310.9	-335.8
344.3	-280.4
377.6	-235.9
444.3	-167.0
510.9	-117.2

11. M.L. McGlashan and D.J.B. Potter, Proc. R. Soc. A267 478 (1962). Precision of B values ± 5.

T	B	T	B
295.4	-399	337.8	-299
306.5	-369	347.9	-274
317.6	-339	357.9	-265
327.6	-324	368.2	-244

377.7	-229	400.1	-201
388.5	-213	412.9	-182

12. W. Kapallo, N. Lund, and K. Schafer, Z. phys. Chem. Frankf. Ausg. 37 196 (1963).
    Class II.

T	B
244.0	-610
273.0	-477
297.0	-394
321.0	-340

13. R.N. Lichtenthaler, B. Schramm, and K. Schäfer, Ber. (dtsch.) Bunsenges. phys. Chem. 73 36 (1969).
    Class I.

T	B
296.0	-404.2

14. R.N. Lichtenthaler and K. Schäfer, Ber. (dtsch.) Bunsenges. phys. Chem. 73 42 (1969).
    Estimated absolute error in B ± 1.

T	B
288.2	-428.4
296.0	-404.2
303.2	-383.5
313.2	-361.4
322.8	-338.2

15. S.E. Babb, Jr. and S.L. Robertson, J. chem. Phys. 53 1097 (1970).
    PVT data given: P range 1500 - 11000 atm. T range 308 - 473 K.

16. K. Strein, R.N. Lichtenthaler, B. Schramm, and Kl. Schäfer, Ber. (dtsch.) Bunsenges. phys. Chem. 75 1308 (1971).
    Estimated accuracy of B is ± 1%.

T	B	T	B
296.1	-396.0	413.8	-191.1
308.0	-360.5	433.1	-172.6
332.9	-309.0	453.5	-156.2
353.8	-271.2	474.9	-140.0
373.4	-242.1	493.3	-128.1
394.0	-213.7		

17. R. Hahn, K. Schäfer, and B. Schramm, Ber. (dtsch.) Bunsenges. phys. Chem. <u>78</u> 287 (1974).

B determined assuming B (296.1 K) = -396.

Quoted accuracy of B is ± 2.

T	B
211.3	-844
231.2	-680
251.5	-567
273.8	-471

18. K. Schäfer, B. Schramm, and J.S.U. Navarro, Z. phys. Chem. Frankf. Ausg. <u>93</u> 203 (1974).

Values of B obtained relative to the value at 296K.

T	B	T	B
296[†]	-404	432.3	-182.9
353.2	-281.0	473.0	-149.4
392.9	-225.2	511.8	-122.3

[†] reference value.

19. W. Warowny and J. Stecki, J. chem. Engng Data <u>23</u> 212 (1978).

Burnett method, P → 70 atm.

T	B	
393.18	-212.31,	-214.76
407.49	-194.69,	-198.19
423.00	-179.91,	-181.29

n-PROPANOL (1-PROPANOL) $CH_3 \cdot (CH_2)_2 OH$

1. O.R. Fox, J. Morcillo, and A. Mendez, An. R. Soc. esp. Fis. Quim. <u>17B</u> 23 (1954).

Values read from diagram.

Class II.

T	B
350	-1 500
371	-1 150
402	- 860

2.   J.D. Cox, Trans. Faraday Soc. <u>57</u> 1674 (1961).
     Maximum pressure 560 torr.

T	B
378.2	-890 ± 8
393.2	-763 ± 6
408.2	-655 ± 9
423.2	-596 ± 14

2-PROPANOL (ISO-PROPANOL) $CH_3.CHOH.CH_3$

1.   O.R. Fox, J. Morcillo, and A. Mendez, An. R. Soc. esp. Fis. Quim.
     <u>17B</u> 23 (1954).
     Values read from diagram.
     Class II.

T	B
350	-1 330
371	- 920
380	- 820
402	- 610

2.   C.B. Kretschmer and R. Wiebe, J. Am. chem. Soc. <u>76</u> 2579 (1954).
     Low-pressure vapour density method. Data fitted to 3-term P series
     (P^0, P^1, and P^3). Uncertainty in B is ± 5 at 373.15 K and above,
     but increases to about ± 70 at 333.15 K.

T	B
333.15	-1609
353.15	-1137
373.15	- 890
393.15	- 721

3.   J.D. Cox, Trans. Faraday Soc. <u>57</u> 1674 (1961).
     Maximum pressure 760 torr.

T	B
378.2	-844 ± 11
393.2	-685 ± 11
408.2	-589 ± 9
423.2	-521 ± 9

4. M.P. Moreland, J.J. McKetta, and I.H. Silberberg, J. chem. Engng Data, <u>12</u> 329 (1967).

Errors in B less than 2 per cent; errors in C less than 5 per cent or ± 2000 whichever is greater.

T	B	C
373.15	-870.8	702 000
373.15	-874.2	669 000
398.15	-716.0	439 000
398.15	-714.5	312 000
398.15	-714.5	264 000
423.15	-558.4	83 000
423.15	-557.8	76 000
448.15	-459.9	9 000
448.15	-454.1	17 000
473.15	-390.0	24 000
473.15	-387.5	28 000

5. J.K. Tseng and L.I. Stiel, A. I. Ch. E. Jl. <u>17</u> 1283 (1971).

PVT data given: P range 65 - 530 atm, T range 473 - 573 K.

1-PROPANETHIOL (n-PROPYL MERCAPTAN) $CH_3(CH_2)_2SH$.

1. R.E. Pennington, D.W. Scott, H.L. Finke, J.P. McCullough, J.F. Messerly, I.A. Hossenlopp, and G. Waddington, J. Am. chem. Soc. <u>78</u> 3266 (1956).

Values of B calculated by the authors from measurements of vapour pressure and heats of vaporization.

T	B
303.02	-1 387
320.63	-1 186
340.87	-1 012

2-PROPANE THIOL (ISOPROPYL MERCAPTAN) $CH_3.CH(SH)CH_3$

1. J.P. McCullough, H.L. Finke, D.W. Scott, M.E. Gross, J.F. Messerly, R.E. Pennington, and G. Waddington, J. Am. chem. Soc. <u>76</u> 4796 (1954).

Values of B calculated by the authors from heats of vaporization and vapour pressure measurements.

T	B
290.40	-1 356
306.19	-1 180
325.71	-1 019

METHYL ETHYL SULPHIDE $CH_3.S.C_2H_5$

1.   G.A. Bottomley and I.H. Coopes, Aust. J. Chem. 15 190 (1962).

T	B
299.15	-1 458 ± 10
339.15	-1 015 ± 9

2.   D.W. Scott, H.L. Finke, J.P. McCullough, M.E. Gross, K.D. Williamson, G. Waddington, and H.M. Huffmann, J. Am. chem. Soc. 73 261 (1951).

B values calculated by the authors from heats of vaporization and vapour pressure measurements.

T	B
301.65	-1 568
319.75	-1 366
339.80	-1 246

TRIMETHYLAMINE $(CH_3)_3N$

1.   J.D. Lambert and E.D.T. Strong, Proc. R. Soc. A200 566 (1950)(†). Maximum pressure 600 torr.   Accuracy ± 50.

T	B	T	B
311.2	-672	343.2	-532.5
313.8	-655	351.0	-508
323.8	-605	363.6	-469
335.0	-573	374.7	-437

2.   W.S. Haworth and L.E. Sutton, Trans. Faraday Soc. 67 2907 (1971). Estimated precision of B ± 10.

T	B
298.2	-764

METHYL BORATE $B(OCH_3)_3$

1.   R.G. Kunz and R.S. Kapner, J. chem. Engng Data <u>14</u> 190 (1969).
     B values calculated from PVT data given by R.G. Griskey,
     W.E.Gorgas, and L.N. Canjar, A. I. Ch. E. Jl <u>6</u> 128 (1960).

T	B
498.15	-380.7
523.15	-338.3
548.15	-301.3
573.15	-267.3

TRIFLUOROACETIC ANHYDRIDE $(CF_3.CO)_2O$

1.   D. Wyrzykowska-Stankiewicz and A. Kreglewski, Bull. Acad. pol. Sci.
     Ser. Sci. chim. <u>7</u> 417 (1963).

T	B
313.15	-1 550 ± 40
303.15	-1 980 ± 150
293.15	-2 360 ± 220

PERFLUOROCYCLOBUTANE $\underline{CF_2.CF_2.CF_2.CF}_2$

1.   D.R. Douslin, R.T. Moore, and G. Waddington, J. phys. Chem.,
     Ithaca <u>63</u> 1959 (1959).

Maximum pressure 394 atm. 3-term fit.

T	B	C
373.15	-434	- -
388.37	-391	46 400
398.15	-366	45 300
423.15	-313	41 900
448.15	-268	38 500
473.15	-229	34 700
498.15	-196	31 400
523.15	-167	28 100
548.15	-142	25 800
573.15	-120	23 500
598.15	-100	21 600
623.15	- 82	20 100

PERFLUORO n-BUTANE $CF_3.CF_2.CF_2.CF_3$

1. J.A. Brown and W.H. Mears, J. phys. Chem., Ithaca 62 960 (1958).
    PVT data given; T range 338 - 448 K, P → 53 atm.

2. T.B. Tripp and R.D. Dunlap, J. phys. Chem., Ithaca 66 635 (1962).
    Values obtained from compressibility data by use of (1) linear
    (2) quadratic equations in both (a) pressure and (b) concentration
    (P → 1 atm.)

T	B			
	1(a)	1(b)	2(a)	2(b)
283.16	-1 164 ± 14	-1 098 ± 9	-1 018 ± 45	-1 030 ± 40
303.04	- 942 ± 19	- 900 ± 18	- 894 ± 120	- 899 ± 110
323.21	- 800 ± 6	- 770 ± 6	- 772 ± 39	- 771 ± 37

2-CHLOROTHIOPHENE Cl.C:CH.CH:CH.S

1. C. Eon, C. Pommier and G. Guiochon, J. chem. Engng Data 16 408(1971).
    Isoteniscopic method: estimated uncertainty in B ± 4%.

T	B
333.45	-2060
343.45	-1930
353.45	-1830
363.45	-1730
373.45	-1650

FURAN (1,4-EPOXY-1,3-BUTADIENE, FURFURAN) CH:CH.CH:CH.O

1. G.B. Guthrie, Jr., D.W. Scott, W.N. Hubbard, C. Katz, J.P. McCullough,
    M.E. Gross, K.D. Williamson, and G. Waddington, J. Am. chem. Soc.
    74 4662 (1952).

    Values of B calculated by the authors from measurements of vapour
    pressures and heats of vaporization.

T	B
279.15	-975
293.15	-851
304.51	-789

2. C. Eon, C. Pommier and G. Guiochon, J. chem. Engng Data <u>16</u> 408 (1971).

Isoteniscopic method: estimated uncertainty in B ± 4%.

T	B
333.45	-645
343.45	-605
353.45	-570
363.45	-535
373.45	-500

THIOPHENE (THIOFURAN) SCH:CH.CH:CH

1. G. Waddington, J.W. Knowlton, D.W. Scott, G.D. Oliver, S.S. Todd, W.N. Hubbard, J.C. Smith, and H.M. Huffman, J. Am. chem. Soc. <u>71</u> 797 (1949).

Values of B calculated by the authors from measurements of vapour pressure and heats of vaporization.

T	B
318.51	-1 173
336.23	-1 025
357.31	- 957

2. C. Eon, C. Pommier and G. Guiochon, J. chem. Engng Data <u>16</u> 408 (1971).

Isoteniscopic method: estimated uncertainty in B ± 4%.

T	B
333.45	-1020
343.45	- 970
353.45	- 915
363.45	- 865
373.45	- 820

PYRROLE CH:CH.CH:CH.NH

1. D.W. Scott, W.T. Berg, I.A. Hossenlopp, W.N. Hubbard, J.F. Messerly, S.S. Todd, D.R. Douslin, J.P. McCullough, and G. Waddington, J. phys. Chem.,Ithaca <u>71</u> 2263 (1967).

B calculated from heats of vaporization and vapour pressure measurements.

T	B
362.11	-1759
381.21	-1456
402.91	-1185

2. C. Eon, C. Pommier and G. Guiochon, J. chem. Engng Data **16** 408 (1971).
   Isoteniscopic method: estimated uncertainty in B ± 4%.

T	B
333.45	-1270
343.45	-1160
353.45	-1070
363.45	- 990
373.45	- 880

1-BUTYNE $CH\dot{:}C.CH_2.CH_3$

1. J.G. Aston, S.V.R. Mastrangelo, and G.W. Moessen, J. Am. chem. Soc. **72** 5287 (1950).

   Accuracy ± 10 percent. Values of B calculated by the authors from measurements of vapour pressure and heats of vaporization.

T	B	T	B
262	- 980	278	- 980
266	-1 001	282	- 940
270	-1 100	298	- 748
274	-1 000		

2,5-DIHYDROFURAN $CH_2.CH.CH.CH_2O$

1. C. Eon, C. Pommier, and G. Guiochon, J. chem. Engng Data **16** 408 (1971).

   Isoteniscopic method: estimated uncertainty in B ± 4%.

T	B
333.45	-670
343.45	-630
353.45	-595
363.45	-565
373.45	-535

BUT-1-ENE(1-BUTENE) $CH_3.CH_2.CH:CH_2$

1. E.E. Roper, J. phys. Chem., Ithaca <u>44</u> 835 (1940)(*)
    Maximum pressure 851 torr.
    Class II.

T	B
243.37	-1 188.6
273.15	- 793.4
273.15	- 793.8
294.25	- 669.0
333.16	- 507.7

2. M.L. McGlashan and C.J. Wormald, Trans. Faraday Soc. <u>60</u> 646 (1964).
    Class I.

T	B	T	B
304.2	-606	363.5	-400
313.5	-559	373.7	-381
323.2	-533	383.6	-363
333.1	-494	392.7	-334
343.8	-457	403.5	-321
353.5	-432	420.1	-295

3. R.H. Olds, B.H. Sage, and W.N. Lacey, Ind. Engng Chem. ind. Edn
    <u>38</u> 301 (1946).
    PVT data given: P range 1 - 680 atm, T range 311 - 444 K.

4. J.A. Beattie and S. Marple, Jr., J. Am. chem. Soc. <u>72</u> 4143 (1950).
    PVT data given: P range 25 - 250 atm, T range 423 - 523 K.

2-METHYL PROPENE (ISOBUTYLENE) $CH_2:C(CH_3)_2$

1. E.E. Roper, J. phys. Chem., Ithaca <u>44</u> 835 (1940)(*).
    Maximum pressure 850 torr.
    Class II.

T	B
243.28	-1 190.6
273.15	- 803.7
273.15	- 815.7
333.14	- 508.4

CIS-2-BUTENE (CIS-β-BUTYLENE) $CH_3.CH:CH.CH_3$

1.   E.E. Roper, J. phys. Chem., Ithaca <u>44</u> 835 (1940)(*).
     Maximum pressure 822 torr.
     Class II.

T	B	T	B
250.92	-1 243.2	273.15	- 968.5
259.93	-1 069.1	308.16	- 656.0
273.15	- 869.6	333.13	- 557.8
273.15	- 902.5	343.09	- 503.2
273.15	- 936.2		

TRANS-2-BUTENE (TRANS-β-BUTYLENE) $CH_3.CH:CH.CH_3$

1.   E.E. Roper, J. phys. Chem.,Ithaca <u>44</u> 835 (1940)(*).
     Maximum pressure 839 torr.
     Class II.

T	B	T	B
243.29	-1 230.2	273.15	- 943.9
250.91	-1 130.4	308.16	- 665.9
273.15	- 885.0	308.16	- 656.8
273.15	- 900.0	308.16	- 665.2
273.15	- 908.3	333.14	- 593.2
273.15	- 911.4	333.14	- 545.2

TETRAHYDROFURAN $\underset{\llcorner\quad\quad\quad\lrcorner}{OCH_2CH_2CH_2CH_2}$

1.   C. Treiner, J.F. Bocquet and M. Chemla, J. Chim. phys. <u>70</u> 72 (1973).
     B determined from mass and pressure measurements on the vapour in a
     cell of known value.

T	B
304.58	-1092 ± 25

METHYL ETHYL KETONE (2-BUTANONE) $CH_3.CO.C_2H_5$

1.   J.K. Nickerson, K.A. Kobe, and J.J. McKetta, J. phys. Chem.,
     Ithaca, 65 1037 (1961).

     B determined from vaporization data.   Accuracy ± 2.

T	B
314.61	-1 968
338.69	-1 589
352.54	-1 362
362.58	-1 228
370.57	-1 130

BUTYRIC ACID (BUTANOIC ACID, ETHYLACETIC ACID) C_3H_7COOH

1.   R.E. Lundin, F.E. Harris, and L.K. Nash, J. Am. chem. Soc. 74 743
     (1952).

     PVT data given: P less than 1 atm., T range 415 - 473 K.

     Vapour density method.

n-PROPYL FORMATE $H.COO.CH_2.CH_2.CH_3$

1.   J.D. Lambert, J.S. Clarke, J.F. Duke, C.L. Hicks, S.D. Lawrence,
     D.H. Morris, and M.G.T. Shone, Proc. R. Soc. A249 414 (1959).

     Accuracy ± 50.

T	B	T	B
328.6	-1 530	353.2	-1 190
332.2	-1 430	361.1	-1 140
333.2	-1 440	368.2	-1 060
334.7	-1 420	368.8	-1 030
337.6	-1 390	377.1	- 960
342.3	-1 370	385.0	- 900
345.1	-1 300	391.6	- 880
347.4	-1 270	397.1	- 860
351.4	-1 190	400.1	- 845

ETHYL ACETATE $CH_3.COO.C_2H_5$

1.   J.D. Lambert, J.S. Clarke, J.F. Duke, C.L. Hicks, S.D. Lawrence,
     D.M. Morris, and M.G.T. Shone, Proc. R. Soc. A249 414 (1959).

     Accuracy ± 50.

T	B	T	B
330.2	-1 550	351.2	-1 250
333.5	-1 490	353.2	-1 240
336.8	-1 470	358.6	-1 160
337.7	-1 360	367.5	-1 060
338.2	-1 440	368.2	-1 080
343.2	-1 300	375.1	-1 040
345.7	-1 300	382.6	- 950
346.4	-1 300	390.2	- 930
348.2	-1 260	398.7	- 875

METHYL PROPIONATE $CH_3.CH_2.COO.CH_3$

1.   J.D. Lambert, J.S. Clarke, J.F. Duke, C.L. Hicks, S.D. Lawrence,
     D.M. Morris, and M.G.T. Shone, Proc. R. Soc. A249 414 (1959).

     Accuracy ± 50.

T	B	T	B
328.6	-1 610	353.2	-1 240
333.2	-1 540	360.8	-1 230
335.2	-1 530	368.2	-1 110
339.6	-1 455	370.0	-1 130
343.1	-1 390	381.7	-1 070
347.0	-1 340	391.4	- 940
351.6	-1 310	398.9	- 905

THIACYCLOPENTANE (TETRAHYDROTHIOPHENE) $CH_2.CH_2CH_2CH_2.S$

1.   W.N. Hubbard, H.L. Finke, D.W. Scott, J.P. McCullough, C. Katz,
     M.E. Gross, J.F. Messerly, R.E. Pennington, and G. Waddington,
     J. Am. chem. Soc. 74 6025 (1952).

     B calculated from heats of vaporization and vapour pressure
     measurements.

T	B
349.86	-1540
370.16	-1181
394.28	-1074

2. C. Eon, C. Pommier and G. Guiochon, J. chem. Engng Data 16 408 (1971). Isoteniscopic method: estimated uncertainty in B ± 4%.

T	B
333.45	-910
343.45	-875
353.45	-840
363.45	-760
373.45	-710

2-BROMO BUTANE $CH_3.CHBr.CH_2.CH_3$

1. A.R. Paniego, J.A.B. Lluna, and J.E. Garcia, An. Quim. 71 349 (1975). Relative compressibility method using nitrogen. Class II.

T	B
363.4	-1197
378.9	-1082
392.7	-1021
407.6	- 930

1-BROMO-2-METHYL PROPANE $CH_2Br.CH(CH_3)_2$

1. A.R. Paniego, J.A.B. Lluna, and J.E.H. Garcia, An. Quim. 71 349 (1975).
Relative compressibility method using nitrogen.
Class II.

T	B
368.5	-1063
384.1	- 985
397.5	- 898
412.6	- 843

n-BUTYL CHLORIDE (1-CHLORO BUTANE) $CH_3.CH_2.CH_2.CH_2Cl$

1. A. Perez Masia and M. Diaz Pena, An. R. Soc. esp. Fis. Quim. 54B
    661 (1958).

    Class II.

T	B
358.15	-1 025.9
373.15	- 924.4
423.15	- 672.9

2. K. Bohmhammel and W. Mannchen, Z. phys. Chem. 248 230 (1971).

T	B	T	B
329.6	-1227.3	459.9	- 528.8
331.2	-1212.3	478.7	- 472.8
358.9	- 977.4	488.1	- 451.8
367.9	- 980.9	501.2	- 424.1
377.2	- 853.6	519.8	- 392.1
398.5	- 738.2	539.4	- 358.1
407.3	- 706.1	544.5	- 343.2
412.4	- 677.1	573.3	- 305.3
444.5	- 568.3		

ISOBUTYL CHLORIDE $(CH_3)_2CHCH_2Cl$

1. A. Perez Masia and M. Diaz Pena, An. R. Soc. esp. Fis. Quim. 54B
    661 (1958).

    Class II.

T	B
348.15	-1 021.3
373.15	- 953.7

2-CHLORO 2-METHYL PROPANE (TERT-BUTYL CHLORIDE) $(CH_3)_3CCl$

1. A. Perez Masia, M. Diaz Pena, and J.A. Burriel Lluna, An. R. Soc.
    esp. Fis. Quim. 60B 229 (1964).

T	B	C
323.11	-995 ± 25	-19 000
348.18	-871 ± 15	- 3 000
373.26	-773 ± 17	+ 6 000
396.86	$^-$701 ± 13	13 000

2.   M. Rätzsch, Z. phys. Chem. <u>238</u> 321 (1968).

T	B
313.2	-1504
333.2	-1051

PYRROLIDINE (1-AZACYCLOPENTANE, TETRAHYDROPYRROLE) $NH.CH_2CH_2.CH_2.CH_2$

1.   J.P. McCullough, D.R. Douslin, W.N. Hubbard, S.S. Todd, J.F.
Messerly, I.A. Hossenlopp, F.R. Frow, J.P. Dawson, and
G. Waddington, J. Am. chem. Soc. <u>81</u> 5884 (1959).

Values of B calculated by the authors from measurements of vapour
pressure and heats of vaporization.

T	B
321.90	-1 399
339.50	-1 180
359.72	-1 001

n-BUTANE $CH_3(CH_2)_2CH_3$

The reported second virial coefficients of n-butane generally agree to
within the estimated experimental uncertainties, though the values
reported by Schafer, Schramm, and Navarro (17) are significantly more
negative than the results of other workers. The following values are
recommended.

T	B	T	B
250	-1170 ± 30	380	- 419 ± 20
260	-1050 ± 30	400	- 370 ± 20
270	- 950 ± 20	420	- 332 ± 15
280	- 862 ± 20	440	- 298 ± 15
290	- 788 ± 20	470	- 256 ± 10
300	- 722 ± 20	500	- 219 ± 10
320	- 620 ± 20	530	- 188 ± 10
340	- 535 ± 20	560	- 164 ± 10
360	- 472 ± 20		

1. B.H. Sage, D.C. Webster, and W.N. Lacey, Ind, Engng Chem. ind. Edn 29 1188 (1937).

   PVT data given: P range 1 - 200 atm, T range 294 - 377 K.

2. F.W. Jessen and J.H. Lightfoot, Ind. Engng Chem. ind. Edn 30 312 (1938)(*).

T	B
303.15	-789

3. J.A. Beattie and W.H. Stockmayer, J. chem. Phys. 10 473 (1942). 2-term fit of PV data. Maximum pressure 350 atm. Class I.

T	B	T	B
423.15	-328.7	523.15	-198.1
448.15	-287.3	548.15	-176.0
473.15	-254.2	573.15	-157.4
498.15	-224.5		

4. R.H. Olds, H.H. Reamer, B.H. Sage, and W.N. Lacey, Ind. Engng Chem. ind. Edn 36 282 (1944).

   PVT data given: P range 1 - 680 atm, T range 311 - 511 K.

5. H.W. Prengle, Jr., L.R. Greenhaus, and R. York, Jr., Chem. Engng Prog. 44 863 (1948).

   PVT data given: P range 0 - 300 atm, T range 277 - 556 K.

6. C.B. Kretschmer and R. Wiebe, J. Am. chem. Soc. 73 3778 (1951)(*). Estimated error in B less than 12.

T	B
303.15	-761

7. R.D. Gunn, M.S. Thesis, University of California, Berkeley (1958). Values of B given by J.A. Huff and T.M. Reed, J. chem. Engng Data 8 306 (1963). Class I.

T	B	T	B
344.3	-505.7	444.3	-293.4
377.6	-424.9	460.9	-272.2
410.9	-353.6	477.6	-245.9
427.6	-322.1	510.9	-199.9

8. D. McA. Mason and B.E. Eakin, J. chem. Engng Data 6 499 (1961)(*).

T	B
288.70	-817.5 ± 5

9. M.L. McGlashan and D.J.B. Potter, Proc. R. Soc. A267 478 (1962).
    Errors in B less than ± 20.

T	B	T	B
296.4	-720	358.4	-466
307.5	-667	368.4	-440
318.2	-619	377.9	-410
328.9	-568	387.6	-383
337.8	-533	400.4	-353
348.4	-501	413.4	-322

10. T.B. Tripp amd R.D. Dunlap, J. phys. Chem., Ithaca 66 635 (1962).
     Values of B obtained from compressibility data by use of (1) linear and
     (2) quadratic equations in both (a) pressure and (b) concentration.
     Pressure range 100 - 700 torr.

T	1(a)	1(b)	B 2(a)	2(b)
283.15	-881 ± 11	-846 ± 10	-862 ± 68	-862 ± 63
303.03	-745 ± 6	-715 ± 5	-691 ± 27	-695 ± 25
323.20	-641 ± 12	-619 ± 11	-599 ± 72	-602 ± 69

11. W. Kapallo, N. Lund, and K. Schafer, Z. phys. Chem. Frankf. Ausg.
     37 196 (1963).
     Class II.

T	B	T	B
244.0	-1 230	305.6	- 718
273.4	- 923	312.0	- 674
282.3	- 862	321.0	- 635
297.0	- 758		

12. G.A. Bottomley and T.H. Spurling, Aust. J. Chem. <u>17</u> 501 (1964).
    Estimated error ± 4, possibly greater at 273.06 K.

T	B	T	B
273.06	-897	346.46	-522
297.14	-735	370.86	-449
323.16	-606	397.34	-389
325.68	-595	426.37	-331

13. A.E. Jones and W.B. Kay, A. I. Ch. E. Jl <u>13</u> 720 (1967).
    Errors in B ± 2.0.

T	B	T	B
368.25	-444.2	423.14	-326.1
373.22	-429.5	448.18	-284.8
378.18	-418.0	473.21	-256.3
398.14	-376.0	498.20	-228.7

14. R.N. Lichtenthaler and K. Schäfer, Ber. (dtsch.) Bunsenges. phys.
    Chem. <u>73</u> 42 (1969)
    Estimated error in B is ± 1, ± 0.5%.

T	B
288.2	-816.9
296.0	-765.8
303.2	-727.8
313.2	-669.2
323.1	-621.2

15. K. Strein, R.N. Lichtenthaler, B. Schramm, and Kl. Schäfer,
    Ber. (dtsch.) Bunsenges. phys. Chem. <u>75</u> 1308 (1971).
    Estimated accuracy of B is ± 1%.

T	B	T	B
296.1	-743.0	393.8	-385.9
309.5	-661.0	413.0	-345.9
334.6	-555.7	433.3	-314.3
353.1	-489.2	472.8	-253.3
374.2	-431.8	498.0	-220.3

16. T.R. Das, C.O. Reed, Jr., and P.T. Eubank, J. chem. Engng Data
    <u>18</u> 244 (1973).

    Values of B calculated by authors from analysis of available PVT
    data.

T	B	T	B
280	-902	400	-371
290	-825	420	-329
300	-757	440	-294
320	-644	470	-250
340	-554	500	-215
360	-481	530	-186
380	-421	560	-161

17. K. Schäfer, B. Schramm, and J.S.U. Navarro, Z. phys. Chem. Frankf.
    Ausg. <u>93</u> 203 (1974).

    Values of B obtained relative to the value at 296K.

T	B	T	B
296[†]	-766	432.2	-365
353.1	-525	472.5	-310
393.3	-427	511.1	-247

[†] reference value.

18. G.A. Bottomley and D.B. Nairn, Aust. J. Chem. <u>30</u> 1645 (1977).
    B determined relative to B (316.18 K) = -613.
    Precision of B is between ± 5 and ± 15.

T	B	T	B
316.18	-612.3	427.88	-303.8
341.75	-512.0	427.88	-306.5
341.83	-513.3	462.69	-248.6
341.54	-503.6	462.86	-246.6
341.49	-518.9	463.32	-243.9
367.13	-432.3	498.81	-213.9
367.83	-435.3	498.98	-201.4
367.99	-429.7	537.16	-167.1
367.09	-432.9	537.32	-160.5
396.39	-368.6	537.38	-158.1
396.46	-367.6	579.46	-125.9
396.60	-363.8	580.48	-124.6

ISOBUTANE (2-METHYL PROPANE) $CH_3 \cdot CH(CH_3)_2$

1. F.W. Jessen and J.H. Lightfoot, Ind. Engng Chem. ind. Edn. 30 312 (1938) (*).

T	B
273.15	-889 ± 25
303.15	-699 ± 10

2. B.H. Sage and W.N. Lacey, Ind. Engng Chem. ind. Edn. 30 673 (1938).
    PVT data given: P range 1 - 200 atm, T range 294 - 394 K.

3. W.M. Morris, B.H. Sage, and W.N. Lacey, Trans. Am. Inst. Min. Metall. Engrs. 136 158 (1940).
    PVT data given: P range 7 - 340 atm, T range 311 - 511 K.

4. J.A. Beattie, S. Marple, Jr., and D.G. Edwards, J. chem. Phys. 18 127 (1950).
    PVT data given: P range 25 - 300 atm, T range 423 - 573 K.

5. C.B. Kretschmer and R. Wiebe, J. Am. chem. Soc. 73 3778 (1951) (*).

T	B
303.16	-644 ± 11

6. R.D. Gunn, M.S. Thesis, University of California, Berkeley (1958).
    Values of B given by J.A. Huff and T.M. Reed, J. chem. Engng Data 8 306 (1963).
    Class I.

T	B	T	B
344.3	-414.0	444.3	-267.8
377.6	-358.0	477.6	-230.2
410.9	-310.6	510.9	-191.6

7. D.McA. Mason and B.E. Eakin, J. chem. Engng Data 6 499 (1961) (*).

T	B
288.70	-714.3 ± 5

8. K. Strein, R.N. Lichtenthaler, B. Schramm, and Kl. Schäfer, Ber. (dtsch.) Bunsenges. phys. Chem. <u>75</u> 1308 (1971).

Estimated accuracy of B is ± 1%.

T	B	T	B
296.1	-691.0	413.8	-349.9
308.1	-634.0	433.8	-320.0
333.9	-532.5	453.6	-291.8
353.9	-476.7	470.2	-267.2
373.9	-427.6	494.0	-243.4
394.6	-384.0		

9. T.R. Das, C.O. Reed, Jr., and P.T. Eubank, J. chem. Engng Data <u>18</u> 253 (1973).

Values of B calculated by the authors from analysis of available PVT data.

T	B	T	B
273.16	-899.9	406.87	-326.5
303.16	-687.4	410.93	-318.5
344.26	-497.2	444.26	-261.9
360.93	-441.2	477.6	-218.1
377.59	-393.9	510.9	-183.7
394.26	-353.4		

n-BUTANOL (1-BUTANOL) $CH_3 \cdot (CH_2)_3 OH$

1. O.R. Fox, J. Morcillo, and A. Mendez, An. R. Soc. esp. Fis. Quim. <u>17B</u> 23 (1954).

Values read from diagram.

Class II.

T	B
350	-1 670
371	-1 500
380	-1 220

2. J.D. Cox, Trans. Faraday Soc. $\underline{57}$ 1674 (1961).
    Maximum pressure 560 torr.

T	B
393.2	-1 053 ± 16
408.2	- 903 ± 7
423.2	- 773 ± 13
439.2	- 624 ± 9

ISO-BUTANOL (2-METHYL-1-PROPANOL) $(CH_3)_2.CH.CH_2OH$

1. J.D. Cox, Trans. Faraday Soc. $\underline{57}$ 1674 (1961).
    Maximum pressure 570 torr.

T	B
393.2	-1 039 ± 11
408.2	- 904 ± 9
423.2	- 789 ± 6
439.2	- 631 ± 18

SEC-BUTANOL (2-BUTANOL) $CH_3.CH_2CHOH.CH_3$

1. J.D. Cox, Trans. Faraday Soc. $\underline{57}$ 1674 (1961).
    Maximum pressure 570 torr.

T	B
378.2	-1 131 ± 8
393.2	- 969 ± 10
408.2	- 831 ± 8
423.2	- 692 ± 12

TERT-BUTANOL (2-METHYL-2-PROPANOL) $(CH_3)_3.OH$

1. J.D. Cox, Trans. Faraday Soc. $\underline{57}$ 1674 (1961).
    Maximum pressure 580 torr.

T	B
378.2	-935 ± 6
393.2	-810 ± 7
408.2	-660 ± 5
423.2	-540 ± 14

DIETHYL ETHER (ETHYL ETHER, ETHOXY ETHANE) $(C_2H_5)_2O$

The various sets of data for diethyl ether are in fairly good agreement. Values taken from a smooth line drawn through the data are given below.

T	B	T	B
280	-1550 ± 50	360	- 640 ± 30
300	-1200 ± 50	380	- 520 ± 50
320	- 950 ± 20	400	- 430 ± 50
340	- 780 ± 20		

1. J.D. Lambert, G.A.H. Roberts, J.S. Rowlinson, and V.J. Wilkinson, Proc. R. Soc. A196 113 (1949)(†).

Maximum pressure 600 torr. Accuracy ± 50.

T	B	T	B
320.0	-907.5	351.0	-714.5
336.2	-794.9	382.3	-564.0
336.8	-806.6	405.2	-480.4
350.4	-713.1		

2. G.A. Bottomley and C.G. Reeves, J. chem. Soc. 3794 (1958). Accuracy ± 3.

T	B
295.2	-1 226
308.2	-1 084
323.2	- 950

3. Sh.D. Zaalishvili and L.E. Kolysko, Zh. fiz. Khim. 34 2596 (1960). Maximum pressure 1200 torr. Accuracy ± 3 per cent.

T	B
339.15	-790
358.15	-720

4. Sh.D. Zaalishvili and L.E. Kolysko, Zh. fiz. Khim. 36 846 (1962). Class II.

T	B	T	B
313.15	-1 046	343.15	- 768
328.15	- 900	358.15	- 682

5. M. Rätzsch and H.-J. Bittrich, Z. phys. Chem. <u>228</u> 81 (1965).

    Class II.

T	B
293.1	-1 386
313.2	-1 107

6. R. Stryjek and A. Kreglewski, Bull. Acad. pol. Sci. Ser. Sci. chim. <u>13</u> 201 (1965).

    Maximum pressure 400 torr.

    Class II.

T	B
298.16	-1 165
323.26	- 955
348.16	- 770
368.16	- 630

7. D.H. Knoebel and W.C. Edmister, J. chem. Engng Data <u>13</u> 312 (1968).

    Low pressure PVT measurements.

T	B
333.2	-895
353.2	-677
373.2	-525

8. J.F. Counsell, D.A. Lee, and J.F. Martin, J. chem. Soc.(A) 313 (1971).

    B calculated from measured vapour pressures and heats of vaporization.

T	B
280.7	-1542
289.4	-1406
295.6	-1300
307.6	-1146

124

DIETHYL SULPHIDE (ETHYL SULPHIDE) $C_2H_5.S.C_2H_5$

1.   G.A. Bottomley and I.H. Coopes, Aust. J. Chem. 15 190 (1962).

T	B
319.16	-1 772 ± 12
339.16	-1 505 ± 9

2.   D.W. Scott, H.L. Finke, W.N. Hubbard, J.P. McCullough, G.D. Oliver,
M.E. Gross, C. Katz, K.D. Williamson, G. Waddington, and M. Kuffman,
J. Am. chem. Soc. 74 4656 (1952).

Values of B were calculated by the authors from heats of vaporization and vapour pressure measurements.

T	B
324.70	-1 524
344.16	-1 389
365.26	-1 240

1-BUTANETHIOL (BUTYL MERCAPTAN) $CH_3(CH_2)_2CH_2SH$

1.   D.W. Scott, H.L. Finke, J.P. McCullough, J.F. Messerly,
R.E. Pennington, I.A. Hossenlopp, and G. Waddington,
J. Am. chem. Soc. 79 1062 (1957).

Values of B calculated by the authors from measurements of vapour
pressure and heats of vaporization.

T	B
330.62	-1 615
349.70	-1 431
371.62	-1 243

2-BUTANETHIOL $CH_3CH_2CHSHCH_3$

1.   J.P. McCullough, H.L. Finke, D.W. Scott, R.E. Pennington, M.E. Gross,
J.F. Messerly, and G. Waddington, J. Am. chem. Soc. 80 4786 (1958).

Values of B calculated by the authors from measurements of vapour
pressure and heats of vaporization.

T	B
318.03	-1 540
328.64	-1 485
336.68	-1 410
358.14	-1 211

2-METHYL 1-PROPANETHIOL (ISOBUTYL MERCAPTAN) $(CH_3)_2CHCH_2SH$

1. D.W. Scott, J.P. McCullough, J.F. Messerly, R.E. Pennington,
    I.A. Hossenlopp, H.L. Finke, and G. Waddington, J. Am. chem. Soc.
    80 55 (1958).

    Values of B calculated by the authors from measurements of vapour
    pressure and heats of vaporization.

T	B
321.31	-1 707
340.07	-1 418
361.65	-1 172

2-METHYL 2-PROPANETHIOL $(CH_3)_3CSH$

1. J.P. McCullough, D.W. Scott, H.L. Finke, W.N. Hubbard, M.E. Gross,
    C. Katz, R.E. Pennington, J.F. Messerly, and G. Waddington,
    J. Am. chem. Soc. 75 1818 (1953).

    Values of B calculated by the authors from measurements of vapour
    pressure and heats of vaporization.

T	B
298.15	-1 588
317.15	-1 316
337.37	-1 160

3-METHYL 2-THIABUTANE (METHYLISOPROPYL SULPHIDE) $CH_3.S.CH(CH_3)_2$

1. J.P. McCullough, H.L. Finke, J.F. Messerly, R.E. Pennington,
    I.A. Hossenlopp, and G. Waddington, J. Am. chem. Soc. 77 6119 (1955).

    Values of B calculated by the authors from measurements of vapour
    pressure and heats of vaporization.

T	B
318.05	-1 575
336.62	-1 366
357.97	-1 173

2-THIAPENTANE (METHYL n-PROPYL SULPHIDE) $CH_3.(CH_2)_2.S.CH_3$

1. D.W. Scott, H.L. Finke, J.P. McCullough, J.F. Messerly, R.E.
    Pennington, I.A. Hossenlopp, and G. Waddington, J. Am. chem. Soc.
    79 1062 (1957).

Values of B calculated by the authors from measurements of vapour pressure and heats of vaporization.

T	B
328.11	-1 663
347.01	-1 452
368.69	-1 233

DIETHYLAMINE $C_2H_5.NH.C_2H_5$

1.   J.D. Lambert and E.D.T. Strong, Proc. R. Soc. A200 566 (1950)(†).
     Maximum pressure 600 torr. Accuracy ± 50.

T	B	T	B
313.6	-1 295	364.2	- 907
322.8	-1 202	373.8	- 850
333.4	-1 118	383.6	- 792
342.8	-1 041	394.4	- 741
351.4	- 970	405.2	- 658
353.6	- 948		

TETRAMETHYLSILANE $(CH_3)_4Si$

1.   S.D. Hamann, J.A. Lambert, and R.B. Thomas, Aust. J. Chem. 8 149 (1955).
     Class II.

T	B	T	B
323.15	-956	363.15	-737
333.15	-886	383.15	-649
343.15	-831	403.15	-580
353.15	-784		

2.   J. Bellm, W. Reineke, K. Schäfer, and B. Schramm, Ber. (dtsch.) Bunsenges. phys. Chem. 78 282 (1974).
     Estimated accuracy of B is ± 30.

T	B	T	B
300	-1347	430	- 719
320	-1199	460	- 651
340	-1069	490	- 595
370	- 921	520	- 547
400	- 805	550	- 508

3.   G.A. Bottomley and D.B. Nairn, Aust. J. Chem. <u>30</u> 1645 (1977).
B determined relative to B (343 K) = -829.
Precision of B is ± 6.

T	B	T	B
294.64	-1191.7	398.38	- 587.6
294.66	-1191.3	430.53	- 491.9
294.83	-1190.0	430.59	- 489.5
318.99	- 978.6	429.83	- 492.4
319.12	- 995.4	429.88	- 491.4
318.66	- 982.4	429.80	- 493.8
318.64	- 984.8	462.90	- 404.3
318.57	- 976.6	463.11	- 408.6
343.46	- 828.0	463.46	- 409.5
343.57	- 829.0	500.46	- 334.1
368.24	- 713.6	500.72	- 334.7
369.00	- 705.2	538.71	- 272.0
369.13	- 703.1	538.65	- 270.0
398.40	- 586.0		

NEOPENTANE-d_{12} $(CD_3)_4 \cdot C$

1.   I. Gainar, Kl. Schäfer, B. Schmeiser, B. Schramm, and K.Strein, Ber. (dtsch.) Bunsenges. phys. Chem. <u>77</u> 372 (1973).

Values of ΔB ($B_{deuterated\ form} - B_{neopentane}$) determined on the basis that ΔB (297 K) = 20.0. Repeatability of ΔB ± 0.3.

T	B
313	18.2
375	13.0
422	9.3
475	8.0
510	6.9

PERFLUORO-n-PENTANE $CF_3.(CF_2)_3.CF_3$

1. M.D.G. Garner and J.C. McCoubrey, Trans. Faraday Soc. <u>55</u> 1524 (1959). Class II.

T	B	T	B
307.6	-1 360	350.9	- 934
329.2	-1 125	372.8	- 811
337.4	-1 036	383.1	- 717

NEOPENTANE-d_9 $(CD_3)_3.C.CH_3$

1. I. Gainar, Kl. Schäfer, B. Schmeiser, B. Schramm, and K. Strein, Ber. (dtsch.) Bunsenges. phys. Chem. <u>77</u> 372 (1973).

Values of ΔB ($B_{deuterated\ form} - B_{neopentane}$) determined on the basis that ΔB (297 K) = 15.0. Repeatability of $\Delta B \pm 0.3$.

T	ΔB	T	ΔB
345	12.7	422	8.0
375	10.9	475	6.7
395	9.7	570	6.1

PYRIDINE (AZINE) CH:CH.CH:CH.CH:N

1. J.D. Cox and R.J.L. Andon, Trans. Faraday Soc. <u>54</u> 1622 (1958); R.J.L. Andon, J.D. Cox, E.F.G. Herington, and J.F. Martin, Trans. Faraday Soc. <u>53</u> 1074 (1957).

Maximum pressure 800 torr.

T	B	T	B
349.1	-1 263 ± 19	409.0	- 823 ± 12
363.0	-1 132 ± 15	424.2	- 759 ± 15
376.9	-1 054 ± 20	437.8	- 650 ± 8
393.1	- 973 ± 11		

2. J.P. McCullough, D.R. Douslin, J.F. Messerly, I.A. Hossenlopp, T.C. Kinchelor, and G. Waddington, J. Am. chem. Soc. <u>79</u> 4289 (1957).

Values of B calculated by the authors from measurements of vapour pressure and heats of vaporization.

T	B
346.64	-1 292
366.10	-1 157
388.39	- 990

NEOPENTANE-d_6 $(CD_3)_2C.(CH_3)_2$

1. I. Gainar, Kl. Schäfer, B. Schmeiser, B. Schramm, and K. Strein, Ber. (dtsch.) Bunsenges. phys. Chem. <u>77</u> 372 (1973).

   Values of ΔB ($B_{deuterated\ form} - B_{neopentane}$) determined on the basis that ΔB (297 K) = 10.0. Repeatability of $\Delta B \pm 0.3$.

T	ΔB	T	ΔB
345	7.9	422	5.0
375	7.3	475	4.2
395	5.8	510	3.8

2-METHYL FURAN $CH_3.C:CH.CH:CH.O$

1. C. Eon, C. Pommier and G. Guiochon, J. chem. Engng Data <u>16</u> 408 (1971). Isoteniscopic method: estimated uncertainty in B ± 4%.

T	B
333.45	-810
343.45	-760
353.45	-725
363.45	-690
373.45	-670

2-METHYL THIOPHENE (α-THIOTOLUENE) $S.C(CH_3):CH.CH:CH$

L. R.E. Pennington, H.L. Finke, W.N. Hubbard, J.F. Messerly, F.R. Frow, I.A. Hossenlopp, and G. Waddington, J. Am. chem. Soc. <u>78</u> 2055 (1956).

   Values of B calculated by the authors from measurements of vapour pressure and heats of vaporization.

T	B
343.49	-1 780
363.14	-1 400
385.72	-1 160

2.   C. Eon, C. Pommier and G. Guiochon, J. chem. Engng Data <u>16</u> 408 (1971).

Isoteniscopic method: estimated uncertainty in B ± 4%.

T	B
333.45	-1700
343.45	-1610
353.45	-1490
363.45	-1390
373.45	-1310

3-METHYL THIOPHENE (β-THIOTOLUENE) S.CH:C(CH$_3$).CH:CH

1.   J.P. McCullough, S. Sunner, H.L. Finke, W.N. Hubbard, M.E. Gross, R.E. Pennington, J.F. Messerly, W.D. Good, and G. Waddington, J. Am. chem. Soc. <u>75</u> 5075 (1953).

Values of B calculated by the authors from measurements of vapour pressure and heats of vaporization.

T	B
328.68	-1 640
346.10	-1 510
365.89	-1 405
388.59	-1 215

2.   C. Eon, C. Pommier and G. Guiochon, J. chem. Engng Data <u>16</u> 408 (1971).

Isoteniscopic method: estimated uncertainty in B ± 4%.

T	B
333.45	-1830
343.45	-1680
353.45	-1550
363.45	-1430
373.45	-1320

1-METHYL PYRROLE   CH$_3$C:CH.CH:CH.NH

1.   C. Eon, C. Pommier and G. Guiochon, J. chem. Engng Data <u>16</u> 408 (1971).

Isoteniscopic method: estimated uncertainty in B ± 4%.

T	B
333.45	-1370
343.45	-1200
353.45	-1190
363.45	-1120
373.45	-1050

SPIROPENTANE $CH_2.CH_2.C.CH_2.CH_2$

1. D.W. Scott, H.L. Finke, W.N. Hubbard, J.P. McCullough, M.E. Gross, K.D. Williamson, G. Waddington, and H.M. Huffman, J. Am. chem. Soc. 72 4664 (1950); Correction J. Am. chem. Soc. 74 6313 (1952).

    Values of B calculated by the authors from measurements of vapour pressure and heats of vaporization.

T	B
283.15	-1 217
398.15	-1 042
312.13	- 943

CYCLOPENTENE $CH:CH.CH_2.CH_2.CH_2$

1. C. Eon, C. Pommier, and G. Guiochon, J. chem. Engng Data 16 408 (1971).

    Isoteniscopic method: estimated uncertainty in B ± 4%.

T	B
333.45	-820
343.45	-755
353.45	-710
363.45	-665
373.45	-630

NEOPENTANE-d$_3$ $CD_3.C.(CH_3)_3$

1. I. Gainar, Kl. Schäfer, B. Schmeiser, B. Schramm, and K. Strein, Ber. (dtsch.) Bunsenges. phys. Chem. 77 372 (1973).

    Values of ΔB ($B_{deuterated\ form} - B_{neopentane}$) determined on the basis that ΔB (297 K) = 5.0. Repeatability of ΔB ± 0.3.

T	ΔB
345	4.0
375	3.3
422	1.6
475	1.2
510	1.0

CYCLOPENTANE $CH_2(CH_2)_3.CH_2$

1. J.P. McCullough, R.E. Pennington, J.C. Smith, I.A. Hossenlopp, and G. Waddington, J. Am. chem. Soc. <u>81</u> 5880 (1959).

Values of B calculated by the authors from heats of vaporization and vapour pressure measurements.

T	B
298.15	-1 066
310.15	- 972
322.41	- 907

PENT-1-ENE (1-PENTYLENE, PROPYLETHYLENE, 1-PENTENE) $CH_3.CH_2.CH_2.CH:CH_2$

1. M.L. McGlashan and C.J. Wormald, Trans. Faraday Soc. <u>60</u> 646 (1964).

Class I.

T	B	T	B
308.0	-982	363.9	-662
315.3	-930	374.0	-628
324.1	-870	383.2	-585
334.1	-810	393.8	-552
344.3	-756	403.4	-519
353.3	-712	410.4	-486

2. D.W. Scott, G. Waddington, J.C. Smith, and H.M. Huffman, J. Am. chem. Soc. <u>71</u> 2767 (1949).

Values of B calculated by the authors from heats of vaporization and vapour pressure measurements.

T	B
283.95	-1 265
298.15	-1 143
303.12	-1 099

2-METHYL-1-BUTENE $CH_2:C(CH_3).CH_2.CH_3$

1.   D.W. Scott, G.Waddington, J.C. Smith, and H.M. Huffman,
     J. Am. chem. Soc. <u>71</u> 2767 (1949).

     Calculated from vaporization data.

T	B
277.95	-1 258
298.15	-1 163
304.31	-1 117

2-METHYL-2-BUTENE $(CH_3)_2C:CH.CH_3$

1.   D.W. Scott, G.Waddington, J.C. Smith and H.M. Huffman,
     J. Am. chem. Soc. <u>71</u> 2767 (1949).

     Calculated from vaporization data.

T	B
289.89	-1 365
298.15	-1 276
311.72	-1 143

METHYL n-PROPYL KETONE $CH_3.CO.CH_2.CH_2CH_3$

1.   J.K. Nickerson, K.A. Kobe, and J.J. McKetta, J. phys. Chem.,
     Ithaca <u>65</u> 1037 (1961).

     B determined from vaporization data.   Accuracy ± 2.

T	B
334.87	-2 635
360.56	-1 760
375.33	-1 540
386.04	-1 377
394.57	-1 280

METHYL ISOPROPYL KETONE $CH_3.CO.CH(CH_3)_2$

1.   J.L. Hales, E.B. Lees, and D.J. Ruxton, Trans. Faraday Soc. <u>63</u> 1876
     (1967).

     B calculated from measured vapour pressures and heats of
     vaporization.

T	B
327.69	-2020
346.22	-1720
367.48	-1430

DIETHYL KETONE $CH_3.CH_2.CO.CH_2.CH_3$

1. J.L. Hales, E.B. Lees, and D.J. Ruxton, Trans. Faraday Soc. <u>63</u> 1876 (1967).

   B calculated from measured vapour pressures and heats of vaporization.

T	B
335.01	-2269
353.70	-1858
375.11	-1511

TRIMETHYLACETIC ACID (2,2-DIMETHYL PROPANOIC ACID, PIVALIC ACID) $C(CH_3)_3COOH$

1. E.W. Johnson and L.K. Nash, J Am. chem. Soc. <u>72</u> 547 (1950).

   PVT data given: P range up to 1.5 atm, T range 353-473 K.

   Vapour density method.

THIACYCLOHEXANE (PENTAMETHYLENESULPHIDE) $CH_2.(CH_2)_4.S$

1. J.P. McCullough, H.L. Finke, W.N. Hubbard, W.D. Good, R.E. Pennington, J.F. Messerly, and G. Waddington, J. Am. chem. Soc. <u>76</u> 2661 (1954).

   Values of B calculated by the authors from measurements of vapour pressure and heats of vaporization.

T	B
351.44	-1 700
368.91	-1 600
390.31	-1 460
404.26	-1 370
414.91	-1 290

CYCLOPENTANETHIOL (CYCLOPENTYL MERCAPTAN) $CH_2.(CH_2)_3.CH.SH$

1. W.T. Berg, D.W. Scott, W.N. Hubbard, S.S. Todd, J.F. Messerly, I.A. Hossenlopp, A. Osborn, D.R. Douslin, and J.P. McCullough, J. phys. Chem., Ithaca <u>65</u> 1425 (1961).

   B calculated from heats of vaporization and vapour pressure measurements.

T	B
360.62	-1674
381.41	-1442
405.31	-1212

2-CHLORO-2-METHYLBUTANE $CH_3.C(CH_3)Cl.CH_2.CH_3$

1. A.R. Paniego, J.A.B. Lluna, and J.E.H. Garcia, An. Quim. <u>71</u> 349 (1975).

    Relative compressibility method using nitrogen.

    Class II.

T	B
363.3	-1249
378.8	-1148
392.6	-1043
408.2	- 969

n-PENTANE $CH_3(CH_2)_3CH_3$

There is close agreement between the different sets of data for the second virial coefficient of n-pentane obtained prior to 1963 and the measurements of Hajjar, Kay and Leverett (10), though the results of Kapallo, Lund, and Schafer (8) and of Ratzsch and Bittrich (9) at temperatures below 325 K are more negative by about 100 cm^3mol^{-1}. It is not possible to say which set of values is the more accurate. The following values are recommended.

T	B	T	B
300	-1250 ± 60	400	- 575 ± 15
310	-1135 ± 60	425	- 515 ± 15
320	-1030 ± 50	450	- 452 ± 15
340	- 860 ± 30	475	- 399 ± 15
360	- 750 ± 25	500	- 347 ± 15
380	- 660 ± 25	550	- 276 ± 15

1. F.W. Jessen and J.H. Lightfoot, Ind. Engng Chem. ind. Edn <u>30</u> 312 (1938)(*).

T	B
303.15	-1277

2. B.H. Sage and W.N. Lacey, Ind. Engng Chem. ind. Edn <u>34</u> 730 (1942).

    PVT data given: P range 1 - 670 atm, T range 311 - 511 K.

3. J.A. Beattie, S.W. Levine, and D.R. Douslin, J. Am. chem. Soc. 74 4778 (1952).

Class I.

T	B
473.16	-405
498.16	-350
523.16	-311
548.16	-274
573.16	-244

4. Sh.D. Zaalishvili, Usp. Khim. 24 759 (1955).

B values given by J.A. Huff and T.M. Reed, 111, J. chem. Engng Data 8 306 (1963).

T	B
377.5	-666
410.9	-547
444.2	-460
477.5	-396
510.0	-343

5. M.D.G. Garner and J.C. McCoubrey, Trans. Faraday Soc. 55 1524 (1959). Standard error in quoted B values varies between 15 and 45.

T	B	T	B
307.6	-1 082	353.0	- 747
329.3	- 896	372.1	- 662
337.7	- 851	383.7	- 623
351.2	- 783		

6. D.McA. Mason and B.E. Eakin, J.chem. Engng Data 6 499 (1961)(*).

T	B
288.70	-1376 ± 17

7. M.L. McGlashan and D.J.B. Potter, Proc.R. Soc. A267 478 (1962). Precision of B values better than ± 20.

T	B	T	B
298.2	-1 194	329.0	- 923
306.1	-1 117	339.0	- 863
318.1	-1 011	349.0	- 800

358.0	- 758	388.4	- 612
368.6	- 701	401.0	- 578
378.9	- 652	413.6	- 517

8. W. Kapallo, N. Lund, and K. Schafer, Z. phys. Chem. Frankf. Ausg. 37 196 (1963).
    Class II.

T	B
273.4	-1 680
283.0	-1 533
297.0	-1 363
312.7	-1 205
321.4	-1 121

9. M. Ratzsch and H.-J. Bittrich, Z. phys. Chem. 228 81 (1965).
    Class II.

T	B
293.1	-1 386
313.2	-1 187

10. R.F. Hajjar, W.B. Kay, and G.F. Leverett, J. chem. Engng Data 14 377 (1969).
    Low pressure PVT measurements.

T	B	T	B
313.15	-1050	398.15	- 575
323.15	- 960	413.15	- 535
336.15	- 875	427.15	- 500
353.15	- 785	453.15	- 445
368.15	- 715	473.15	- 400
382.15	- 655		

11. T.R. Das, C.O. Reed, Jr., and P.T. Eubank, J. chem. Engng Data 22 3 (1977).
    B calculated from compressibility data given by different authors.

T	B	T	B
310	-1173	425	- 526
320	-1082	450	- 455
340	- 926	475	- 396
360	- 801	500	- 347
380	- 698	550	- 270
400	- 613		

ISOPENTANE (2-METHYLBUTANE) $(CH_3)_2 \cdot CH.CH_2 \cdot CH_3$

The values derived from Das, Reed, and Eubank (5) are recommended.   The
following values are given with estimated uncertainties at rounded
temperatures.

T	B	T	B
280	-1265 ± 40	375	- 645 ± 20
300	-1075 ± 35	400	- 555 ± 20
325	- 900 ± 30	450	- 425 ± 20
350	- 760 ± 25		

1. I.H. Silberberg, J.J. McKetta, and K.A. Kobe, J. chem. Engng Data
   4 323 (1959).

   Class II.

T	B	T	B
323.15	-960	423.15	-494
348.15	-763	448.15	-434
373.15	-645	461.65	-407
398.15	-570	473.15	-376

2. D.W. Scott, J.P. McCullough, K.D. Williamson, and G. Waddington,
   J. Am. chem. Soc. 73 1707 (1951).

   Values of B calculated by the authors from heats of vaporization and
   vapour pressure measurements.

T	B
279.47	-1 313
298.15	-1 155
301.00	-1 114

3. D.McA. Mason and B.E. Eakin, J. chem. Engng Data 6 499 (1961)(*).

T	B
288.70	-1259 ± 12

4. I.H. Silberberg, D.C.K. Lin, and J.J. McKetta, J. chem. Engng Data
   12 226 (1967).

   Values of B, C, and D calculated by the authors from the PVT data of
   I.H. Silberberg, J.J. McKetta, and K.A. Kobe, J. chem. Engng Data 4
   323 (1959).

   The errors in B are less than 2 per cent, errors in C are less than
   5 per cent or 2000 (whichever is the greater).

T	B	C
273.15	(-1 370.6)	
298.15	(-1 149.7)	
323.15	- 954.0	506 000
348.15	- 775.2	364 000
373.15	- 642.9	239 000
398.15	- 555.7	151 000
423.15	- 482.5	97 000
448.15	- 419.1	67 000
461.65	- 388.4	56 000
473.15	- 366.4	49 000

5. T.R. Das, C.O. Reed, Jr., and P.T. Eubank, J. chem. Engng Data 22 9 (1977).

B calculated from experimental compressibility data given by various authors.

T	B	T	B
273.15	-1323	398.15	- 569
298.15	-1087	423.15	- 495
323.15	- 909	448.15	- 433
348.15	- 770	461.65	- 403
373.15	- 659	473.15	- 380

NEOPENTANE $(CH_3)_4C$

Values reported for the second virial coefficient of neopentane prior to 1970 generally agree to within the estimated error limits. The results given by Silberberg et al. (6) from PVT data should be used in preference to those given by Heichelheim and McKetta (4). Since 1970, measurements made by Strein et al.(7) and Bellm et al. (8) give values which are significantly more negative than the earlier values. The following values are recommended.

T	B	T	B
300	-920 ± 35	400	-468 ± 15
310	-840 ± 25	425	-405 ± 10
330	-720 ± 20	450	-354 ± 10
350	-630 ± 20	500	-276 ± 10
375	-540 ± 15	550	-220 ± 10

140

1. J.A. Beattie, D.R. Douslin, and S.W. Levine, J. chem. Phys. <u>20</u> 1619 (1952).

    2-term fit of PV data. Maximum pressure 300 atm.

T	B	T	B
433.75	-383	498.15	-273
448.15	-354	523.15	-243
473.15	-312	548.15	-216

2. S.D. Hamann, and J.A. Lambert, Aust. J. Chem. <u>7</u> 1 (1954); S.D. Hamann, J.A. Lambert, and R.B. Thomas, Aust. J. Chem. <u>8</u> 149 (1955).

    Class II.

T	B	T	B
303.15	-842	363.15	-566
323.15	-734	373.15	-536
333.15	-686	383.15	-507
343.15	-643	398.15	-472
348.15	-626	403.15	-452
353.15	-602		

3. H.M. Ashton and E.S. Halberstadt, Proc. R. Soc. <u>A245</u> 373 (1958).

    Values obtained from refractive index measurements.

T	B (average)
300.35	-940
313.57	-902
322.69	-776
342.29	-679

4. H.R. Heichelheim and J.J. McKetta, Jr., Chem. Engng Prog. Symp. Ser. (Thermodynamics). <u>59</u> 23 (1963).

    Burnett method: Maximum pressure 70 atm.

    Class II.

T	B	T	B
303.15	-933	423.15	-444
323.15	-809	434.65	-417
348.15	-676	448.15	-379
373.15	-566	473.15	-349
398.15	-503		

5. A. Perez Masia, M. Diaz Pena, and J.A. Burriel Lluna, An. R. Soc. esp. Fis. Quim. 60B 229 (1964).

T	B	C
303.22	-881 ± 14	8 600
323.23	-737 ± 9	1 300
348.11	-618 ± 12	- 1 500
373.21	-552 ± 13	4 400
397.70	-476 ± 14	1 500

6. I.H. Silberberg, D.C.K. Lin, and J.J. McKetta, J. chem. Engng Data 12 226 (1967).

Values of B and C calculated by the authors from PVT data of H.R. Heichelheim, K.A. Kobe, I.H. Silberberg, and J.J. McKetta, J. chem. Engng Data 7 507 (1962).

Errors in B are stated to be less than 2 per cent and errors in C are less than 5 per cent or ± 2000, whichever is the greater.

T	B	C
303.15	-900.4	282 000
323.15	-764.0	151 000
348.15	-637.2	108 000
373.15	-548.7	91 000
398.15	-477.5	77 000
423.15	-413.5	65 000
434.65	-386.4	59 000
448.15	-357.7	54 000
473.15	-318.0	44 000

7. K. Strein, R.N. Lichtenthaler, B. Schramm, and Kl. Schäfer, Ber. (dtsch.) Bunsenges. phys. Chem. 75 1308 (1971).

Estimated accuracy of B is ± 1%.

T	B	T	B
296.1	-975.6	413.5	-518.0
308.9	-892.0	433.0	-480.5
334.1	-767.7	453.2	-442.5
353.9	-688.5	473.9	-408.9
373.5	-622.1	492.6	-375.2
393.1	-568.0		

8. J. Bellm, W. Reineke, K. Schäfer, and B. Schramm, Ber. (dtsch.) Bunsenges. phys. Chem. 78 282 (1974).

Estimated accuracy of B is ± 10.

T	B	T	B
300	-956.0	430	-494.0
320	-849.0	460	-441.5
340	-755.0	490	-397.0
370	-644.5	520	-360.0
400	-560.0	550	-331.0

9. P.P. Dawson, Jr., I.H. Silberberg and J.J. McKetta, J. chem. Engng Data 18 7 (1973).

PVT data P 5 - 310 atm., T 343.15 - 498.15 K.

Estimated uncertainty in B ± 4, in C ± 2000.

T	B	C
348.15	-615	13000
373.15	-530	41300
398.15	-460	46200
423.15	-401	47100
433.75	-383	49800
434.65	-381	49800
448.15	-354	46400
473.15	-311	41500
498.15	-275	37600

10. T.R. Das, C.O. Reed, Jr., and P.T. Eubank, J. chem. Engng Data 22 16 (1977).

B calculated from experimental compressibility factors given by various authors.

T	B	T	B
303.15	-941.6	398.15	-480.1
323.15	-803.5	423.15	-412.7
333.15	-745.2	448.15	-357.5
348.15	-668.4	473.15	-311.4
363.15	-602.4	498.15	-272.7
373.15	-563.4	523.15	-239.6
383.15	-527.9	548.15	-211.6

1-PENTANETHIOL (AMYL MERCAPTAN) $CH_3(CH_2)_4SH$

1. H.L. Finke, I.A. Hossenlopp, and W.T. Berg, J. phys. Chem., Ithaca
    <u>69</u> 303 (1965).

    Values of B calculated by the authors from measurements of vapour
    pressure and heats of vaporization.

T	B
356.08	-1 910
376.43	-1 670
399.80	-1 406

CHLOROPENTAFLUOROBENZENE C_6F_5Cl

1. R.J.L. Andon, J.F. Counsell, J.L. Hales, E.B. Lees, and J.F. Martin,
    J. chem. Soc. (A) 2357 (1968).

    B calculated from measured vapour pressures and heats of vaporization.

T	B
349.41	-2598
368.86	-2131
391.10	-1780

HEXAFLUOROBENZENE C_6F_6

1. J.F. Counsell, J.H.S. Green, J.L. Hales and J.F. Martin, Trans.
    Faraday Soc. <u>61</u> 212 (1965).

    B determined from measured vapour pressures and heats of vaporization.

T	B
315.96	-2104
333.41	-1792
353.41	-1498

2. D.R. Douslin, R.H. Harrison and R.T. Moore, J. chem. Thermodyn.
    <u>1</u> 305 (1969).

    a. Calculated from enthalpies of vaporization and vapour pressure
    measurements.

T	B	C
300.57	-2 469	
315.94	-2 039	
333.39	-1 725	
353.39	-1 417	
376.52	-1 161	

b. PVT data given: P range 20 - 350 atm, T range 225 - 350°K.

T	B	C
498.20	-532.2	110 100
516.74	-482.7	101 000
523.22	-466.9	98 100
548.23	-411.2	86 900
573.25	-363.2	76 600
598.26	-321.4	67 400
623.26	-285.0	59 700

3. R.J. Powell, Ph. D. Thesis, University of Strathclyde (1969).
   Class II.

T	B
364.988	-1314
389.573	-1059
414.709	- 880.5
443.958	- 745

4. D. Ambrose, J.H. Ellender, C.H.S Sprake, and R. Townsend, J. Chem.
   Soc. Faraday 1 71 35 (1975).

   B calculated from measured vapour pressures and enthalpy of vaporization.

   Class III.

T	B
353.41	-1500

'ERFLUORO-c-HEXANE cC_6F_{12}

l. Z.L. Taylor, Jr., and T.M. Reed III, A.I. Ch.E. Jl. 16 738 (1970).
    B ± 1% ; C ± 10%

T	B	C
350.32	-1152	
378.36	- 941	
395.00	- 851	150 000
410.81	- 770	150 000
433.17	- 669	130 000
451.01	- 599	120 000

PERFLUORO-n − HEXANE $CF_3 \cdot (CF_2)_4 \cdot CF_3$

1. M.D.G. Garner and J.C. McCoubrey, Trans. Faraday Soc. 55 1524 (1959).
    Class II.

T	B	T	B
307.7	-1 922	350.8	-1 416
329.8	-1 676	372.3	-1 145
338.0	-1 579	383.9	-1 049

2. Z.L. Taylor, Jr., and T.M. Reed III, A.I. Ch.E. Jl. 16 738 (1970).
    B ± 1% T > 350, B ± 5% T < 350 ; C ± 10%.

   a. Data of R.R. Cecil, Doctoral dissertation Univ. Fla. (1964)
       reported in above reference.

T	B	C
303.2	-2320	
323.5	-1922	
341.1	-1571	
351.3	-1528	

   b.

T	B	C
341.10	-1612	
351.85	-1484	
374.78	-1209	
395.56	-1051	250 000
415.48	- 920	230 000
432.68	- 818	210 000
451.55	- 725	200 000

PERFLUORO 2-METHYLPENTANE $(CF_3)_2 \cdot CF \cdot (CF_2)_2 \cdot CF_3$

1. Z.L. Taylor, Jr., and T.M. Reed III, A.I. Ch.E. Jl. <u>16</u> 738 (1970).
   B ± 1% T > 350, B ± 5% T < 350; C ± 10%.

   a. Data of R.R. Cecil, Doctoral dissertation Univ. Fla. (1964) reported in above reference.

T	B	C
303.8	-2142	
304.0	-2098	
322.5	-1813	
334.8	-1601	
341.3	-1500	
352.0	-1444	

	T	B	C
b.	333.23	-1837	
	351.65	-1412	
	370.65	-1179	
	417.39	- 882	200 000
	433.13	- 794	180 000
	441.70	- 756	180 000
	454.63	- 691	170 000

PERFLUORO-3-METHYLPENTANE $CF_3 \cdot CF_2 \cdot CF(CF_3) \cdot CF_2 \cdot CF_3$

1. Z.L. Taylor, Jr., and T.M. Reed III, A.I. Ch.E. Jl. <u>16</u> 738 (1970).
   B ± 1% T > 350; B ± 5% T < 350; C ± 10%

   a. Data of R.R. Cecil, Doctoral dissertation Univ. Fla. (1964) reported in above reference.

T	B	C
303.9	-1700	
321.0	-1716	
323.2	-1642	
332.0	-1438	
342.0	-1434	
350.4	-1364	

b.	351.45	-1443	
	376.72	-1113	
	394.73	-1003	
	417.05	- 883	190 000
	432.99	- 795	180 000
	450.49	- 715	170 000

PERFLUORO-2,3-DIMETHYLBUTANE $(CF_3)_2CF.CF(CF_3)_2$

1. Z.L. Taylor, Jr., and T.M. Reed III, A.I. Ch.E. Jl. <u>16</u> 738 (1970).
   B ± 1% T > 350;   B ± 5% T < 350; C ± 10%.

   a. Data of R.R. Cecil, Doctoral dissertation Univ. Fla. (1964)
      reported in above reference.

T	B	C
303.6	-2670	
323.1	-1610	
335.4	-1582	
345.3	-1391	

b.	351.21	-1372	
	374.63	-1130	
	394.65	-1012	
	414.62	- 906	190 000
	433.22	- 786	160 000
	450.36	- 702	140 000

PENTAFLUOROBENZENE C_6HF_5

1. J.F. Counsell, J.L. Hales, and J.F. Martin, J. Chem. Soc. (A) 2042
   (1968).

   B calculated from measured vapour pressures and heats of vaporization.

T	B
320.71	-2147
338.51	-1768
358.90	-1455

1,2-DIFLUOROBENZENE $C_6H_4F_2$

1. D.W. Scott, J.F. Messerly, S.S. Todd, I.A. Hossenlopp, A. Osborn, and J.P. McCullough, J. chem. Phys. 38 532 (1963).

    B calculated from heats of vaporization and vapour pressure measurements.

T	B
326.90	-1625
345.61	-1423
367.07	-1186

FLUOROBENZENE C_6H_5F

1. D.W. Scott, J.P. McCullough, W.D. Good, J.F. Messerly, R.E. Pennington, T.C. Kincheloe, I.A. Hossenlopp, D.R. Douslin, and G. Waddington, J. Am. chem. Soc. 78 5457 (1956).

    Values of B calculated by the authors from vapour pressure and heat of vaporization measurements.

T	B
318.40	-1 446
336.78	-1 275
357.89	-1 086
382.37	- 909

2. D.R. Douslin, R.T. Moore, J.P. Dawson, and G. Waddington, J. Am. chem. Soc. 80 2031 (1958).

    Maximum pressure 400 atm.

    Class I.

T	B	C
548.15	-378	49 500
560.08	-360	48 000
573.15	-341	45 800
598.15	-307	41 700
623.15	-279	39 800

3. Sh.D. Zaalishvili, Z.S. Belousova and V.P. Verkhova, Russ. J. phys. Chem. 46 291 (1972).

    Constant volume piezometer: P 6 → 12 atm.

    Class II.

T	B	T	B
443.2	-664	473.2	-560
453.2	-627	483.2	-530
463.2	-593	493.2	-502

4. D. Ambrose, J.H. Ellender, C.H. Sprake, and R. Townsend, J.C.S. Faraday 1 71 35 (1975).

B calculated from measured vapour pressures and enthalpy of vaporization.

Class III.

T	B
358.2	-1120

5. B. Kaussman, R. Matzky, G. Opel, and E. Vogel, Z. phys. Chem. 258 730 (1977).

B determined from pressure measurements at constant volume of a known mass of fluorobenzene at different temperatures. Pressure range → 1.6 atm.

a. Virial equation in terms of molar volume.

T	B	T	B
373.15	-957 ± 23	523.15	-418 ± 4
398.15	-818 ± 9	548.15	-372 ± 3
423.15	-705 ± 8	573.15	-332 ± 2
448.15	-613 ± 8	598.15	-298 ± 4
473.15	-536 ± 8	623.15	-269 ± 6
498.15	-472 ± 7		

b. Virial equation in terms of pressure.

T	B	T	B
373.15	-979 ± 30	523.15	-422 ± 3
398.15	-835 ± 15	548.15	-375 ± 2
423.15	-718 ± 6	573.15	-335 ± 3
448.15	-622 ± 7	598.15	-300 ± 5
473.15	-542 ± 8	623.15	-270 ± 7
498.15	-477 ± 5		

BENZENE $CH(CH)_4CH$

Much of the controversy that surrounded the second virial coefficient of benzene has now been removed by recent measurements which are, with the exception of Hajjar et al (27), all consistent with the function of Al-Bizreh and Wormald (31). The recommended values given below are close to values calculated from this function.

T	B	T	B
290	1590 ± 30	400	710 ± 20
300	1450 ± 20	440	570 ± 10
310	1340 ± 20	480	470 ± 10
320	1230 ± 20	520	390 ± 10
340	1050 ± 20	560	340 ± 10
360	920 ± 20	600	290 ± 10
380	810 ± 20		

1. S. Young, Proc. R. Ir. Acad. 12 374 (1910).

   Following values taken from P.W. Allen, D.H. Everett, and M.F. Penney, Proc. R. Soc. A212 149 (1952).

T	B
313	-1 440
323	-1 400
333	-1 260
343	-1 250
353	-1 040

2. A. Eucken and L. Mayer, Z. phys. Chem. B5 452 (1929).

   Following values read from the authors B - T graph.

   Class III.

T	B	T	B
335	-1 295	362	- 981
342	-1 135	366	- 893
348	-1 128	372	- 836
354	-1 052	375	- 807
357	-1 010		

3. E. Steurer and K.L. Wolf, Z. phys. Chem. B39 101 (1938).

   B calculated from vapour density measurements and given by H.G. David, S.D. Hamann, and R.B. Thomas, Aust. J. Chem. 12 309 (1959).

T	B
373.2	-873

4. E.J. Gornowski, E.H. Amick, Jr., and A.N. Hixson, Ind. Engng. Chem.
    ind. Edn <u>39</u> 1348 (1947).

    PVT data given: P range 25 - 65 atm, T range 513 - 628 K.

    Values of B given by A.E. Sherwood and J.M. Prausnitz, J. chem. Phys.
    <u>41</u> 429 (1964).

T	B	T	B
553.2	-330	603.2	-279
563.2	-319	613.2	-268
573.2	-307	623.2	-256
583.2	-298	628.2	-251
593.2	-288		

5. D.C. Scott, G. Waddington, J.C. Smith, and H.M. Huffman, J. chem.
    Phys. <u>15</u> 565 (1947).

    Calculated by the authors from heat of vaporization data and vapour
    pressure measurements.

T	B
298.1	-1 570
314.8	-1 272
323.1	-1 213
334.1	-1 138
353.3	- 994
383.2	- 847

6. J.D. Lambert, G.A.H. Roberts, J.S. Rowlinson, and V.J. Wilkinson,
    Proc. R. Soc. <u>A196</u> 113 (1949) (†).

    Maximum pressure 600 torr. Accuracy ± 50.

T	B	T	B
319.1	-1 208.9	350.7	- 980.3
326.9	-1 109.4	359.1	- 936.0
334.8	-1 071.6	359.1	- 923.4
344.9	-1 048.1		

7. J.H. Baxendale, B.V. Enustun, and J. Stern, Phil. Trans. R. Soc. <u>A243</u>
    176 (1950).

    a. Values recalculated from PVT data assuming $PV=RT+BP$. Errors
        given are mean deviations.

T	B
303.15	-1 880 ± 40
313.15	-1 540 ± 28
323.15	-1 370 ± 8
333.15	-1 220 ± 2
343.15	-1 110 ± 2.5
353.15	-1 020 ± 2.5

b. Values calculated by J.H. Baxendale from his compressibility data assuming $PV=RT+BP+CP^2$.

T	B
303.15	-2 170
313.15	-1 885
323.15	-1 660
333.15	-1 430
343.15	-1 250
353.15	-1 095

8. F.L. Casado, D.S. Massie, and R. Whytlaw-Gray, Proc. R. Soc. A207 483 (1951)(*).

Maximum pressure 69.2 torr.

T	B
295.15	-1 465 ± 20

9. W. Mund, A. Gerbaux, and J. Mornigny, Bull. Acad. r. Belg. Cl. Sci. 37 706 (1951).

B calculated from listed apparent molecular weight data by H.G. David, S.D. Hamann, and R.B. Thomas, Aust. J. Chem. 12 309 (1959).

T	B
338.3	-1018
373.2	- 904

10. D.H. Everett (ref. (1) above). B values calculated by the author from an analysis of best available heat of vaporization data.

Accuracy ± 50.

T	B
298.2	-1 440
308.2	-1 280
318.2	-1 230
328.2	-1 170
338.2	-1 110
348.2	-1 020

11. Calculated from D.H. Everett's empirical formula (ref. (1): B = 70-1320 X $10^5/T^2$.

T	B	T	B
280	-1 614	350	-1 007
290	-1 500	360	- 949
300	-1 397	370	- 896
310	-1 303	380	- 844
320	-1 219	390	- 798
330	-1 142	400	- 755
340	-1 072		

12. P.G. Francis, M.L. McGlashan, S.D. Hamann, and W.J. McManamey, J. chem. Phys. 20 1341 (1952).

Class II.

a.

T	B	T	B
316.25	-1 300	345.25	-1 075
325.05	-1 198	348.25	-1 090
331.25	-1 137	359.05	-1 004
333.15	-1 064	364.95	-1 006
336.05	-1 112	367.35	- 947
338.85	-1 099	372.25	- 916

b.

T	B
343.15	-1 035
353.15	- 971
373.15	- 852
373.15	- 839
398.15	- 733

13. F.G. Waelbroek, J. chem. Phys. 23 749 (1955); J. Chim. phys. 54 710 (1957).

T	B	T	B
331.15	-1 125 ± 6.5	341.15	-1 054 ± 10.5
333.15	-1 117 ± 11.5	343.15	-1 035 ± 10.7
335.15	-1 103	348.15	-1 011 ± 8.6
338.15	-1 088 ± 2.1		

14. R. Whytlaw-Gray and G.A. Bottomley, Nature, Lond. 180 1252 (1957).

T	B
295.2	-1 525
295.2	-1 537

15. R.J.L. Andon, J.D. Cox, E.F.G. Herington, and J.F. Martin, Trans. Faraday Soc. 53 1074 (1957).
Maximum pressure 800 torr.

T	B
340.3	-1 046 ± 16
349.1	- 969 ± 12
363.0	- 900 ± 12
376.9	- 830 ± 8
393.1	- 729 ± 11

16. J.D. Cox and R.J.L. Andon, Trans. Faraday Soc. 54 1622 (1958).
Maximum pressure 570 torr.

T	B
409.0	-668 ± 18
424.2	-572 ± 14
437.8	-516 ± 22

17. G.A. Bottomley, C.G. Reeves, and R. Whytlaw-Gray, Proc. R. Soc. A246 504 (1958).
Maximum pressure 400 torr. Accuracy ± 10.

T	B
295.2	-1 525
308.2	-1 349
323.2	-1 202
343.2	-1 035

18. G.A. Bottomley, C.G. Reeves, and R. Whytlaw-Gray, Proc. R. Soc. A246 514 (1958).

T	B
295.2	-1 528 ± 12
308.2	-1 352 ± 22

19. J.D. Cox and D. Stubley, Trans. Faraday Soc. 56 484 (1960).

T	B
373.2	-814 ± 6

20. J.F. Connolly and G.A. Kandalic, Physics Fluids 3 463 (1960).

3-term fit. Maximum pressure 25 atm. Standard deviation in B of ± 0.2.

Values of B calculated from their equation.

T	B	C
493.2	-442	39 000
513.2	-405	41 000
533.2	-372	40 500
553.2	-341	39 200
573.2	-315	36 800

21. Sh. D. Zaalishvili and Z.S. Belousova, Russ. J. phys. Chem. 38 269 (1964); Zh. fiz. Khim. 38 503 (1964).

Class II.

T	B
353.2	-991
363.2	-897
373.2	-840
383.2	-814

22. Sh. D. Zaalishvili, Z.S. Belousova, and L.E. Kolysko, Russ. J. phys. Chem. 39 232 (1965); Zh. fiz. Khim. 39 447 (1965).

Errors do not exceed 3.5 per cent.

T	B
353.2	-944
363.2	-891
373.2	-833
383.2	-808

23. G.A. Bottomley and T.H. Spurling, Aust. J. Chem. <u>19</u> 1331 (1966).

(e) indicates low-pressure results for which the initial pressure of the experiments was 99 torr. The remaining results were obtained with an initial pressure of 312 torr.

For high pressure results error ± 12. Low pressure results somewhat less accurate.

T	B	T	B
308.37	-1 394(e)	380.47	- 785
331.12	-1 123	405.00	- 724(e)
337.64	-1 065(e)	416.08	- 641
343.16	-1 023, - 1 023(e)	416.50	- 638
354.48	- 938(e)	429.70	- 599(e)
354.96	- 939	457.67	- 521

24. Z.S. Belousova and Sh. D. Zaalishvili, Russ. J. phys. Chem. <u>41</u> 1290 (1967); Zh. fiz. Khim. <u>41</u> 2388 (1967).

T	B	C
433.2	-637	
443.2	-610	
453.2	-584	
463.2	-555	
473.2	-532	
483.2	-506	
493.2	-483	
503.2	-458	

25. D.H. Knoebel and W.C. Edmister, J. chem. Engng. Data <u>13</u> 312 (1968).
Low pressure PVT measurements.

T	B
313.2	-1303
333.2	-1107
353.2	- 944
373.2	- 815

26. P.G. Francis, M.L. McGlashan, and C.J. Wormald, J. chem. Thermodyn. <u>1</u> 441 (1969).

T	B	C
329.54	-1143	
330.78	-1118	
331.77	-1124	
332.96	-1114	
335.60	-1076	
337.84	-1038	
340.52	-1048	
343.65	-1023	
344.15	-1007	
347.69	- 975	
354.00	- 928	
358.87	- 915	
368.28	- 847	
374.50	- 825	
380.02	- 806	
383.92	- 764	
398.75	- 726	
403.07	- 688	
408.25	- 678	
412.68	- 658	
422.99	- 611	

27. R.F. Hajjar, W.B. Kay, and G.F. Leverett, J. chem. Engng. Data 14 377 (1969).

Low pressure PVT measurements.

T	B	T	B
313.15	-1565	382.15	- 765
323.15	-1350	398.15	- 720
331.15	-1215	413.15	- 645
335.15	-1150	427.15	- 610
353.15	- 945	453.15	- 550
368.15	- 840	473.15	- 510

28. R.J. Powell, Ph.D. Thesis, University of Strathclyde (1969).

Class II.

T	B
364.065	-915
389.497	-749
413.856	-681
443.778	-581.5
474.805	-469

29. C. Eon, C. Pommier and G. Guiochon, J. chem. Engng Data <u>16</u> 408 (1971).
Isoteniscopic method: estimated uncertainty in B ± 4%.

T	B
333.45	-1110
343.45	-1030
353.45	- 960
363.45	- 900
373.45	- 850

30. C.J. Wormald, J.C.S. Faraday 1 <u>71</u> 726 (1975).

The following equation was determined by analysis of available
second virial coefficient data, isothermal Joule-Thomson coefficients
and pressure coefficients of the heat capacity.

$$B = 1109 \quad -1.734 \times 10^6 T^{-1} + 8.731 \times 10^8 T^{-2}$$
$$-2.248 \times 10^{11} T^{-3} + 1.491 \times 10^{13} T^{-4}$$

for T in the range 295 to 630 K.

31. N. Al-Bizreh and C.J. Wormald, J. chem. Thermodyn. <u>9</u> 749 (1977).

The following equation was determined by analysis of available
second virial coefficent data, pressure coefficients of the heat
capacity and isothermal Joule-Thomson coefficients, including new
data given in this paper.

Standard percentage deviation in B is 2.6%; T range 295 - 630 K.

$$B = 41.5 \quad -1.118 \times 10^5 T^{-1} - 3.850 \times 10^7 T^{-2}$$
$$-1.289 \times 10^9 T^{-3} - 5.158 \times 10^{12} T^{-4}$$

Calculated values:

T	B	T	B
298.15	-1468	473.15	- 482
323.15	-1184	498.15	- 432
348.15	- 979	523.15	- 391
373.15	- 825	548.15	- 355
398.15	- 708	573.15	- 325
423.15	- 616	598.15	- 299
448.15	- 541	623.15	- 276

32. M. Gehrig and H. Lentz, J. chem. Thermodyn. $\underline{9}$ 445 (1977).

PVT data given; P range 50 - 3000 atm, T → 683 K.

PHENOL C_6H_5OH

1. R.J.L. Andon, D.P. Biddiscombe, J.D. Cox, R. Handley, D. Harrop, E.F.G. Herington, and J.F. Martin, J. chem. Soc. 5246 (1960).

B determined from (a) pressure series and (b) volume series.

T	B	
	(a)	(b)
428.5	-1495 ± 30	-1454 ± 28
437.8	-1373 ± 25	-1338 ± 25
453.7	-1169 ± 18	-1135 ± 16

BENZENETHIOL (MERCAPTOBENZENE) $C_6H_5.SH$

1. D.W. Scott, J.P. McCullough, W.N. Hubbard, J.F. Messerly, I.A. Hossenlopp, F.R. Frow, and G. Waddington, J. Am. chem. Soc. $\underline{78}$ 5463 (1956).

Values of B were calculated by the authors from heats of vaporization and vapour pressure measurements.

T	B
375.23	-2 100
394.73	-1 905
407.33	-1 730
416.89	-1 609

α-PICOLINE (2-METHYLPYRIDINE, 2-PICOLINE) $CH_3.C_5H_4N$

1. J.D. Cox and R.J.L. Andon, Trans. Faraday Soc. <u>54</u> 1622 (1958); R.J.L. Andon, J.D. Cox, E.F.G. Herington, and J.F. Martin, Trans. Faraday Soc. <u>53</u> 1074(1957).

    Maximum pressure 560 torr.

T	B
363.0	-1 608 ± 19
376.9	-1 449 ± 33
393.1	-1 272 ± 29
409.0	-1 119 ± 17
424.2	- 998 ± 3
437.8	- 933 ± 18

2. D.W. Scott, W.N. Hubbard, J.F. Messerly, S.S. Todd, I.A. Hossenlopp, W.D. Good, D.R. Douslin, and G. Waddington, J. phys. Chem., Ithaca <u>67</u> 680 (1963).

    Values of B calculated by the authors from measurements of vapour pressure and heats of vaporization.

T	B
359.34	-1 673
379.47	-1 399
402.53	-1 204

β-PICOLINE (3-METHYLPYRIDINE, 3-PICOLINE) $CH_3.C_5H_4N$

1. J.D. Cox and R.J.L. Andon, Trans. Faraday Soc. <u>54</u> 1622 (1958); R.J.L. Andon, J.D. Cox, E.F.G. Herington, and J.F. Martin, Trans. Faraday Soc. <u>53</u> 1074 (1957).

    Maximum pressure 550 torr.

T	B
376.9	-1 909 ± 33
393.1	-1 514 ± 33
409.0	-1 377 ± 12
424.2	-1 193 ± 20
437.8	-1 132 ± 27

2. D.W. Scott, W.D. Good, G.B. Guthrie, S.S. Todd, I.A. Hossenlopp, A.G. Osborn, and J.P. McCullough, J. phys. Chem., Ithaca <u>67</u> 685 (1963).

    Values of B calculated by the authors from measurements of vapour pressure and heats of vaporization.

T	B
372.44	-1 799
393.35	-1 600
417.28	-1 355

γ-PICOLINE (4-METHYLPYRIDINE) $CH_3.C_5H_4N$

1. J.D. Cox and R.J.L. Andon, Trans. Faraday Soc. 54 1622 (1958); R.J.L. Andon, J.D. Cox, E.F.G. Herington, and J.F. Martin, Trans. Faraday Soc. 53 1074 (1957).

Maximum pressure 480 torr.

T	B
376.9	-1 880 ± 25
393.1	-1 482 ± 16
409.0	-1 324 ± 18
424.2	-1 222 ± 9
437.8	-1 122 ± 27

DIHYDROBENZENE C_6H_8

1. J.D. Lambert and J.S. Rowlinson (1950)(†).

Class II.

T	B
329.2	-1 560
351.8	-1 310

2-ETHYL THIOPHENE $CH_2.CH_2.C : CH.CH:CH.S$

1. C. Eon, C. Pommier and G. Guiochon, J. chem. Engng Data 16 408 (1971).

Isoteniscopic method: estimated uncertainty in B ± 4%.

T	B
333.45	-2210
343.45	-2070
353.45	-1970
363.45	-1890
373.45	-1790

2,5-DIMETHYL THIOPHENE $CH_3 \cdot \underline{C : CH.CH : C(CH_3)} \cdot S$

1. C. Eon, C. Pommier, and G. Guiochon, J. chem. Engng Data <u>16</u> 408 (1971).

    Isoteniscopic method: estimated uncertainty in B ± 4%.

T	B
333.45	-2410
343.45	-2280
353.45	-2160
363.45	-2040
373.45	-1920

CYCLOHEXANE (HEXAHYDROBENZENE, HEXAMETHYLENE) $CH_2 \cdot \underline{(CH_2)_4} CH_2$

The results of Al-Bizreh and Wormald (10) and Rotinjanz and Nagornow (1) are in good agreement above 430K. Below 310K there is a considerable discrepancy between the determinations of Al-Bizreh and Wormald (10) and Bottomley and Remmington (4). The results of Kerns et al. (9) Hajjar et al. (7) and Lambert et al. (2) have been ignored in establishing the recommended values given below.

T	B	T	B
300	1700 ± 100	380	880 ± 50
310	1525 ± 50	400	790 ± 50
320	1390 ± 50	440	640 ± 50
330	1280 ± 50	480	540 ± 50
340	1175 ± 50	520	440 ± 50
350	1080 ± 50	560	370 ± 50
360	1000 ± 50		

1. L. Rotinjanz and N. Nagornow, Z. phys. Chem. <u>A169</u> 20 (1934).

    PVT data given; P range 2-20 atm, T range 433-574 K.

    B values calculated by H.G. David, S.D. Hamann, and R.B. Thomas, Aust. J. Chem. <u>12</u> 309 (1959). The following values are read from their graph.

T	B	T	B
433.5	-680	544.1	-398
453.4	-614	549.1	-388
473.5	-588	550.7	-388
493.7	-500	554.2	-385
513.5	-446	558.6	-375
534.1	-431	573.8	-351

2. J.D. Lambert, G.A.H. Roberts, J.S. Rowlinson, V.J. Wilkinson, Proc. R. Soc. A196 113 (1949)(†).

Maximum pressure 600 torr. Accuracy ± 50.

T	B	T	B
324.3	-1 320	350.9	-1 100
331.2	-1 250	362.2	- 930
334.2	-1 270	370.3	- 840
337.5	-1 120	380.4	- 900
341.6	-1 140	393.4	- 750
344.0	-1 080	404.2	- 800

3. F.G. Waelbroek, J. chem. Phys. 23 749 (1955); J. Chim. phys. 54 710 (1957).

T	B	T	B
315.15	-1 518	333.15	-1 268 ± 13.3
320.15	-1 435 ± 8.5	338.15	-1 236 ± 10.7
323.15	-1 382 ± 13.5	343.15	-1 180 ± 10.0
328.15	-1 355 ± 16.0	348.15	-1 171 ± 3.1
331.15	-1 301 ± 2.5		

4. G.A. Bottomley and T.A. Remmington, J. chem. Soc. 3800 (1958).

Errors ~ 2 per cent in B.

T	B
295.2	-1 600
295.2	-1 663(a)
308.2	-1 515
308.2	-1 523(a)

(a) These values are considered less reliable by the authors.

5. J.D. Cox and D. Stubley, Trans. Faraday Soc. 56 484 (1960).

T	B
373.2	-910 ± 8

6. G.A. Bottomley and I.H. Coopes, Nature, Lond. 193 268 (1962).

T	B
308.2	-1 457 ± 10
323.2	-1 309 ± 10
343.2	-1 121 ± 10

7. R.F. Hajjar, W.B. Kay, and G.F. Leverett, J. chem. Engng Data 14 377 (1969).

Low pressure PVT measurements.

T	B	T	B
316.58	-1625	396.62	- 795
326.07	-1365	411.05	- 690
336.40	-1225	428.15	- 625
353.15	-1045	453.15	- 540
368.15	- 940	473.15	- 490
382.00	- 885		

8. R.J. Powell, Ph.D. Thesis, University of Strathclyde (1969).

Class II.

T	B
364.883	-965
389.712	-809
414.668	-701
443.873	-595
475.125	-512

9. W.J. Kerns, R.G. Anthony, and P.T. Eubank, A. I. Ch. E. Symp. Ser. 70 14 (1974).

Burnett method.

T	B	C
423	- 643	-2 330 000
448	- 535	-1 090 000
473	- 475	- 650 000
498	- 420	- 290 000
523	- 377	- 270 000

10. N. Al-Bizreh and C.J. Wormald, J. chem. Thermodyn. <u>9</u> 749 (1977).

The following equation was determined by analysis of available second virial coefficient data, and measured isothermal Joule-Thomson coefficients.

Standard percentage deviation in B is 2.2%; T range 273 to 573 K.

$B = 545.4 - 360.7 \exp (527.6/T)$.

Calculated values:

T	B	T	B
273.15	-1943	448.15	- 625
298.15	-1571	473.15	- 555
323.15	-1300	498.15	- 495
348.15	-1096	523.15	- 443
373.15	- 938	548.15	- 399
398.15	- 812	573.15	- 360
423.15	- 710		

METHYLCYCLOPENTANE (METHYLPENTAMETHYLENE) $\underbrace{CH.(CH_3).(CH_2)_3.CH_2}$

1. J.P. McCullough, R.E. Pennington, J.C. Smith, I.A. Hossenlopp, and G. Waddington, J. Am. chem. Soc. <u>81</u> 5880 (1959).

Values calculated from vaporization data.

T	B
304.08	-1 456
325.97	-1 263
344.96	-1 118

HEX-1-ENE(BUTYLETHYLENE) $CH_2:CH.(CH_2)_3.CH_3$

1. M.L. McGlashan and C.J. Wormald, Trans. Faraday Soc. <u>60</u> 646 (1964). Class I.

T	B	T	B
313.8	-1 493	373.1	- 947
325.7	-1 340	383.4	- 886
333.7	-1 262	392.6	- 845
342.6	-1 194	403.2	- 779
353.8	-1 087	410.3	- 763
364.3	-1 009		

2, 3-DIMETHYL 2-BUTENE (TETRAMETHYLETHYLENE) $(CH_3)_2C:C(CH_3)_2$

1. D.W. Scott, H.L. Finke, J.P. McCullough, M.E. Gross, J.F. Messerly, R.E. Pennington, and G. Waddington, J. Am. chem. Soc. <u>77</u> 4993 (1955).

    Values of B calculated by the authors from measurements of vapour pressure and heats of vaporization.

T	B
292.12	-2 179
307.86	-1 818
325.78	-1 596
346.36	-1 362

ETHYL PROPYL KETONE $CH_3.CH_2.CO.(CH_2)_2.CH_3$

1. J.L. Hales, E.B. Lees, and D.J. Ruxton, Trans. Faraday Soc. <u>63</u> 1876 (1967).

    B calculated from measured vapour pressures and heats of vaporization.

T	B
354.38	-2436
374.07	-1982
396.65	-1670

n-HEXANE $CH_3.(CH_2)_4.CH_3$

Values obtained by different workers for the second virial coefficient of n-hexane generally agree to within the estimated experimental uncertainty although some values given by Lambert et al (2) are less negative and values given by Hajjar, Kay, and Leverett (7) below 350 K are more negative, than expected. Values taken from a smooth curve through the data are given below.

T	B	T	B
300	-1925 ± 40	380	- 980 ± 20
310	-1720 ± 35	400	- 855 ± 20
320	-1555 ± 35	425	- 725 ± 20
340	-1310 ± 30	450	- 620 ± 20
360	-1130 ± 25		

1. E.A. Kelso and W.A. Felsing, J. Am. chem. Soc. <u>62</u> 3132 (1940).

    PVT data given: P range 5 - 312 atm, T range 373 - 498 K.

2. J.D. Lambert, G.A.H. Roberts, J.S. Rowlinson, and V.J. Wilkinson, Proc. R. Soc. A196 113 (1949).

Values of B read from the graph. Accuracy of B values ± 50.

T	B	T	B
303	-1 600	350	-1 135
320	-1 560	350	-1 130
322	-1 500	364	-1 000
335	-1 265	395	- 865

3. G.A. Bottomley and C.G. Reeves, J. chem. Soc. 3794 (1958).

T	B
298.2	-1 984 ± 3
313.2	-1 620 ± 3
328.2	-1 406 ± 3

4. R.G. Griskey and L.N. Canjar, A. I. Ch. E. Jl 5 29 (1959).

PVT data given: P range 25 - 230 atm, T range 513 - 573 K.

5. M.L. McGlashan and D.J.B. Potter, Proc. R. Soc. A267 478 (1962).

Precision of B values better than ± 20.

Sample (1)

T	B	T	B
313.5	-1 676	354.0	-1 194
318.1	-1 637	360.2	-1 147
324.0	-1 496	368.0	-1 075
332.1	-1 419	378.4	- 981
338.5	-1 326	387.2	- 913
339.4	-1 310	398.3	- 895
342.4	-1 289	413.1	- 771
347.3	-1 239		

Sample (2)

T	B	T	B
318.0	-1 598	370.6	-1 050
329.0	-1 419	377.7	- 992
337.7	-1 338	386.7	- 928
348.0	-1 235	395.9	- 880
358.5	-1 144	410.1	- 800

6. S.F. Di Zio, M.M. Abbott, D. Zibello, and H.C. Van Ness, Ind. Engng Chem. Fundamen. 5 569 (1966).

Density balance. Values of B read from graph.

T	B	T	B
318.2	-1580	333.2	-1420
318.2	-1560	348.2	-1200
318.2	-1620	348.2	-1220
333.2	-1320	348.2	-1230
333.2	-1340	363.2	-1100
333.2	-1380		

7. R.F. Hajjar, W.B. Kay, and G.F. Leverett, J. chem. Engng Data 14 377 (1969).

Low pressure PVT measurements.

T	B	T	B
315.15	-1750	382.15	- 995
323.15	-1590	403.15	- 850
335.15	-1450	427.15	- 730
353.15	-1205	453.15	- 615
368.15	-1090	473.15	- 530

8. Sh. D. Zaalishvili, Z.S. Belousova, and V.P. Verkhova, Russ. J. phys. Chem. 45 149 (1971); Zh. fiz. Khim. 45 268 (1971).

T	B	C
433.2	-714	
443.2	-677	
453.2	-644	
463.2	-613	
473.2	-583	
478.2	-570	

9. N. Al-Bizreh and C.J. Wormald, J. chem. Thermodyn. 10 231 (1978).

B values calculated from the authors best fit to second virial coefficient and Joule-Thomson coefficient data.

T	B	T	B
300	-1822	450	- 650
325	-1466	475	- 572
350	-1207	500	- 508
375	-1014	550	- 406
400	- 864	600	- 331
425	- 746	700	- 228

2-METHYLPENTANE $(CH_3)_2CH(CH_2)_2CH_3$

1. E.A. Kelso and W.A. Felsing, J. Am. chem. Soc. **62** 3132 (1940).

    PVT data given: P range 5 - 312 atm, T range 373 - 498 K.

2. G. Waddington, J.C. Smith, D.W. Scott, and H.M. Huffman, J. Am. chem. Soc. **71** 3902 (1949).

    Values of B calculated by the authors from measurements of vapour pressure and heats of vaporization.

T	B
298.15	-1 712
298.15	-1 792†
318.15	-1 487
333.42	-1 332

    † This point was determined from the data of Osborne and Ginnings, J. Res. natn. Bur. Stand. **39** 453 (1947).

3-METHYLPENTANE(DIETHYLMETHYLMETHANE) $C_2H_5CH(CH_3)C_2H_5$

1. G. Waddington, J.C. Smith, D.W. Scott, and H.M. Huffman, J. Am. chem. Soc. **71** 3902 (1949).

    Values of B calculated by the authors from measurements of vapour pressure and heats of vaporization.

T	B
298.15	-1 718†
303.25	-1 592
323.70	-1 418
336.44	-1 285

    † This point was determined from the data of Osborne and Ginnings, J. Res. natn. Bur. Stand. **39** 453 (1947).

2,2-DIMETHYL BUTANE (NEOHEXANE, ETHYLTRIMETHYLMETHANE) $(CH_3)_3CC_2H_5$

1. W.A. Felsing and G.M. Watson, J. Am. chem. Soc. $\underline{65}$ 1889 (1943).
    PVT data given: P range 10 - 300 atm, T range 373 - 548 K.

2,3-DIMETHYLBUTANE $(CH_3)_2.CH.CH.(CH_3)_2$

1. E.A. Kelso and W.A. Felsing, Ind. Engng Chem. ind. Edn $\underline{34}$ 161 (1942).
    PVT data given: P range 35 - 290 atm, T values 523.16 and 548.16 K.

2. G. Waddington, J.C. Smith, D.W. Scott, and H.M. Huffman, J. Am. chem. Soc. $\underline{71}$ 3902 (1949).
    Values of B calculated by the authors from measurements of vapour pressure and heats of vaporization.

T	B
295.95	-1 554
298.15	-1 569+
303.02	-1 496
313.12	-1 400
331.14	-1 256

    + Data of Osborne and Ginnings, J. Res. natn. Bur. Stand. $\underline{39}$ 453 (1947).

ISOPROPYL ETHER $(CH_3)_2CH.O.CH(CH_3)_2$

1. R.J.L. Andon, J.F. Counsell, D.A. Lee, and J.F. Martin, J.C.S. Faraday 1 $\underline{70}$ 1914 (1974).
    B calculated from measured vapour pressures and enthalpy of vaporization.
    Class III.

T	B
310.85	-1815
329.51	-1531
341.73	-1407

TRIETHYLAMINE $(C_2H_5)_3.N$

1. J.D. Lambert and E.D.T. Strong, Proc. R. Soc. A200 566 (1950)(†).
    Maximum pressure 600 torr. Accuracy ± 50.

T	B	T	B
323.2	-1 679	351.6	-1 328
325.3	-1 593	362.4	-1 224
329.6	-1 582	374.2	-1 159
332.8	-1 566	383.2	-1 069
331.5	-1 490	393.6	-1 003
342.0	-1 421	405.0	- 966

HEXAMETHYLDISILOXANE $(CH_3)_3Si.O.Si(CH_3)_3$

1. D.W. Scott, J.F. Messerly, S.S. Todd, G.B. Guthrie, I.A. Hossenlopp,
    R.T. Moore, A. Osborn, W.T. Berg, and J.P. McCullough, J. phys. Chem.,
    Ithaca 65 1320 (1961).

    B calculated from heats of vaporization and vapour pressure
    measurements.

T	B
332.31	-2493
351.50	-2191
373.67	-1856

PERFLUOROMETHYLCYCLOHEXANE $C_6F_{11}.CF_3$

1. F.L. Casado, D.S. Massie, and R. Whytlaw-Gray, Proc. R. Soc. A214
    466 (1952)(*).

    Maximum pressure 22 torr.

T	B
295.2	-1 671

2. R.F. Hajjar, W.B. Kay, and G.F. Leverett, J. chem. Engng Data 14 377
    (1969).

    Low pressure PVT measurements.

T	B	T	B
313.15	-2610	398.15	- 905
323.15	-2140	413.15	- 815
337.15	-1595	427.15	- 730
353.15	-1370	453.15	- 600
368.15	-1145	473.15	- 505
382.15	-1065		

1-HYDRO-PENTADECAFLUORO-n-HEPTANE $CHF_2 \cdot (CF_2)_5 \cdot CF_3$

1. G.A. Bottomley.

Values given by J.L. Carson, R.C. Stewart, and A.G. Williamson, J. chem. Engng   Data 11 231 (1966).

T	B
323.16	-299.0
326.46	-289.0
346.61	-237.2
380.00	-181.3
398.71	-156.8
431.56	-124.3

2,3,4,5,6-PENTAFLUOROTOLUENE (METHYLPENTAFLUOROBENZENE) $C_6F_5 \cdot CH_3$

1. J.F. Counsell, J.L. Hales, E.B. Lees, and J.F. Martin, J. chem. Soc. (A) 2994 (1968).

B calculated from measured vapour pressures and heats of vaporization.

T	B
349.16	-2765
368.52	-2253
390.65	-1896

2. D. Ambrose, J.H. Ellender, C.H. Sprake, and R. Townsend, J.C.S. Faraday 1 71 35 (1975).

B calculated from measured vapour pressures and enthalpy of vaporization.

Class III.

T	B
390.649	-1890

BENZOTRIFLUORIDE(α,α,α-TRIFLUOROTOLUENE) $C_6H_5 \cdot CF_3$

1.   D.W. Scott, D.R. Douslin, J.F. Messerly, S.S. Todd, I.A. Hossenlopp, T.C. Kincheloe, and J.P. McCullough, J. Am. chem. Soc. <u>81</u> 1015 (1951).

Values of B calculated by the authors from heats of vaporization and vapour pressure measurements.

T	B
334.21	-1 960
353.30	-1 735
375.20	-1 450

o-FLUOROTOLUENE $C_6H_4 \cdot (CH_3) \cdot (F)$

1.   L.M. Mozhginskaya and L.E. Kolysko, Russ. J. phys. Chem. <u>48</u> 881 (1974); Zh. fiz. Khim. <u>48</u> 1506 (1974).

Constant volume piezometer.

T	B	T	B
373.2	-1485 ± 6	413.2	-1110 ± 5
383.2	-1370 ± 5	423.2	-1042 ± 5
393.2	-1274 ± 5	433.2	- 980 ± 4
403.2	-1188 ± 5	443.2	- 930 ± 4

m-FLUOROTOLUENE $C_6H_4 \cdot (CH_3) (F)$

1.   L.V. Mozhginskaya and L.E. Kolysko, Russ. J. phys. Chem. <u>48</u> 881 (1974); Zh. fiz. Khim. <u>48</u> 1506 (1974).

Constant volume piezometer.

T	B	T	B
373.2	-1585 ± 8	423.2	-1110 ± 7
383.2	-1460 ± 5	433.2	-1045 ± 6
393.2	-1350 ± 5	443.2	- 994 ± 6
403.2	-1260 ± 7	453.2	- 940 ± 8
413.2	-1180 ± 7		

p-FLUOROTOLUENE $C_6H_4 \cdot (CH_3) (F)$

1.   L.V. Mozhginskaya and L.E. Kolysko, Russ. J. phys. Chem. <u>48</u> 881 (1974); Zh. fiz. Khim. <u>48</u> 1506 (1974).

Constant volume piezometer.

T	B	T	B
373.2	-1440 ± 8	413.2	-1175 ± 8
383.2	-1360 ± 9	423.2	-1128 ± 7
393.2	-1290 ± 9	433.2	-1082 ± 7
403.2	-1230 ± 8	443.2	-1047 ± 6

TOLUENE (METHYLBENZENE, PHENYLMETHANE) $C_6H_5.CH_3$

1. Huggett and M.L. McGlashan, (1953) (†).

    Class II.

T	B	T	B
373.15	-1 449	383.15	-1 297
373.15	-1 532	383.15	-1 284
373.15	-1 506	393.15	-1 249
383.15	-1 364	393.15	-1 273

2. R.J.L. Andon, J.D. Cox, E.F.G. Herington, and J.F. Martin, Trans. Faraday Soc. 53 1074 (1957).

    Maximum pressure 700 torr.

T	B
349.1	-1 660 ± 21
363.0	-1 453 ± 16
376.9	-1 335 ± 9
393.1	-1 189 ± 9

3. J.D. Cox and R.J.L. Andon, Trans. Faraday Soc. 54 1622 (1958).

    Maximum pressure 560 torr.

T	B
409.0	-1 013 ± 12
424.2	- 940 ± 8
437.8	- 864 ± 17

4. D.W. Scott, G.B. Guthrie, J.F. Messerly, S.S. Todd, W.T. Berg, I.A. Hossenlopp, and G. Waddington, J. phys. Chem., Ithaca 66 911 (1962).

    Values of B calculated by the authors from measurements of vapour pressure and heats of vaporization.

T	B
341.26	-1 718
361.05	-1 497
483.76	-1 280
410.10	-1 057

5. L.E. Kolysko, Z.S. Belousova, T.D. Sulimova, L.V. Mozhginskaya, and
V.M. Prokhorov, Russ. J. phys. Chem. 47 1067 (1973); Zh. fiz. Khim.
47 1890 (1973).

Constant volume piezometer. Standard error in B given as ± 5.

T	B	T	B
373.2	-1376	453.2	- 850
388.2	-1227	463.2	- 812
403.2	-1098	473.2	- 775
423.2	- 970	483.2	- 746
443.2	- 886	493.2	- 704

6. G. Opel, B. Zorn, and K.-D. Zwerschke, Z. phys. Chem. 255 997 (1974).

B determined from measurements of pressure at constant volume of a
known mass of toluene at different temperatures.

Pressure → 1.9 atm.

(a) Virial equation in terms of molar volume.

T	B	T	B
413.15	-1010 ± 5	513.15	- 594 ± 5
433.15	- 896 ± 6	533.15	- 541 ± 7
453.15	- 803 ± 3	553.15	- 495 ± 8
473.15	- 723 ± 2	573.15	- 450 ± 6
493.15	- 654 ± 3	583.15	- 432 ± 6

(b) Virial equation in terms of pressure.

T	B	T	B
413.15	-1011 ± 17	513.15	- 587 ± 11
433.15	- 911 ± 8	533.15	- 531 ± 13
453.15	- 810 ± 2	553.15	- 483 ± 15
473.15	- 724 ± 4	573.15	- 457 ± 21
493.15	- 651 ± 8	583.15	- 437 ± 22

METHYL PHENYL ETHER $C_6H_5OCH_3$

1. J.L. Hales, E.B. Lees, and D.J. Ruxton, Trans. Faraday Soc. <u>63</u> 1876 (1967).

    B calculated from measured vapour pressures and heats of vaporization.

T	B
366.91	-2593
381.67	-1988
402.70	-1698
426.73	-1475

o-CRESOL $C_6H_4 \cdot (OH)(CH_3)$

1. R.J.L. Andon, D.P. Biddiscombe, J.D. Cox, R. Handley, D. Harrop, E.F.G. Herington, and J.F. Martin, J. chem. Soc. 5246 (1960).

    B determined from (a) pressure series and (b) volume series.

T	B	
	(a)	(b)
437.8	-1748 ± 24	-1693 ± 24
453.7	-1378 ± 35	-1343 ± 31

c-HEPTATRIENE(1,3,5-CYCLOHEPTATRIENE,TROPILIDENE) $\underline{CH_2 \cdot CH:CH_2 \cdot CH:CH_2 \cdot CH:CH}$

1. H.L. Finke, D.W. Scott, M.E. Gross, J.F. Messerly, and G. Waddington, J. Am. chem. Soc. <u>78</u> 5469 (1956).

T	B
298.15	-2 300

    Uncertainty in B is about 20 per cent. The value for B was estimated from a correlation published in the paper by D.W. Scott, H.L. Finke, M.E. Gross, G.B. Guthrie, and H.M. Huffman, J. Am. chem. Soc. <u>72</u> 2424 (1950).

2,3-LUTIDINE $(CH_3)_2 \cdot C_5H_3N$

1. J.D. Cox, Trans. Faraday Soc. <u>56</u> 959 (1960).

    B calculated from compressibility data using the virial equation linear in (a) pressure and (b) concentration.

T	B		
	(a)		(b)
408.15	-1830 ± 24		-1776 ± 23
423.15	-1692 ± 13		-1614 ± 10
438.95	-1494 ± 9		-1428 ± 8

2,4-LUTIDINE $(CH_3)_2 \cdot C_5H_3N$

1. J.D. Cox, Trans. Faraday Soc. <u>56</u> 959 (1960).

    B calculated from compressibility data using the virial equation
    linear in (a) pressure and (b) concentration.

T	B		
	(a)		(b)
392.9	-1949 ± 34		-1896 ± 34
408.2	-1779 ± 24		-1716 ± 22
423.2	-1693 ± 16		-1627 ± 16
439.0	-1496 ± 8		-1433 ± 8

2,5-LUTIDINE $(CH_3)_2 \cdot C_5H_3N$

1. J.D. Cox, Trans. Faraday Soc. <u>56</u> 959 (1960).

    B calculated from compressibility data using the virial equation
    linear in (a) pressure and (b) concentration.

T	B		
	(a)		(b)
408.2	-1784 ± 15		-1724 ± 14
423.2	-1687 ± 8		-1606 ± 7
439.0	-1484 ± 12		-1420 ± 10

2,6-LUTIDINE(2,6-DIMETHYLPYRIDINE) $(CH_3)_2 \cdot C_5H_3N$

1. J.D. Cox and R.J.L. Andon, Trans. Faraday Soc. <u>54</u> 1622 (1958); R.J.L.
    Andon, J.D. Cox, E.F.G. Herington, and J.F. Martin, Trans. Faraday
    Soc. <u>53</u> 1074 (1957).

T	B	T	B
376.9	-1 898 ± 33	424.2	-1 423 ± 4
393.1	-1 662 ± 16	437.8	-1 237 ± 21
409.0	-1 544 ± 15		

178

3,4-LUTIDINE $(CH_3)_2.C_5H_3N$

1. J.D. Cox, Trans. Faraday Soc. 56 959 (1960).

B calculated from compressibility data using the virial equation linear in (a) pressure and (b) concentration.

T	B	
	(a)	(b)
423.3	-2113 ± 17	-2040 ± 16
439.0	-1783 ± 10	-1703 ± 10

3,5-LUTIDINE $(CH_3)_2.C_5H_3N$

1. J.D. Cox, Trans. Faraday Soc. 56 959 (1960).

B calculated from compressibility data using the virial equation linear in (a) pressure and (b) concentration.

T	B	
	(a)	(b)
423.2	-1992 ± 12	-1908 ± 11
439.0	-1763 ± 13	-1670 ± 9

HEPT-1-ENE $CH_2:CH.(CH_2)_4.CH_3$

1. M.L. McGlashan and C.J. Wormald, Trans. Faraday Soc. 60 646 (1964). Class I.

T	B	T	B
333.8	-1 847	374.2	-1 370
338.2	-1 809	383.8	-1 288
343.6	-1 760	393.3	-1 196
353.8	-1 614	404.3	-1 120
364.0	-1 482	411.4	-1 069

CYCLOHEPTANE $CH_2.(CH_2)_5.CH_2$

1. H.L. Finke, D.W. Scott, M.E. Gross, J.F. Messerly, and G. Waddington, J. Am. chem. Soc. 78 5469 (1956).

T	B
298.15	-2 700

Uncertainty in B is about 20 per cent. This value was estimated
from a correlation published in the paper by D.W. Scott, H.L. Finke,
M.E. Gross, G.B. Guthrie, and H.M. Huffman, J. Am. chem. Soc. <u>72</u> 2424
(1950).

1-cis-3-DIMETHYLCYCLOPENTANE $CH_2.CH(CH_3).CH_2.CH(CH_3).CH_2$

1. J.P. McCullough, R.E. Pennington, J.C. Smith, I.A. Hossenlopp, and
    G. Waddington, J. Am. chem. Soc. <u>81</u> 5880 (1959).

    Values calculated from vaporization data.

T	B
322.61	-1 826
341.81	-1 590
363.92	-1 351

HEPTANOIC ACID (ENANTHIC/ENANTHYLIC/OENANTHIC/n-HEPTOIC/n-HEPTYLIC ACID)
$CH_3(CH_2)_5COOH$

1. R.E. Lundin, F.E. Harris, and L.K. Nash, J. Am. chem. Soc. <u>74</u> 743
    (1952).

    PVT data given: P less than 1 atm, T range 463 - 500 K.

n-HEPTANE $CH_3.(CH_2)_5.CH_3$

1. L.B. Smith, J.A. Beattie, and W.C. Kay, J. Am. chem. Soc. <u>59</u> 1587
    (1937).

    PVT data (P→315 atm).

    Values of B given by J.O. Hirschfelder, F.T. McClure, and I.F. Weeks,
    J. chem. Phys. <u>10</u> 201 (1942).

T	B
548.15	- 569
573.15	- 503
598.15	- 449
623.15	- 399

2. M.L. McGlashan and D.J.B. Potter, Proc. R. Soc. <u>A267</u> 478 (1962).
   Precision of B values better than ± 20.
   Sample (i)

T	B	T	B
349.4	-1 819	389.1	-1 325
358.3	-1 691	400.8	-1 212
368.5	-1 560	413.7	-1 108
378.1	-1 446		

   Sample (ii)

T	B
370.8	-1 520
378.2	-1 436
389.2	-1 334

3. Z.S. Belousova and Sh.D. Zaalishvili, Russ. J. phys. Chem. <u>41</u> 1290
   (1967); Zh. fiz. Khim. <u>41</u> 2388 (1967).

T	B	C
473.2	-825	
483.2	-766	
493.2	-703	
503.2	-655	

4. R.F. Hajjar, W.B. Kay, and G.F. Leverett, J. chem. Engng Data <u>14</u>
   377 (1969).
   Low pressure PVT measurements.

T	B	T	B
313.15	-3735	382.15	-1385
323.15	-2765	398.15	-1265
336.15	-2240	427.15	-1045
353.15	-1760	453.15	- 880
368.15	-1555	473.15	- 775

5. N. Al-Bizreh and C.J. Wormald, J. chem. Thermodyn. <u>10</u> 231 (1978).
   B values calculated from the authors best fit to second virial
   coefficient and Joule-Thomson coefficient data.

T	B	T	B
300	-2784	450	- 912
325	-2192	475	- 796
350	-1774	500	- 701
375	-1467	550	- 555
400	-1235	600	- 449
425	-1055	700	- 306

o-XYLENE (1,2-DIMETHYLBENZENE) $C_6H_4 \cdot (CH_3)_2$

1. R.J.L. Andon, J.D. Cox, E.F.G. Herington, and J.F. Martin, Trans.
    Faraday Soc. 53 1074 (1957);   J.D. Cox and R.J.L. Andon, Trans.
    Faraday Soc. 54 1622 (1958).

    Maximum pressure 550 torr.

T	B
376.9	-2 128 ± 40
393.1	-1 769 ± 17
409.0	-1 547 ± 24
424.2	-1 437 ± 17
437.8	-1 252 ± 21

m-XYLENE (1,3-DIMETHYLBENZENE) $C_6H_4 \cdot (CH_3)_2$

1. J.D. Cox and R.J.L. Andon, Trans. Faraday Soc. 54 1622 (1958); R.J.L.
    Andon, J.D. Cox, E.F.G. Herington, and J.F. Martin, Trans. Faraday
    Soc. 53 1074 (1957).

    Maximum pressure 550 torr.

T	B
376.9	-2 167 ± 18
393.1	-1 781 ± 16
409.0	-1 559 ± 19
424.2	-1 343 ± 22
437.8	-1 200 ± 44

p-XYLENE (1,4-DIMETHYLBENZENE) $C_6H_4 \cdot (CH_3)_2$

1. J.D. Cox and R.J.L. Andon, Trans. Faraday Soc. 54 1622 (1958); R.J.L.
    Andon, J.D. Cox, E.F.G. Herington, and J.F. Martin, Trans. Faraday
    Soc. 53 1074 (1957).

Maximum pressure 550 torr.

T	B
376.9	-2 117 ± 15
393.1	-1 777 ± 17
409.0	-1 543 ± 21
434.2	-1 371 ± 7
437.8	-1 178 ± 31

OCT-1-ENE $CH_2:CH.(CH_2)_5.CH_3$

1. M.L. McGlashan and C.J. Wormald, Trans. Faraday Soc. <u>60</u> 646 (1964).
    Class I.

T	B	T	B
358.8	-2 162	383.8	-1 818
363.2	-2 100	389.9	-1 728
368.1	-2 000	395.4	-1 666
374.5	-1 941	403.7	-1 552
375.3	-1 914	411.6	-1 465
383.3	-1 823		

c-OCTANE $CH_2(CH_2)_6CH_2$

1. H.L. Finke, D.W. Scott, M.E. Gross, J.F. Messerly, and G. Waddington,
    J. Am. chem. Soc. <u>78</u> 5469 (1956).

T	B
298.15	-4 200

Uncertainty in B is about 20 per cent. This value for B was
estimated from a correlation published in the paper by D.W. Scott,
H.L. Finke, M.E. Gross, G.B. Guthrie, and H.M. Huffmann, J. Am. chem.
Soc. <u>72</u> 2424 (1950).

n-OCTANE $CH_3.(CH_2)_6.CH_3$

1. M.L. McGlashan and D.J.B. Potter, Proc. R. Soc. <u>A267</u> 478 (1972).
    Maximum pressure 150 torr. Precision of B values better than ± 20.

T	B	T	B
373.0	-2 122	389.2	-1 828
378.2	-2 048	393.5	-1 778
383.2	-1 939	394.2	-1 764
384.3	-1 905	403.5	-1 641
388.1	-1 869	413.5	-1 518

2. J.F. Connolly and G.A. Kandalic, Physics Fluids 3 463 (1960).

3-term fit. Maximum pressure ~25 atm. Standard deviation in B ± 0.6. B values calculated from their equation.

T	B	C
493.2	-971	170 000
513.2	-876	180 000
533.2	-793	179 000
553.2	-719	170 000
573.2	-653	156 000

3. W.A. Felsing and G.M. Watson, J. Am. chem. Soc. 64 1822 (1942).

PVT data given. (P range 5 - 300 atm, T range 373 - 548 K).

4. Sh. D. Zaalishvili, Z.S. Belousova, and V.P. Verkhova, Russ. J. phys. Chem. 45 902 (1971); Zh. fiz. Khim. 45 1589 (1971).

T	B	C
478.2	-1022	
483.2	- 990	
488.2	- 960	
493.2	- 930	
498.2	- 900	

5. N. Al-Bizreh and C.J. Wormald, J. chem. Thermodyn. 10 231 (1978).

B values calculated from the authors best fit to second virial coefficient and Joule-Thomson coefficient data. *calc. from square-well eq. extrapolated above 400 K*

T	B	T	B
300	-4047	450	-1237
325	-3139	475	-1071
350	-2508	500	- 935
375	-2050	550	- 727
400	-1708	600	- 578
425	-1445	700	- 380

CAESIUM Cs

1. J.P. Stone, C.T. Ewing, J.R. Spann, E.W. Steinkuller, D.D. Williams, and R.R. Miller, J. chem. Engng Data $\underline{11}$ 309 (1966); C.T. Ewing, J.P. Stone, J.R. Spann, and R.R. Miller, J. chem. Engng Data $\underline{11}$ 468 (1966).

    PVT data given (P $\rightarrow$32 atm, T 980 - 1655 K).

2. C.T. Ewing, J.R. Spann, J.P. Stone, and R.R. Miller, J. chem. Engng Data $\underline{16}$ 27 (1971).

    PVT data given: T 1905 - 2135 K, P 71 - 124 atm.

DEUTERIUM CHLORIDE DCl

1. B. Schramm and U. Leuchs (1979)(†).

    Values of B agree within experimental error with B values of HCl.

DEUTERIUM D_2

    All the second virial coefficients reported for deuterium agree within the estimated error limits.

1. A. Michels and M. Goudeket, Physica,'s Grav. $\underline{8}$ 353 (1941)(*).
    Class I.

    (a) 3-term fit of PV data. Maximum pressure 50 atm.

T	B	C
273.15	12.98	706
298.15	13.48	775
323.15	13.79	793
348.15	14.17	815
373.15	14.41	829
398.15	14.75	848
423.15	14.83	870

    (b) 4-term fit of PV data. Maximum pressure 50 atm.

T	B	C
273.15	12.99	2 250
298.15	13.22	948
323.15	12.91	2 139
348.15	13.19	2 338
373.15	13.64	1 898

T	B	C
398.15	13.89	1 815
423.15	14.17	1 727

(c) 3-term fit of PV data. Maximum pressure 230 atm.

T	B	C
273.15	13.25	395
298.15	13.80	390
323.15	14.25	383
348.15	14.66	370
373.15	14.93	364
398.15	15.14	358
423.15	15.39	345

2. H.J. Hoge and J.W. Lassiter, J. Res. natn. Bur. Stand. 47 75 (1951).
PVT data given in the region of the critical point.

3. H.G. David and S.D. Hamann, Trans. Faraday Soc. 49 711 (1953).
PVT data given: P range 150 - 900 atm, T values 64.5 and 78.9 K.

4. F.H. Varekamp and J.J.M. Beenakker, Physica,'s Grav. 25 889 (1959).
Class I.

T	B
18	- 222
19	- 203
20	- 187
21	- 173

A differential method is used: the non-ideality of the gas is compared with that of helium.

5. J.J.M. Beenakker, F.H. Varekamp, and A. Van Itterbeek, Physica,'s Grav. 25 9 (1959).

T	B
20.4	- 181 ± 2

6. A. Michels, W. de Graaff, T. Wassenaar, J.M.H. Levelt, and P. Louwerse, Physica,'s Grav. 25 25 (1959)(*).
PVT data: 6-term fit in density series. P range 5 - 3000 atm.
Class I.

T	B	C
98.15	-4.34	426
103.15	-2.86	447
113.15	-0.09	349
123.15	+1.85	494
138.15	4.43	463
153.15	6.40	450
173.15	8.39	439
198.15	10.16	442
223.15	11.51	410
248.15	12.45	419
273.15	13.32	363
298.15	13.93	349
323.15	14.45	330
348.15	14.85	322
373.15	15.10	335
398.15	15.44	304
423.15	15.55	329

7. H.F.P. Knaap, M. Knoester, C.M. Knobler, and J.J.M. Beenakker, Physica,'s Grav. 28 21 (1962)(*).

Low pressure differential method using helium.

Class I.

T	B
20.51	-179.1
21.55	-165.2

8. R.L. Mills, D.H. Liebenberg, and J.C. Bronson, J. chem. Phys. 68 2663 (1978).

PVT data given: T range 75 - 300 K, P range 2 - 20 kbar.

DEUTERIUM OXIDE D_2O

1. G.S. Kell, G.E. McLaurin, and E. Whalley, J. chem. Phys. 49 2839 (1968).

T	B	C
423.15	-343 ± 13	
423.15	-338 ± 4	
448.14	-269 ± 2	

T	B	C
448.14	-269 ± 2	
473.17	-206 ± 4	-35000 ± 7000
473.16	-202 ± 4	-40000 ± 8000
498.14	-177.8 ± 1.6	-13400 ± 1700
498.14	-174.6 ± 1.2	-18000 ± 1300
523.14	-153.1 ± 1.0	$- 6600 \pm 900$
523.15	-152.1 ± 0.6	$- 7400 \pm 400$
548.14	-135.6 ± 0.8	$- 1000 \pm 400$
573.14	-117.1 ± 0.5	$- 400 \pm 300$
573.14	-118.3 ± 0.3	$+ 300 \pm 100$
598.09	-103.8 ± 0.5	700 ± 200
598.09	-104.1 ± 0.5	900 ± 200
623.21	$- 92.0 \pm 0.3$	1000 ± 100
623.21	$- 92.3 \pm 0.2$	1100 ± 100
748.14	$- 81.5 \pm 0.2$	900 ± 100
648.14	$- 82.4 \pm 0.1$	1200 ± 100
673.16	$- 73.9 \pm 0.2$	1300 ± 100
673.16	$- 73.5 \pm 0.3$	1100 ± 100
698.23	$- 66.3 \pm 0.2$	1200 ± 100
698.23	$- 66.3 \pm 0.2$	1100 ± 100
723.19	$- 59.9 \pm 0.2$	1100 ± 100
723.20	$- 59.8 \pm 0.1$	1000 ± 100
748.17	$- 54.2 \pm 0.2$	1000 ± 100
748.17	$- 54.2 \pm 0.2$	1000 ± 100
773.17	$- 49.0 \pm 0.2$	900 ± 100
773.17	$- 49.1 \pm 0.2$	900 ± 100

AMMONIA-d_3 ND_3

1. G. Adam and B. Schramm, Ber. (dtsch.) Bunsenges. phys. Chem. **81** 442 (1977).

Values of ΔB ($B_{deuterated\ form} - B_{NH_3}$) determined assuming that ΔB (511K) = 0. Maximum error in ΔB estimated to be ± 0.5.

T	ΔB	T	ΔB
298	-9.3	403	-1.1
313	-7.1	433	-0.5
333	-4.9	473	0
363	-2.8		

FLUORINE F_2

1. D. White, J.-H. Hu, and H.L. Johnston, J. chem. Phys. <u>21</u> 1149 (1953).

   Values of B determined using a constant volume gas thermometer.
   Results were obtained in the form $(B_T - B_{300 \ K})$; the latter quantity
   was taken as per tables - 19.0.

T	B (average)	T	B (average)
80	-385	125	-137
85	-310	150	- 98
90	-256	175	- 67
95	-224	200	- 49
100	-202	250	- 26
110	-175		

2. R. Prydz and G.C. Stratz, J. Res. natn. Bur. Stand. <u>74A</u> 747 (1970).
   3 term fit to PV data at densities up to 6.0 mol.ℓ^{-1}.

T	B	C
80	-239.6 ± 40	-22557 ± 20000
85	-213.2	-13548
90	-191.0	- 7748
95	-172.2	- 4016
100	-156.1 ± 10	- 1624 ± 3000
105	-142.2	- 106
110	-130.1	+ 838
115	-119.4	1409
120	-110.0	1736
125	-101.7 ± 2	1905 ± 500
130	- 94.2	1973
135	- 87.5	1979
140	- 81.5	1947
145	- 75.9	1893
150	- 70.9 ± 0.3	1828 ± 30
155	- 66.3	1758
160	- 62.1	1689
165	- 58.2	1621
170	- 54.6	1557
175	- 51.2	1498

T	B	C
180	- 48.1	1442
185	- 45.2	1391
190	- 42.5	1344
195	- 40.0	1301
200	- 37.6 ± 0.3	1261 ± 30
205	- 35.4	1224
210	- 33.2	1190
215	- 31.3	1159
220	- 29.4	1130
225	- 27.6	1104
230	- 25.9	1080
235	- 24.3	1057
240	- 22.8	1037
245	- 21.4	1019
250	- 20.0 ± 0.3	1003 ± 30
255	- 18.7	988
260	- 17.5	976
265	- 16.3	966
270	- 15.2	957
275	- 14.1	951
280	- 13.1	947
285	- 12.2	946
290	- 11.2	946
295	- 10.4	949
300	- 9.5 ± 0.3	955 ± 40

SILICON TETRAFLUORIDE SiF_4

1. S.D. Hamann, W.J. McManamey, and J.F. Pearse, Trans. Faraday Soc. 49
    351 (1953).

    Reproducibility ± 4.

T	B	T	B
293.15	- 145.6	323.15	- 109.4
303.15	- 132.3	333.15	- 100.4
303.15	- 132.6	343.15	- 92.5
313.15	- 121.3	353.15	- 83.7

2. B. Schramm and R. Gehrmann (1979)(†).
    Estimated error in B is ± 6.

T	B	T	B
205.9	-272.2	242.1	-216.0
212.0	-260.4	260.1	-188.8
224.4	-242.9	295.2	-145.0

3. B. Schramm and H. Schmiedel (1979)(†).
    Estimated error in B is ± 10.

T	B	T	B
295	-145.0	425	- 45.0
330	-107.0	450	- 34.5
365	- 79.0	475	- 26.0
400	- 57.0		

IODINE PENTAFLUORIDE IF_5

1. D.W. Osborne, F. Schreiner, and H. Selig, J. chem. Phys. <u>54</u> 3790
    (1971).

    Vapour density measurements.

T	B
329.96	-2813
371.99	-1232

2. A. Heintz and R.N. Lichtenthaler, Ber. (dtsch.) Bunsenges. phys. Chem.
    <u>80</u> 962 (1976).

    Estimated accuracy of B ± 5%.

T	B	T	B
319.8	-2574	365.2	-1760
331.6	-2222	379.4	-1641
333.3	-2305	392.2	-1638
338.4	-2203	399.3	-1459
348.2	-2077	409.1	-1383
360.5	-1975	411.2	-1511

PHOSPHOROUS PENTAFLUORIDE PF_5

1. A. Heintz and R.N. Lichtenthaler, Ber. (dtsch.) Bunsenges. phys.
    Chem. 80 962 (1976).

    Estimated accuracy of B ± 5%.

T	B	T	B
319.8	-164	411.4	- 88
326.1	-156	418.3	- 84
339.6	-148	422.2	- 88
353.4	-128	437.1	- 86
367.1	-119	445.0	- 72
377.4	-116	453.4	- 63
381.3	-106	466.3	- 61
401.5	- 98		

MOLYBDENUM HEXAFLUORIDE MoF_6

1. D.W. Osborne, F. Schreiner, J.G. Malm, H. Selig, and L. Rochester, J.
    chem. Phys. 44 2802 (1966).

T	B
298.15	- 923

2. P. Morizot, J. Ostorero, and P. Plurien, J. Chim. phys. 70 1582
    (1973).

    Density measurements: P → 1.7 atm. Probable uncertainty in B ± 5%.

T	B	T	B
313.2	-790	393.2	-450
333.2	-690	413.2	-400
353.2	-600	433.2	-330
373.2	-530	453.2	-320

3. A. Heintz and R.N. Lichtenthaler, Ber. (dtsch.) Bunsenges. phys.
    Chem. 80 962 (1976).

    Estimated accuracy of B ± 3%.

T	B	T	B
320.0	-713	365.7	-550
329.3	-679	374.4	-514
338.1	-647	387.5	-486
350.6	-588	394.6	-486

SULPHUR HEXAFLUORIDE SF_6

The second virial coefficients of sulphur hexafluoride are generally in good agreement. The following values were taken from a smooth curve drawn through all the points.

T	B	T	B
200	-685 ± 15	350	-190 ± 5
210	-615 ± 12	375	-159 ± 5
220	-555 ± 12	400	-135 ± 4
240	-455 ± 10	425	-113 ± 2
260	-380 ± 8	450	- 97 ± 2
280	-323 ± 7	475	- 81 ± 2
300	-277 ± 5	500	- 67 ± 2
325	-226 ± 5	525	- 56 ± 2

1. K.E. MacCormack and W.G. Schneider, J. chem. Phys. <u>19</u> 845, 849 (1951). (*).

    4-term fit of PV data. Maximum pressure 50 atm. Error in B ± 12 at 273.15 K increasing to ± 1.5 per cent at 473.15 K.

T	B	C
273.15	-333.7	101 310
323.15	-216.9	41 075
373.15	-159.2	23 800
423.15	-118.5	
423.15	-118.3	
473.15	- 78.9	
473.15	- 79.9	
523.15	- 53.8(a)	
523.15	- 51.5	

(a) (The authors give B = -54.8 at 523.15 K.)

2. (a) S.D. Hamann, W.J. McManamey, and J.F. Pearse, Trans. Faraday Soc. <u>49</u> 351 (1953).

    Reproducibility of results ± 5.

T	B	T	B
293.15	-294.1	333.15	-223.1
303.15	-274.8	343.15	-206.7
313.15	-252.6	348.15	-198.6
323.15	-236.4	348.15	-195.6
323.15	-232.4	353.15	-191.6

T	B
373.15	-162.8
398.15	-139.9
423.15	-119.2
448.15	- 94.1
448.15	- 96.8

(b) S.D. Hamann, J.A. Lambert, and R.B. Thomas, Aust. J. Chem. <u>8</u> 149 (1955).

T	B
393.15	- 145

3. H.P. Clegg, J.S. Rowlinson, and J.R. Sutton, Trans. Faraday Soc. <u>51</u> 1327 (1955).

3-term fit of PV data. Maximum pressure 100 atm. Reproducibility of B values ± 3.

T	B	C
307.34	- 259	19 920
323.01	- 230	18 710
348.01	- 191	15 720
370.33	- 164	13 910
404.45	- 131	12 390

4. J.H. Dymond and E.B. Smith, (1962)(†).

Estimated accuracy of B ± 6.

T	B
273.15	-343.3
283.15	-316.8
293.15	-292.0
298.15	-278.5
303.15	-269.5
313.15	-248.8
323.15	-230.0

5. M. Rigby, (1963)(†).

T	B
298.2	- 286 ± 3
323.2	- 230 ± 3

6. W.H. Mears, E. Rosenthal, and J.V. Sinka, J. phys. Chem. <u>73</u> 2254 (1969).

    PVT data given: P range 10 → 80 atm, T range 298 → 363 K. B and C calculated using their own data together with other available PVT data. Standard errors given.

T	B	C
273.15	-341.5 ± 2.4	
293.15	-285.3 ± 0.8	
303.15	-262.7 ± 0.6	
307.34	-254.1 ± 0.5	15702 ± 226
313.15	-242.9 ± 0.5	
323.01	-225.5 ± 0.5	14545 ± 205
323.15	-225.3 ± 0.5	
333.15	-209.5 ± 0.4	
343.15	-195.3 ± 0.4	
348.01	-188.9 ± 0.4	13201 ± 170
348.15	-188.7 ± 0.4	
353.15	-182.4 ± 0.3	
370.33	-162.7 ± 0.3	12333 ± 144
373.15	-159.8 ± 0.3	
398.15	-135.9 ± 0.3	
404.45	-130.5 ± 0.2	11359 ± 112
423.15	-115.6 ± 0.2	
448.15	- 98.0 ± 0.2	
473.15	- 82.5 ± 0.3	
523.15	- 56.3 ± 0.5	

7. R.F. Hajjar and G.E. MacWood, J. chem. Engng Data <u>15</u> 3 (1970).

T	B
343.43	-193.5
373.15	-161.0

8. R.D. Nelson, Jr. and R.H. Cole, J. chem. Phys. <u>54</u> 4033 (1971).

T	B
323.2	-243.3

9. J. Bellm, W. Reineke, K. Schäfer, and B. Schramm, Ber. (dtsch.)
    Bunsenges. phys. Chem. 78 282 (1974).

    Estimated accuracy of B is ± 2.

T	B	T	B
300	-273.7	430	-108.5
320	-233.0	460	- 90.0
340	-200.7	490	- 72.9
370	-161.2	520	- 57.7
400	-131.3	550	- 44.5

10. P.M. Sigmund, I.H. Silberberg, and J.J. McKetta, J. chem. Engng
    Data 17 168 (1972).

    Burnett method : standard errors given.

T	B	C
271.61	-338.92 ± 0.44	18640 ± 190
308.12	-255.87 ± 0.23	19120 ± 360
323.55	-230.00 ± 0.72	17870 ± 230
348.10	-194.29 ± 0.54	15380 ± 330
373.15	-163.41 ± 0.23	12120 ± 130
423.15	-113.61 ± 1.53	

11. R. Hahn, K. Schäfer, and B. Schramm, Ber. (dtsch.) Bunsenges. phys.
    Chem. 78 287 (1974).

    B determined assuming B (296.0 K) = -286.

    Quoted accuracy of B is ± 2.

T	B	T	B
199.6	-693	308.9	-260
210.8	-609	373.9	-167
230.9	-494	432.8	-114
251.8	-409	472.3	- 87
273.2	-342		

12. C. Hosticka and T.K. Bose, J. chem. Phys. 60 1318 (1974).

T	B
323.3	-218.80 ± 1.12
348.3	-198.71 ± 3.52
373.9	-156.40 ± 0.92

13. J. Santafe, J.S. Urieta and C. Gutierrez, Revta Acad. Cienc. exact. fis. - quim. nat. Zaragoza 31 63 (1976).

B values also given by J. Santafe, J.S. Urieta, and C.G. Losa, Chem. phys. 18 341 (1976).

Compressibility measurements. Accuracy of B estimated to be ± 3.

T	B
273.2	-335.5
283.2	-315.7
293.2	-295.2
303.2	-276.3
313.2	-252.4
323.2	-235.2

URANIUM HEXAFLUORIDE UF_6

1. P. Morizot, J. Ostorero, and P. Plurien, J. Chim. phys. 70 1582 (1973).

Density measurements: $P \to 1.7$ atm.

Class II.

T	B	T	B
328.2	-842	413.2	-477
343.2	-778	433.2	-410
358.2	-713	453.2	-362
373.2	-624	463.2	-360
393.2	-557		

2. A. Heintz, E. Meisinger, and R.N. Lichtenthaler, Ber. (dtsch.) Bunsenges. phys. Chem. 80 163 (1976).

Estimated accuracy of B ± 20.

T	B	T	B
320.6	-1032	375.3	- 736
332.2	- 939	385.2	- 710
342.2	- 902	398.7	- 664
347.2	- 870	410.2	- 636
350.2	- 853	417.2	- 615
351.7	- 843	430.2	- 582
357.7	- 815	435.2	- 570
367.2	- 763	445.2	- 548

T	B
451.2	- 530
464.2	- 523
469.2	- 504

TUNGSTEN HEXAFLUORIDE WF_5 ← ↦ √°?

1. P. Morizot, J. Ostorero, and P. Plurien, J. Chim. phys. 70 1582 (1973).

   Density measurements: $P \to 1.7$ atm.

   Class II.

T	B
313.2	-664
333.2	-596
353.2	-490
373.2	-427
393.2	-373
413.2	-330
433.2	-277
453.2	-271

2. A. Heintz and R.N. Lichtenthaler, Ber. (dtsch.) Bunsenges. phys. Chem. 80 962 (1976).

   Estimated accuracy of B ± 3%.

T	B	T	B
320.0	-653	401.2	-411
334.4	-582	411.4	-400
342.8	-584	427.2	-381
351.6	-531	435.0	-377
361.7	-516	438.8	-345
374.0	-486	447.9	-332
382.2	-480	458.1	-326
391.8	-440	461.9	-311

XENON HEXAFLUORIDE XeF_6

1. F. Schreiner, D.W. Osborne, J.G. Malm, and G.N. McDonald, J. chem. Phys. 51 4838 (1969).

    Vapour density measurements.

T	B
346.00	-955

HYDROGEN CHLORIDE HC1

1. G. Glockler, C.P. Roe, and D.L. Fuller, J. chem. Phys. 1 703 (1933) (*).

    PVT data given: P → 100 atm.

    Class II.

T	B
328.7	-100.2
368.7	- 69.5

2. G. Glockler, D.L. Fuller, and C.P. Roe, J. chem. Phys. 1 709 (1933) (*).

    PVT data given: P → 100 atm.

    Class II.

T	B
368.7	- 89.6

3. K. Mangold and E.U. Franck, Z. Elektrochem. 66 260 (1962).

    PVT data given: P range 200 - 2000 atm, T range 333 - 673 K.

4. B. Schramm and U. Leuchs (1979)(†).

    Estimated error in B ± 5 at 480 K, ± 7 at 200 K.

T	B	T	B
190	-456.0	330	-114.0
200	-392.0	370	- 90.0
225	-287.0	400	- 76.0
250	-221.0	420	- 68.5
275	-175.0	450	- 59.0
295	-147.1	480	- 53.0
300	-142.0		

HYDROGEN DEUTERIDE HD

1. H.J. Hoge and J.W. Lassiter, J. Res. natn. Bur. Stand. 47 75 (1951).
    PVT data given in the region of the critical point.

2. F.H. Varekamp and J.J.M. Beenakker, Physica,'s Grav. 25 889 (1959).
    Class I.

T	B	T	B
16	-230	19	-180
17	-214	20	-167
18	-196	21	-154

    The purity of the gas is about 92 per cent, the remainder being
    hydrogen. Differential method: the non-ideality of the gas was
    compared with that of helium.

3. J.J.M. Beenakker, F.H. Varekamp, and A. Van Itterbeek, Physica,'s
    Grav. 25 9 (1959).

T	B
20.4	-162 ± 1

    Gas expansion method. The gas was 92 per cent pure, the rest
    hydrogen.

4. H.F.P. Knaap, M. Knoester, C.M. Knobler, and J.J.M. Beenakker,
    Physica,'s Grav. 28 21 (1962).
    Class I.

T	B
20.55	-160.5
20.67	-158.9

    Low pressure differential method, with helium as reference gas.

HYDROGEN H_2

The different sets of values for the second virial coefficient of
hydrogen all agree within the estimated error limits. The following
values were obtained from a smooth curve through all the data.

T	B	T	B
14	-254 ± 5	50	- 33 ± 2
15	-230 ± 5	75	- 12 ± 1
17	-191 ± 5	100	- 1.9 ± 1
19	-162 ± 5	150	+ 7.1 ± 0.5
22	-132 ± 5	200	11.3 ± 0.5
25	-110 ± 3	300	14.8 ± 0.5
30	- 82 ± 3	400	15.2 ± 0.5
40	- 52 ± 2		

1. H. Kamerlingh Onnes and C. Braak, Communs phys. Lab. Univ. Leiden
    100b (1907)(*).
    Class I.

T	B	C
55.74	-26.95	876.5
60.33	-22.60	819.3
68.45	-15.50	681.9
77.88	- 9.69	558.6
90.34	- 5.39	594.7
109.01	+ 0.41	523.2
133.27	5.14	416.2
169.58	8.82	429.8
273.15	13.02	338
373.15	14.17	223

2. W.J. de Hass, Communs phys. Lab. Univ. Leiden 127a,c(1912)(*).
    Class II.

T	B
15.89	-244.9
17.69	-168.6
20.52	-143.0

3. L. Holborn, Annln Phys. 63 674 (1920)(*).
    3-term fit of PV data (P range 20 - 100 atm).
    Class I.

T	B	C
273.15	14.00	304.6
293.15	14.47	279.3
323.15	15.17	
373.15	15.58	

4. L. Holborn and J. Otto, Z. Phys. 33 1 (1925); 38 359 (1926)(*).
   5-term fit of PV data (P series; terms in P^0, P^1, P^2, P^4, and P^6).
   Maximum pressure 100 atm.
   Class I.

T	B	C
65.25	-18.36	505.4
90.15	- 5.54	663.5
123.15	+ 2.95	463.0
173.15	9.16	369.4
223.15	12.10	350.0
273.15	14.00	304.6
323.15	15.17	
373.15	15.58	
473.15	15.71	

5. F.P.G.A.J. van Agt and H.K. Onnes, Communs phys. Lab. Univ. Leiden
   176b (1925)(*).
   Class II.

T	B	T	B
14.57	-245.6	20.60	-141.4
15.71	-216.8	20.62	-140.2
16.72	-194.0	69.93	- 17.01
18.23	-170.5	90.30	- 8.16
18.29	-170.1		

6. T.T.H. Verschoyle, Proc. R. Soc. A111 552 (1926)(*).
   3-term fit of PV data. Maximum pressure 210 atm.
   Class I.

T	B	C
273.15	14.05	281.8
293.15	14.59	250.1

7. C.W. Gibby, C.C. Tanner, and I. Masson, Proc. R. Soc. A122 283 (1928) (*).

2-term fit of PV data. Maximum pressure 125 atm.

Class I.

T	B	T	B
298.15	14.71	398.35	15.74
323.15	15.05	423.25	15.54
348.15	15.39	448.15	15.41
373.53	15.54		

8. G.P. Nijhoff and W.H. Keesom, Communs phys. Lab. Univ. Leiden 188d, e (1928)(*).

Class II.

T	B	T	B
24.84	-108.62	47.62	- 34.54
31.32	- 76.48	273.16	+ 13.58
36.60	- 57.05	293.16	14.16
41.64	- 45.55	373.16	15.39

9. E.P. Bartlett, H.L. Cupples, and T.H. Tremearne, J. Am. chem. Soc. 50 1275 (1928).

PVT data given: P range 1 - 1000 atm, T range 273 - 673 K.

10. G.A. Scott, Proc. R. Soc. A125 330 (1929)(*).

3-term fit of PV data. Maximum pressure 170 atm.

T	B	C
298.15	14.60	231.3

11. C.C. Tanner and I. Masson, Proc. R. Soc A126 268 (1930)(*).

2-term fit of PV data (3-term fit at 298.15 K). Maximum pressure 126 atm.

Class I.

T	B	C
298.15	14.71	274.0
323.15	15.25	
348.15	15.45	
373.15	15.63	
398.15	15.81	
423.15	15.95	
447.15	15.92	

12. E.P. Bartlett, H.C. Hetherington, H.M. Kvalnes, and T.H. Tremearne,
    J. Am. chem. Soc. 52 1363 (1930).
    PVT data given: P range 1 - 1000 atm, T range 200 - 293 K.

13. D.T.A. Townend and L.A. Bhatt, Proc. R. Soc. A134 502 (1932)(*).
    4-term fit of PV data (P series). Maximum pressure 600 atm.
    Class I.

T	B	C
273.15	14.64	260.5
298.15	13.95	347.9

14. E.A. Long and O.L.I. Brown, J. Am. chem. Soc. 59 1922 (1937)(*).
    Class II.

T	B	T	B
20.87	-136.5	37.08	- 56.0
24.11	-110.3	41.64	- 45.6
27.65	- 90.2	46.45	- 37.2
32.43	- 70.1		

    Double constant-volume gas thermometer.

15. R. Wiebe and V.L. Gaddy, J. Am. chem. Soc. 60 2300 (1938).
    PVT data given: P range 25 - 1000 atm, T range 273 - 573 K.

16. A. Michels and M. Goudeket, Physica,'s Grav. 8 347 (1941)(*).
    Class I.
    (a) 3-term fit of PV data. Maximum pressure 50 atm.

T	B	C
273.15	13.71	423
298.15	13.81	662
323.15	14.21	693
348.15	14.76	530
373.15	14.72	713
398.15	14.94	692
423.15	15.10	755

(b) 4-term fit of PV data. Maximum pressure 50 atm.

T	B	C
273.15	13.50	923
298.15	13.58	1 162
323.15	13.86	1 109
348.15	14.14	1 417
373.15	14.52	965
398.15	14.34	1 501
423.15	14.50	1 409

(c) 3-term fit of PV data. Maximum pressure 230 atm.

T	B	C
273.15	13.81	415
298.15	14.12	405
323.15	14.55	397
348.15	14.90	393
373.15	15.16	384
398.15	15.37	371
423.15	15.59	354

17. H.L. Johnston and D. White, Trans. Am. Soc. mech. Engrs 72 785 (1950).

PVT data given: P range 1 - 200 atm, T range 35 - 300 K.

18. H.J. Hoge and J.W. Lassiter, J. Res. natn. Bur. Stand. 47 75 (1951).
PVT data given in the region of the critical point.

19. H.G. David and S.D. Hamann, Trans. Faraday Soc. 49 711 (1953).
PVT data given: P range 300 - 1250 atm, T values 64.5 and 78.9 K.

20. T.L. Cottrell, R.A. Hamilton, and R.P. Taubinger, Trans. Faraday Soc.
52 1310 (1956).

T	B
303.2	14.1 ± 1.7

21. J.J.M. Beenakker, F.H. Varekamp, and A. Van Itterbeek, Physica,'s
Grav. 25 9 (1959).

T	B
20.4	-152 ± 1

22. A. Michels, W. de Graaff, T. Wassenaar, J.M.H. Levelt, and P. Louwerse, Physica,'s Grav. 25 25 (1959)(*).

PVT data: 6-term fit in density series. P range 5 - 3000 atm.

Class I.

T	B	C
98.15	-2.99	503
103.15	-1.60	511
113.15	+0.80	506
123.15	2.68	519
138.15	5.03	516
153.15	6.98	480
173.15	8.93	459
198.15	10.79	414
223.15	12.05	406
248.15	13.03	388
273.15	13.74	389
298.15	14.37	356
323.15	14.92	323
348.15	15.38	295
373.15	15.67	290
398.15	15.86	296
423.15	16.08	280

23. F.H. Varekamp and J.J.M. Beenakker, Physica,'s Grav. 25 889 (1959).

Class I.

T	B	T	B
14	- 255	18	- 181
15	- 232	19	- 168
16	- 212	20	- 156
17	- 196	21	- 146

Differential method - the non-ideality of the gas is compared with that of helium.

24. H.F.P. Knaap, M. Knoester, C.M. Knobler, and J.J.M. Beenakker, Physica,'s Grav. 28 21 (1962)(*).

Low-pressure differential method.

Class I.

T	B	T	B
20.47	- 151.3	36.21	- 67.2
20.53	- 150.8	39.17	- 59.6
20.58	- 150.2	39.36	- 59.4
34.46	- 72.4		

The authors adopt the value B - 149.7 at 20.4 K, and give the following recommended values for B.

T	B	T	B
14	- 253	30	- 85.8
15	- 229	35	- 68.6
16	- 210	40	- 55.6
17	- 193	45	- 45.3
18	- 178	50	- 37.2
19	- 165	55	- 30.0
20	- 154	60	- 23.7
21	- 144	65	- 18.4
25	- 113		

25. Z.E.H.A. El Hadi, J.A. Dorrepaal, and M. Durieux, Physica,'s Grav. 41 320 (1969).

T	B	T	B
19.26	-158.5	22.19	-126.3
20.37	-145.3	22.69	-121.7
21.40	-134.0	23.26	-117.3
21.71	-131.2		

26. R.L. Mills, D.H. Liebenberg, J.C. Bronson, and L.C. Schmidt, J. chem. Phys. 66 3076 (1977).

PVT data given: P range 2 - 20 kbar, T range 75 - 307 K.

27. B. Schramm and H. Schmiedel (1979)(†).

Estimated error in B is ± 4.

T	B
295	14.5
350	15.0
400	15.3
450	15.4

PARAHYDROGEN H_2

1. E.A. Long and O.L.I. Brown, J. Am. chem. Soc. 59 1922 (1937)(*).

    Double constant-volume gas thermometer method.   Purity of gas 99.8 per cent.

    Class II.

T	B	T	B
20.87	- 138.9	41.64	- 46.3
24.11	- 113.4	43.95	- 42.0
27.65	- 90.2	48.45	- 33.5
32.43	- 71.2	52.51	- 27.4
37.08	- 56.7	56.21	- 23.5
41.49	- 46.7		

2. R.D. Goodwin, D.E. Diller, H.M. Roder, and L.A. Weber, J. Res. natn. Bur. Stand. 68A 121 (1964).

    Analysis of data obtained by same authors and published in J. Res. natn. Bur. Stand. 67A 173 (1963).

    3-term fit of PV data.

    Class I.

T	B	C
16	- 204.2	-
20	- 148.8	-
25	- 106.2	1 402
30	- 80.73	1 600
35	- 63.17	1 426
40	- 50.32	1 209
50	- 33.39	964
75	- 12.42	726
100	- 2.52	609
138.15	+ 5.01	540
198.15	10.65	458
248.15	12.97	415
298.15	14.38	370
348.15	15.27	313
373.15	15.60	310
423.15	16.08	302

Values given at selected temperatures only.   For full tables see original paper.

WATER H_2O

The various sets of values for the second virial coefficient of water
are in close agreement down to 573 K, but below this temperature the
more negative results of Kell, McLaurin, and Whalley (5) should be used
in preference to other data.

1. F.G. Keyes, L.B. Smith, and H.T. Gerry, Proc. Am. Acad. Arts Sci. 70
    319 (1936).

    PVT data given: (P - 350 atm, T 470 - 730 K).

    S.C. Collins and F.G. Keyes, Proc. Am. Acad. Arts Sci. 72 283 (1938).

    PVT data given: (P<1 atm, T 312 - 400 K).

    Following values calculated from equations given for B.

    Class I.

T	B	T	B
323.15	- 838.3	573.15	- 112.0
373.15	- 451.0	623.15	- 89.1
423.15	- 283.5	673.15	- 72.6
473.15	- 196.7	723.15	- 59.9
523.15	- 145.3		

2. G.C. Kennedy, Am. J. Sci. 248 540 (1950).

    PVT data given (P → 2500 atm, T 473 - 1273 K).

3. G.S. Kell, G.E. McLaurin, and E. Whalley, Advances in Thermophysical
    Properties at Extreme Temperatures and Pressures, p.104,   Lafayette,
    Indiana, 1965. Am. Soc. mech. Engrs, New York.

    3-term fit of PV data (ρ series). Standard error in B ± 1 and in
    C ± 200 at 523 K.

T	B	C
523.2	- 150.8	-6 931
	- 148.8	-8 558
623.2	- 91.2	+ 739
	- 92.3	1 150
723.2	- 59.93	1 115
	- 59.96	1 112

4. M.P. Vukalovich, M.S. Trakhtengerts, and G.A. Spiridonov,
    Teploenergetika 14 (7) 65 (1967); Heat Pwr Engng, Wash. 14 (7) 86
    (1967)(*).
    Class I.

T	B	C
353.15	- 844.4	
373.15	- 453.6	
423.15	- 283.3	
473.15	- 196.1	
523.15	- 145.4	- 10 175
573.15	- 112.9	- 3 470
623.15	- 90.2	- 520
673.15	- 72.4	+ 690
723.15	- 60.6	1 090
773.15	- 50.4	1 155
823.15	- 42.0	1 070
873.15	- 35.2	950
923.15	- 29.4	820
973.15	- 24.6	705
1 023.15	- 20.5	615
1 073.15	- 17.0	545
1 123.15	- 14.1	485
1 173.15	- 11.6	440

Values of B and C were calculated from PVT data given in the following references.

T<773.15: F.G. Keyes, L.B. Smith, and H.T. Gerry, Proc. Am. Acad. Arts Sci. 70 319 (1936); S.L. Rivkin and T.S. Akhundov, Teploenergetika 9 (1) 57 (1962); 10 (9) 66 (1963).

T>773.15: M.P. Vukalovich, V.N. Zubarev, and A.A. Aleksandrov, Teploenergetika 8 (10) 79 (1961); 9 (1) 49 (1962).

5. G.S. Kell, G.E. McLaurin, and E. Whalley, J. chem. Phys. 48 3805 (1968).

The following values of the virial coefficients were determined by the authors by analysis of their isothermal compressibility data.

T	B	C
423.13	-334 ± 2	
423.14	-326 ± 2	
448.12	-263 ± 2	
448.12	-264 ± 1	
473.17	-209 ± 2	-23 000 ± 6 000
473.18	-209 ± 6	-27 000 ± 15 000
473.12	-215 ± 3	-10 000 ± 7 000
498.16	-178.4 ± 1.3	-11 000 ± 2 000

T	B	C
498.16	-181.7 ± 3.0	- 7 000 ± 5 000
523.17	-152.5 ± 0.2	- 5 800 ± 200
523.18	-151.8 ± 0.3	- 6 500 ± 300
548.18	-133.2 ± 0.2	- 1 730 ± 140
548.18	-133.0 ± 0.1	- 2 020 ± 90
548.19	-133.3 ± 0.1	- 1 710 ± 80
573.14	-117.9 ± 0.3	+ 390 ± 160
573.14	-117.1 ± 0.1	120 ± 70
598.17	-103.5 ± 0.1	750 ± 50
598.17	-103.6 ± 0.1	790 ± 50
623.15	- 92.38 ± 0.27	1 270 ± 170
623.14	- 91.64 ± 0.11	860 ± 70
648.12	- 81.78 ± 0.09	960 ± 60
648.16	- 82.30 ± 0.23	1 280 ± 140
673.15	- 73.47 ± 0.07	1 110 ± 40
673.16	- 73.26 ± 0.09	1 010 ± 60
698.21	- 65.92 ± 0.10	990 ± 60
698.21	- 65.74 ± 0.08	910 ± 50
723.19	- 59.36 ± 0.06	890 ± 40
723.19	- 59.25 ± 0.09	840 ± 50

Standard errors are quoted above. The estimated accuracy of B is ± 10 at 423 K going to ± 0.7 at 723 K.

6. E.J. Le Fevre, M.R. Nightingale, and J.W. Rose, J. mech. Eng. Sci. 17 243 (1975).

The following B values are calculated from the correlation given.

T	B	T	B
293.15	-1251.5	598.15	- 102.9
303.15	-1073.3	648.15	- 82.0
323.15	- 812.2	698.15	- 66.4
348.15	- 599.2	748.15	- 54.5
373.15	- 459.8	798.15	- 45.1
398.15	- 364.0	848.15	- 37.5
423.15	- 295.4	898.15	- 31.3
448.15	- 244.6	948.15	- 26.2
473.15	- 206.0	998.15	- 21.8
498.15	- 175.8	1048.15	- 18.1
548.15	- 132.4	1098.15	- 14.9

T	B
1148.15	- 12.1
1198.15	- 9.7
1248.15	- 7.6

HYDROGEN SULPHIDE H_2S

1. H.H. Reamer, B.H. Sage, and W.N. Lacey, Ind. Engng Chem. ind. Edn 42 140 (1950).

    PVT data given: P range 0 - 670 atm, 5 range 278 - 444 K.

    Values of B calculated by D.R. Pesuit (†).

    Class III.

T	B
277.7	-248.7
310.9	-185.5

Class II.

T	B
344.4	-145.3
377.7	-117.6
410.9	- 95.1
444.4	- 79.2

2. L.C. Lewis and W.J. Fredericks, J. chem. Engng Data 13 482 (1968).

    Constant volume apparatus;   P → 1700 atm.

T	B	C
373	-69.66	
393	-135.4	10260
413	-104.0	6048
433	- 90.02	4912
453	- 72.06	2886
473	- 63.99	2412
493	- 52.56	1524

AMMONIA NH_3

1. C.H. Meyers and R.S. Jessup, Refrig. Engng 11 345 (1925).

    PVT data given. Maximum pressure 30 atm.

Values of B calculated by J.O. Hirschfelder, F.T. McClure, and I.F. Weeks, J. chem. Phys. <u>10</u> 201 (1942).

T	B	T	B
273.15	- 345	423.15	- 101
298.15	- 261	473.15	- 75
323.15	- 209	523.15	- 58
373.15	- 142	573.15	- 44

2. J.A. Beattie and C.K. Lawrence, J. Am. chem. Soc <u>52</u> 6 (1930). PVT data given. P → 130 atm, T 323 - 698 K.

3. J.D. Lambert and E.D.T. Strong, Proc. R. Soc. <u>A200</u> 566 (1950)(†). Accuracy of B values ± 20.

T	B	T	B
293.4	- 288	364.0	- 154
313.6	- 231	372.8	- 139
323.0	- 205	383.0	- 137
333.2	- 187	392.6	- 128
343.7	- 179	395.6	- 118
351.4	- 165		

4. H.Y. Cheh, J.P. O'Connell and J.M. Prausnitz, Can. J. Chem. <u>44</u> 429 (1966).

Values of B calculated from volumetric data from other references.

T	B	T	B
323.2	-202	398.2	-116
348.2	-165	423.2	- 99
373.2	-137		

5. D.S. Tsiklis, A.I. Semenova, and S.S. Tsimmerman, Russ. J. phys. Chem. <u>48</u> 106 (1974); Zh. fix. Khim. <u>48</u> 184 (1974).

PVT data given: P range 2000 - 9500 atm,
T range 373 - 473 K.

6. G. Adam and B. Schramm, Ber. (dtsch.) Bunsenges. phys. Chem. <u>81</u> 442 (1977).

Values of B determined assuming B (298 K) = -262. Maximum error in B estimated to be ± 2.

T	B		T	B
313	-226		433	-101
333	-194		473	- 81
363	-157		523	- 64
392	-130			

PHOSPHINE PH$_3$

1. M. Ritchie, Proc. R. Soc. A128 551 (1930)(*).

T	B
273.16	-204.4

2. E.A. Long and E.A. Gulbransen, J. Am. chem. Soc. 58 203 (1936).
    Measurements made by constant volume gas thermometer.
    Class II.

    (a)

T	B		T	B
189.91	-452.3		252.19	-249.8
195.99	-428.2		260.14	-238.4
200.42	-399.6		267.17	-220.4
205.07	-383.9		273.10	-198.1
210.55	-361.9		273.10	-198.1
219.05	-325.5		273.10	-198.1
229.10	-314.4		284.52	-181.6
237.72	-289.7		296.95	-152.6
243.12	-269.6			

    (b)

T	B		T	B
189.99	-453.5		240.51	-301.1
195.13	-442.3		249.95	-252.6
201.20	-395.6		260.11	-238.1
210.07	-357.9		273.10	-197.0
219.12	-328.5		273.10	-197.0
229.91	-303.3			

214

DIBORANE B_2H_6

1. E.M. Carr, J.T. Clarke, and H.L. Johnston, J. Am. chem. Soc. 71 740 (1949).

T	B
275.15	- 227 ± 11

HELIUM He

Agreement between the various sets of data for the second virial coefficient of helium is good down to 15 K. Below this temperature there is a significant discrepancy between the results of Keesom (8) and Keller (12b, 13b) and the values by Angus, de Reuch, and McCarty (31). The following values are recommended.

T	B	T	B
2.0	-174 ± 8	15.0	- 10.8 ± 1
2.5	-134 ± 5	20.0	- 3.4 ± 0.5
2.75	-120 ± 5	30.0	+ 2.5 ± 0.5
3.0	-109 ± 2	50.0	7.4 ± 0.5
3.5	- 92.6 ± 2	100.0	11.7 ± 0.5
4.0	- 80.2 ± 2	200.0	12.1 ± 0.5
5.0	- 62.7 ± 1	400.0	11.2 ± 0.5
7.0	- 40.9 ± 1	700.0	10.1 ± 0.5
10.0	- 23.1 ± 1		

1. L. Holborn and J. Otto, Z. Phys. 33 1 (1925); 38 359 (1926)(*).
   5-term fit of PV data (P series; terms in P^0, P^1, P^2, P^4, and P^6).
   Class I.

T	B	C
15.15	- 13.59	1062.9
20.35	- 2.80	409.5
65.15	+ 9.39	253.4
90.15	10.45	205.1
123.15	11.42	189.2
173.15	11.92	194.4
223.15	11.93	181.1
273.15	11.85	
323.15	11.74	
373.15	11.39	

T	B	C
473.15	11.07	
573.15	10.50	
673.15	10.13	

2. F.P.G.A.J. Van Agt and H. Kamerlingh Onnes, Communs phys. Lab. Univ. Leiden 176b (1925)(*).

Class I.

T	B	T	B
16.71	- 9.01	20.59	- 2.68
18.28	- 8.21	20.61	- 1.25
20.57	- 2.72	69.92	+ 8.80

3. (a) C.W. Gibby, C.C. Tanner, and I. Masson, Proc. R. Soc. A122 283 (1928)(*).

2-term fit of PV data (P series). Maximum P 125 atm.

T	B	T	B
298.15	11.44	398.35	11.08
323.15	11.37	423.25	10.36
348.15	11.26	448.15	10.92
373.55	10.90		

(b) C.C. Tanner and I. Masson, Proc. R. Soc. A126 268 (1930)(*).

3-term fit of PV data (P series). P range 29 - 127 atm.

Class I.

T	B	T	B
298.15	11.55	398.55	10.97
323.15	11.39	423.45	10.88
348.15	11.24	448.15	10.92
373.55	11.17		

4. R. Wiebe, V.L. Gaddy, and C. Heins, Jr., J. Am. chem. Soc. 53 1721 (1931)(*).

3-term fit of PV data (P series). Maximum P 1000 atm. General accuracy better than 0.1 per cent for B.

T	B	C
203.15	11.94	126
238.15	11.88	122
273.15	11.70	117

T	B	C
323.15	11.45	110
373.15	11.31	104
473.15	10.72	95

5. W.H. Keesom and H.H. Kraak, Physica,'s Grav. <u>2</u> 37 (1935)(*).
3-term fit of PV data (ρ series).
Class II.

T	B	C
2.581	- 117.2	- 32 850
3.095	- 96.6	- 12 350
4.224	- 74.7	- 1 500

6. A. Michels and H. Wouters, Physica,'s Grav. <u>8</u> 923 (1941)(*).
Class I.
(a) 4-term fit of PV data (P → 300 atm).

T	B	C
273.15	11.87	75.5
298.15	11.74	72.1
323.15	11.58	72.3
348.15	11.43	94.8
373.15	11.35	90.5
398.15	11.24	93.8
423.15	11.07	109.6

(b) 3-term fit of PV data (P → 300 atm).

T	B	C
273.15	11.87	91.3
298.15	11.68	96.6
323.15	11.52	97.9
348.15	11.44	88.8
373.15	11.37	80.2
398.15	11.29	72.1
423.15	11.15	69.7

7. W.H. Keesom and his collaborators quoted by W.H. Keesom in Helium, p.34, table 2.02 and table at foot of p.35, Elsevier, Amsterdam, 1942(*).

T	B	C
2.610	- 118.4	- 24 140
3.105	- 107.0	- 5 280
3.721	- 87.9	- 780
4.245	- 75.5	- 131
14.16	- 15.52	
17.30	- 9.02	
20.58	- 3.61	
23.35	- 3.09	
37.4	+ 4.60	
48.2	7.24	
71.6	9.85	
90.1	10.61	
126.5	11.33	
273.15	11.24	
293.15	11.15	
373.15	10.78	

8. Values adopted by W.H. Keesom in Helium, p.49, table 2.14, Elsevier, Amsterdam, 1942(*).

T	B	C
2.6	- 118.9	
3.0	- 109.2	
3.1	- 106.1	- 5 500
3.5	- 94.0	- 1 210
4.0	- 81.0	- 302
4.5	- 70.4	+ 151
5.0	- 62.1	322
6	- 49.1	458
8	- 33.0	513
10	- 23.3	518
12	- 16.9	508
14	- 12.3	498
16	- 8.86	488
18	- 6.19	478
20	- 4.04	468
22	- 2.27	453
30	+ 2.42	417

218

T	B	C
40	5.63	382
50	7.58	352
60	8.86	327
70	9.76	307
80	10.36	292
90	10.76	277
123.15	11.42	241
173.15	11.68	201
223.15	11.59	176
273.15	11.48	156
323.15	11.30	136
373.15	11.08	126
473.15	10.59	101
573.15	10.11	101

9. J. Kistemaker and W.H. Keesom, Physica,'s Grav. 12 227 (1946)(*).
Uncertainty in B is about 15 per cent.

T	B	C
1.59	- 283	- 302 000
1.70	- 200	- 252 000
1.84	- 195	- 181 000
1.93	- 177	- 146 000
2.04	- 177	- 108 000
2.34	- 153	- 53 000
2.73	- 128	- 18 000

10. W.H. Keesom and W.K. Walstra, Physica,'s Grav. 13 225 (1947)(*).
Accuracy of B values ± 0.5.

T	B	T	B
9.62	- 25.1	16.05	- 6.73
11.69	- 17.9	18.04	- 3.88
14.45	- 10.0	20.48	- 0.96

11. (i) W.G. Schneider and J.A.H. Duffie, J. chem. Phys. <u>17</u> 751 (1949).
(ii) J.L. Yntema and W.G. Schneider, J. chem. Phys. <u>18</u> 641 and 646
(1950).

2-term fit of PV date (P → 80 atm). Errors in B increase from 0.5
per cent at 273.15 to 1 per cent at 873.15, $2\frac{1}{2}$ per cent at 1073.15
and 1273.15 and $3\frac{1}{2}$ per cent at 1473.15 K.

T	B	T	B
273.15	11.77	773.15	10.14
373.15	11.42	873.15	9.82
473.15	11.08	1 073.15	9.17
573.15	10.76	1 273.15	8.66
673.15	10.45	1 473.15	8.19

12. W.E. Keller, Phys. Rev. <u>97</u> 1 (1955).

(a) 2-term fit of PV data.

T	B	T	B
2.154	- 176.4	3.348	- 103.4
2.324	- 157.7	3.961	- 83.7
2.862	- 123.6		

(b) 3-term fit of PV data.
Class I.

T	B	C
2.154	- 159.4	- 5 231
2.324	- 140.8	- 3 855
2.862	- 117.8	- 7 282
3.348	- 102.4	- 823
3.961	- 83.3	- 358

13. W.H. Keesom and W.K. Walstra, Communs phys. Lab. Univ. Leiden 260c
(1940); Physica,'s Grav. <u>7</u> 985 (1940).

Coefficients recalculated by W.E. Keller, Phys. Rev. <u>97</u> 1 (1955).

(a) 2-term fit of PV data.

T	B
2.610	- 138.1
3.105	- 114.6
3.721	- 89.7
4.245	- 74.45

220

(b) 3-term fit of PV data.
Class I.

T	B	C
2.610	- 123.8	- 18 300
3.105	- 108.3	- 4 361
3.721	- 85.2	- 1 747
4.245	- 78.25	+ 1 013

14. W.C. Pfefferle, Jr., J.A. Goff, and J.G. Miller, J. chem. Phys. <u>23</u> 509 (1955).
Class I.
Burnett Method. Maximum pressure 55 atm.

T	B	C
303.15	11.84 ± 0.04	105
303.15	11.74 ± 0.02	105
303.15	11.73 ± 0.04	110

15. I.H. Silberberg, K.A. Kobe, and J.J. McKetta, J. chem. Engng Data <u>4</u> 314 (1959)(*).
Class I.

T	B
373.15	11.31
473.15	10.76

16. L. Stroud, J.E. Miller, and L.W. Brandt, J. chem. Engng Data <u>5</u> 51 (1960)(*).
2-term fit of PV data (P → 270 atm).
Class I.

T	B	T	B
249.9	11.48	294.3	11.24
255.4	11.42	311.0	11.10
261.0	11.37	327.6	11.01
277.6	11.27		

17. D. White, T. Rubin, P. Camky, and H.L. Johnston, J. phys. Chem., Ithaca <u>64</u> 1607 (1960).
3-term fit of PV data. P range 1 - 33 atm. Errors in B 0.2 - 0.5.

T	B	T	B
20.58	- 2.62	75.01	10.70
24.65	+ 0.80	80.02	11.01
28.82	2.46	90.04	11.60
33.00	4.00	100.02	11.85
35.10	5.18	125.03	12.18
40.09	6.57	150.04	12.15
45.10	7.48	175.02	12.24
50.09	8.06	200.11	12.23
55.00	8.96	249.99	12.15
60.03	9.55	273.15	12.08
69.00	10.30	299.99	11.99

18. R.J. Witonsky and J.G. Miller, J. Am. chem. Soc. 85 282 (1963)(*).
    Class I.

T	B
448.2	10.89
523.2	10.69
598.2	10.75
673.2	9.67
748.2	9.70

19. F.B. Canfield, T.W. Leland, and R. Kobayashi, Adv. cryogen. Engng
    8 146 (1963).
    Burnett method. Maximum pressure 500 atm.
    Class I.

T	B	C
133.15	12.10	182
143.14	12.20	159
158.15	12.25	156
183.15	12.30	150
223.13	12.46	108
273.15	12.09	116

20. A.E. Hoover, F.B. Canfield, R. Kobayashi, and T.W. Leland, Jr., J. chem. Engng Data 9 568 (1964).

3-term fit of PV data. Full analysis of errors given.

Class I.

T	B	C
133.15	11.97	172
143.14	11.84	175
158.15	11.98	163.8
183.15	12.09	149
223.13	12.23	121.8
273.15	11.96	117.5

21. P.S. Ku and B.F. Dodge, J. chem. Engng Data 12 158 (1967)(*).

3-term fit of PV data.

Class I.

T	B	C
311.65	11.64	108
373.15	11.21	115

22. K.W. Suk and T.S. Storvick, A. I. Ch. E. Jl 13 231 (1967).

Burnett method. Maximum pressure 70 atm.

Errors in B ± 0.1.

T	B
473.15	10.98
573.15	10.59

23. J.A. Sullivan and R.E. Sonntag, Cryogenics 7 13 (1967).

PVT data given: T 70 - 120, P → 690 atm.

Following B values read from graph.

T	B
70	10.4
80	10.8
100	11.6
120	12.4

24. N.K. Kalfoglou and J.G. Miller, J. phys. Chem., Ithaca <u>71</u> 1256 (1967).

T	B	T	B
303.2	11.40	573.2	10.30
373.2	11.27	673.2	9.67
473.2	10.87	773.2	9.45

25. M.E. Boyd, S.Y. Larsen and H. Plumb, J. Res. natn. Bur. Stand. <u>72A</u> 155 (1968).

Values of B calculated from sound velocity measurements using analytical expressions involving temperature to represent B.

Following values obtained by determining the coefficients in the series

$$B = a + \frac{b}{T} + \frac{c}{T^2}$$

(I) T range 2 - 10 K.

T	B
2	-209
4	- 90.0
10	- 23.3

(II) T range 11 - 20 K.

10	- 22.5
16	- 5.8
20	- 1.1

(III) T range 4 - 16 K.

2	-230
4	- 91.6
10	- 22.6
16	- 6.9
20	- 1.9

26. A.L. Blancett, K.R. Hall, and F.B. Canfield, Physica,'s Grav. <u>47</u> 75 (1970).
    Burnett method. Maximum pressure 700 atm.

T	B	C
223.15	12.16 ± .02	118.3 ± 1.6
273.15	11.94 ± .03	111.7 ± 2.5
323.15	11.76 ± .03	102.5 ± 2.9

27. K.R. Hall and F.B. Canfield, Physica,'s Grav. <u>47</u> 219 (1970).
    Burnett method. Maximum pressure 700 atm.

T	B	C
83.15	10.97 ± .04	193.0 ± 4.3
103.15	11.57 ± .04	171.3 ± 3.0
113.15	11.77 ± .03	166.3 ± 2.3

28. A.P. Kudchadker and P.T. Eubank, J. chem. Engng Data <u>15</u> 7 (1970).

T	B
376.16	11.36 ± 0.52
423.16	10.76 ± 0.63
473.16	10.16 ± 0.48

29. J.A. Provine and F.B. Canfield, Physica,'s Grav. <u>52</u> 79 (1971).

T	B	C
143.15	12.13 ± .02	146.8 ± 2.2
158.15	12.66 ± .07	113.0 ± 3.7
183.15	12.27 ± .03	123.7 ± 1.6

30. W.J. Kerns, R.G. Anthony, and P.T. Eubank, A. I. Ch. E. Symp. Ser. <u>70</u> 14 (1974)(*).
    Burnett method.

T	B
323	11.61
373	11.42
423	11.27

31. S. Angus, K.M. de Reuck, and R.D. McCarty, International
Thermodynamic Tables of the Fluid State Helium -4, Pergamon Press,
Oxford, 1977.

T	B	T	B
2.5	-152.9	30.0	4.5
3.0	-122.0	35.0	6.3
3.5	-101.0	40.0	7.6
4.0	- 85.8	45.0	8.6
4.5	- 74.2	50.0	9.3
5.0	- 65.0	55.0	9.9
5.5	- 57.5	60.0	10.4
6.0	- 51.2	65.0	10.7
6.5	- 45.9	70.0	11.0
7.0	- 41.3	75.0	11.3
7.5	- 37.4	80.0	11.5
8.0	- 33.9	90.0	11.8
8.5	- 30.8	100.0	12.0
9.0	- 28.1	125.0	12.3
9.5	- 25.6	150.0	12.4
10.0	- 23.4	175.0	12.4
11.0	- 19.5	200.0	12.3
12.0	- 16.3	225.0	12.3
13.0	- 13.6	250.0	12.2
14.0	- 11.3	273.15	12.1
15.0	- 9.2	275.00	12.1
16.0	- 7.5	298.15	12.0
17.0	- 5.9	300.00	12.0
18.0	- 4.6	350.00	11.7
19.0	- 3.3	400.00	11.5
20.0	- 2.2	450.00	11.3
21.0	- 1.3	500.00	11.1
22.0	- .4	600.00	10.7
23.0	.4	700.00	10.4
24.0	1.2	800.00	10.0
25.0	1.8	900.00	9.7
26.0	2.5	1000.00	9.4
27.0	3.0	1100.00	9.1
28.0	3.6	1200.00	8.9
29.0	4.0	1300.00	8.6

T	B
1400.00	8.4

32. D.D. Dillard, M. Waxman, and R.L. Robinson, Jr., J. chem. Engng Data
    23 269 (1978).
    Burnett Method.

T	B
223.15	12.04
273.15	11.93
323.15	11.69

33. G.S. Kell, G.E. McLaurin, and E. Whalley, J. chem. Phys. 68 2199
    (1978).
    Modified Burnett Method.

T	B	C
273.15	11.89 ± 0.15	127 ± 34
273.15	12.23 ± 0.02	
298.15	11.79 ± 0.06	121 ± 11
298.15	12.13 ± 0.02	
373.15	11.52 ± 0.07	106 ± 24
373.15	11.80 ± 0.02	
423.15	11.26 ± 0.08	143 ± 31
423.15	11.41 ± 0.10	
423.15	11.30 ± 0.04	92.± 13
423.15	11.56 ± 0.05	
523.15	10.87 ± 0.16	143 ± 73
523.15	10.98 ± 0.12	
523.15	10.86 ± 0.04	100 ± 14
523.15	11.19 ± 0.02	
623.15	10.42 ± 0.11	
623.15	10.61 ± 0.01	
623.15	10.79 ± 0.04	
673.15	10.61 ± 0.05	
723.15	10.53 ± 0.04	
773.15	10.26 ± 0.07	

34. M. Prasad and A.P. Kudchadker, J. chem. Engng Data 23 190 (1978).
Burnett method.

T	B	T	B
298.15	11.60	373.15	11.30
333.15	11.48	393.15	11.20
353.15	11.42	413.15	11.15

35. W. Warowny and J. Stecki, J. chem. Engng Data 23 212 (1978).
Burnett method, P → 70 atm.

T	B
393.20	11.33
407.50	11.26
423.00	11.13

HELIUM He^3

1. W.E. Keller, Phys. Rev. 98 1571 (1955).

T	B	C
1.516	- 168.0	
1.818	- 142.3	
2.161	- 117.9	
2.991	- 86.25	1 718
3.786	- 65.37	1 132

POTASSIUM K

1. J.P. Stone, C.T. Ewing, J.R. Spann, E.W. Steinkuller, D.D. Williams,
and R.R. Miller, J. chem. Engng Data 11 309 (1966).
C.T. Ewing, J.P. Stone, J.R. Spann, and R.R. Miller, J. chem. Engng
Data 11 468 (1966).

PVT data given (P → 26 atm, T 1150 - 1655 K).

KRYPTON Kr

There is good agreement between the different sets of data for the
second virial coefficient of krypton at temperatures above 125 K. Below
this temperature, the results of Fender and Halsey (5) are preferred.
The following values are recommended.

T	B	T	B
110	-364 ± 10	200	-116.9 ± 1
115	-333 ± 10	250	- 75.7 ± 1
120	-306 ± 5	300	- 50.5 ± 1
130	-264 ± 5	400	- 22.0 ± 1
140	-229.5 ± 5	500	- 8.1 ± 0.5
150	-200.7 ± 2	600	+ 1.7 ± 0.5
170	-159.0 ± 2	700	8.2 ± 0.5

1. G. Glocker, C.P. Roe, and D.L. Fuller, J. chem. Phys. 1 703 (1933)(*).
   PVT data given: P → 100 atm.
   Class II.

T	B
328.7	-37.0
368.7	-24.9

2. J.A. Beattie, J.S. Brierley, and R.J. Barriault, J. chem. Phys. 20 1615 (1952)(*).
   4-term fit of PV data (V series: terms in V^0, V^{-1}, V^{-2}, and V^{-4}).
   Maximum pressure 410 atm.
   Class I.

T	B	C
273.16	- 62.96	2 757
298.15	- 52.36	2 612
323.15	- 42.78	2 260
348.15	- 35.21	2 076
373.16	- 28.86	1 942
398.17	- 23.47	1 842
423.18	- 18.82	1 759
448.20	- 14.73	1 671
473.21	- 11.11	1 583
498.23	- 8.40	1 637
523.25	- 5.69	1 626
548.26	- 3.17	1 569
573.28	- 1.15	1 613

3. E. Whalley and W.G. Schneider, Trans. Am. Soc. mech. Engng 76 1001 (1954)(*).
   3-term fit of PV data (P series). P range 10 - 80 atm.

T	B	C
273.15	- 62.70 ± 0.49	3 130 ± 125
323.15	- 42.78 ± 0.27	3 000 ± 120
373.15	- 29.28 ± 0.36	2 570 ± 90
423.15	- 18.13 ± 0.36	1 960 ± 85
473.15	- 10.75 ± 0.36	1 755 ± 75
573.15	+ 0.42 ± 0.38	1 355 ± 115
673.15	7.24 ± 0.56	1 260 ± 150
773.15	12.70 ± 0.44	1 055 ± 150
873.15	17.19 ± 0.42	760 ± 170

4. G. Thomaes and R. van Steenwinkel, Nature, Lond. 193 160 (1962).
   Class II.

T	B	T	B
109.95	- 358.4	235.1	- 88.5
134.1	- 255.2	235.1	- 88.2
134.1	- 256.1	270.3	- 67.2
174.4	- 151.3	270.3	- 65.3
174.4	- 151.7		

5. B.E.F. Fender and G.D. Halsey, Jr., J. chem. Phys. 36 1881 (1962).
   3-term fit of PV data (P less than 1 atm). Maximum error in B ± 2
   per cent.

T	B	T	B
107.55	- 386.67	121.47	- 301.59
108.89	- 374.23	121.64	- 297.47
109.94	- 365.03	128.14	- 270.49
112.28	- 349.75	132.13	- 255.62
115.35	- 330.80	138.07	- 236.73
118.50	- 314.83		

6. N.J. Trappeniers, T. Wassenaar, and G.J. Wolkers, Physica,'s Grav. 32
   1503 (1966)(*).
   Class I.
   (a) 7-term fit of PV data (P → 2900 atm).

T	B	C
273.15	- 60.33	1 755
298.15	- 49.79	1 785
323.15	- 41.08	1 745
348.15	- 33.85	1 725
373.15	- 27.42	1 522
398.15	- 22.05	1 430
423.15	- 17.41	1 350

(b) 3-term fit of PV data (P → 80 atm).

T	B	C
273.15	- 61.24	2 455
298.15	- 50.39	2 235
323.15	- 41.44	2 035
348.15	- 33.95	1 855
373.15	- 27.61	1 710
398.15	- 22.22	1 615
423.15	- 17.53	1 515

7. R.D. Weir, I. Wynn Jones, J.S. Rowlinson, and G. Saville, Trans. Faraday Soc. 63 1320 (1967).

3-term fit of PV data. Errors in B ± 10(T 110 - 115 K); ± 3(T 115 - 125 K); and ± 1(T>125 K).

T	B	T	B
110.64	- 370.5	142.81	- 220.1
112.42	- 354.4	155.09	- 188.7
115.15	- 334.1	168.55	- 162.9
118.14	- 315.9	180.75	- 142.8
120.24	- 307.4	200.59	- 116.6
127.79	- 273.6	224.19	- 95.46
133.11	- 251.4		

8. M.A. Byrne, M.R. Jones, and L.A.K. Staveley, Trans. Faraday Soc. 64 1747 (1968).

Class I.

T	B	T	B
117.41	- 326.9	129.15	- 265.4
120.19	- 307.9	134.69	- 247.0
124.01	- 287.5	142.45	- 220.3

T	B	T	B
150.98	- 198.1	186.86	- 132.7
161.12	- 175.2	209.16	- 106.9
172.29	- 154.5	251.94	- 74.8

9. F. Theeuwes and R.J. Bearman, J. chem. Thermodyn. $\underline{2}$ 171 (1970).
   PVT data given: P range - 280 atm, T range - 240 K.

10. C.A. Pollard and G. Saville (unpublished results). See also C.A. Pollard, Ph.D. thesis, University of London (1971).
    Standard deviation of the curve fit for B is 2 except at the highest temperature where σ_B is 3.

T	B	T	B
166.71	-163.1	232.27	- 86.6
175.48	-149.2	242.81	- 81.0
187.20	-131.8	247.44	- 77.1
191.45	-124.4	248.46	- 78.9
199.50	-119.0	254.48	- 71.8
204.45	-113.2	259.02	- 70.3
217.38	-100.7	274.25	- 64.7
221.02	- 98.9		

11. J. Santafe, J.S. Urieta, and C. Gutierrez, Revta. Acad. Cienc. exact. fis. - quim. nat. Zaragoza $\underline{31}$ 63 (1976).
    B values also given by J. Santafe, J.S. Urieta, and C.G. Losa, Chem. phys. $\underline{18}$ 341 (1976).
    Compressibility measurements. Accuracy of B estimated to be ± 3.

T	B	T	B
273.2	-61.7	303.2	-48.2
283.2	-57.2	313.2	-44.9
293.2	-51.9	323.2	-41.7

12. H.-P. Rentschler and B. Schramm, Ber. (dtsch.) Bunsenges. phys. Chem. $\underline{81}$ 319 (1977).
    Maximum error in B estimated to be ± 4.

T	B	T	B
300	-50.6	553	- 2.0
410	-19.7	630	+ 3.4
475	-11.5	715	8.4

13. B. Schramm, H. Schmiedel, R. Gehrmann, and R. Bartl, Ber. (dtsch.) Bunsenges. phys. Chem. 81 316 (1977).

Maximum error in B estimated to be ± 4.

T	B	T	B
202.0	-116.7	332.9	- 39.5
214.6	-103.0	367.5	- 30.0
233.9	- 88.9	402.2	- 21.9
259.4	- 72.4	431.7	- 16.8
278.9	- 61.4	461.7	- 11.5
295.2	- 52.0	497.1	- 7.6

14. D.D. Dillard, M. Waxman, and R.L. Robinson, Jr., J. chem. Engng Data 23 269 (1978).

Burnett Method.

T	B
223.15	-93.05
273.15	-61.53
323.15	-41.72

NITRIC OXIDE NO

1. H.L. Johnston and H.R. Weimer, J. Am. chem. Soc. 56 625 (1934). Measurements made using a constant volume gas thermometer.

Class II.

(a) Gas in bulb 1. Maximum pressure 1.2 atm.

T	B	T	B
121.72	- 224.4	201.09	- 56.1
135.08	- 150.9	210.70	- 50.1
141.30	- 141.7	219.97	- 46.0
146.91	- 120.6	230.64	- 41.3
153.06	- 107.9	240.48	- 36.0
160.12	- 97.8	250.28	- 32.8
172.21	- 82.2	260.25	- 28.9
180.37	- 72.7	274.00	- 22.8
189.90	- 63.9	274.09	- 21.9
200.47	- 56.5		

The experiment was repeated with the gas in bulb 2, and then the series of measurements was made with the gas in each bulb at a lower initial pressure ($\frac{1}{2}$ atm). Values of B at corresponding temperatures agree to within ± 2. For full results consult the original paper.

(b) Low-temperature readings for gas in bulb 2.

T	B
124.95	- 199.3
129.79	- 176.1
135.16	- 149.7
144.32	- 123.8

2. B.H. Golding and B.H. Sage, Ind. Engng Chem. ind. Edn. <u>43</u> 160 (1951) (†).

T	B	C
277.60	- 26.2	1 400
310.94	- 19.0	1 900

NITROGEN DIOXIDE NO_2

1. W.G. Schlinger and B.H. Sage, Ind. Engng Chem. ind. Edn. <u>42</u> 2158 (1950).

    PVT data given: P range 1 - 150 atm, T range 294 - 444 K.

NITROGEN N_2

Values of the second virial coefficient of nitrogen obtained by different sets of workers are in good agreement except for the values derived by Pocock and Wormald (26) which are significantly more negative. The following values are recommended.

T	B	T	B
75	-275 ± 8	200	- 35.2 ± 1
80	-243 ± 7	250	- 16.2 ± 1
90	-197 ± 5	300	- 4.2 ± 0.5
100	-160 ± 3	400	+ 9.0 ± 0.5
110	-132 ± 2	500	16.9 ± 0.5
125	-104 ± 2	600	21.3 ± 0.5
150	- 71.5 ± 2	700	24.0 ± 0.5

1. H. Kamerlingh Onnes and A.T. van Urk, Communs phys. Lab. Univ.
    Leiden 169d, e(1924)(*).
    Class I.

T	B	C
126.83	- 101.8	3 345
128.69	- 98.3	3 469
131.62	- 94.5	3 240
141.88	- 80.7	2 874
151.96	- 69.83	2 694
170.90	- 52.53	2 092
192.05	- 39.12	1 763
222.89	- 24.79	1 345
249.53	- 16.73	1 535
273.15	- 9.19	1 037
293.15	- 5.48	1 196

2. L. Holborn and J. Otto, Z. Phys. 33 1 (1925)(*).
    5-term fit of PV data (P series; terms in P^0, P^1, P^2, P^4, and P^6).
    Maximum pressure 100 atm.
    Class I.

T	B	C
143.15	- 79.78	2 645
173.15	- 51.86	2 125
223.15	- 26.38	2 048
273.15	- 10.34	1 674
323.15	- 0.26	1 302
373.15	+ 6.14	1 286
423.15	11.53	1 094
473.15	15.34	1 028
573.15	20.64	883
673.15	23.51	1 130

3. T.T.H. Verschoyle, Proc. R. Soc. A111 552 (1926)(*).
    3-term fit of PV data. Maximum pressure 205 atm.
    Class I.

T	B	C
273.15	- 11.11	1 798
293.15	- 6.27	1 548

4. E.P. Bartlett, H.L. Cupples, and T.H. Tremearne, J. Am. chem. Soc. 50 1275 (1928).

PVT data given: P range 1 - 1000 atm, T range 273 - 673 K.

5. E.P. Bartlett, H.C. Hetherington, H.M. Kvalnes, and T.H. Tremearne, J. Am. chem. Soc. 52 1363 (1930).

PVT data given: P range 1 - 1000 atm, T range 200 - 293 K.

6. J. Otto, A. Michels, and H. Wouters, Phys. Z. 35 97 (1934)(*).
6-term fit of PV data. Maximum pressure 410 atm.

Class I.

T	B	C
273.15	- 10.14	1 427
298.15	- 4.87	1 405
323.15	- 0.50	1 375
348.15	+ 3.25	1 334
373.15	6.23	1 336
398.15	9.04	1 277
423.15	11.37	1 278

7. A. Michels, H. Wouters, and J. de Boer, Physica,'s Grav. 1 587 (1934). (*). (See also Physica,'s Grav. 3 585 (1936)).

(P → 80 atm).

Class I.

T	B	C
273.15	- 10.27	1 537
298.15	- 4.71	1 315
323.15	- 0.28	1 219
348.15	+ 3.20	1 257
373.15	6.56	1 079
398.15	9.45	963
423.15	12.29	694

8. M. Benedict, J. Am. chem. Soc. 59 2233 (1937).

PVT data given: P range 980 - 5800 atm, T range 100 - 473 K.

9. M. Benedict, J. Am. chem. Soc. 59 2224 (1937).

PVT data given: P range 1 - 1550 atm, T range 90 - 273 K.

10. A. Michels, R.J. Lunbeck, and G.J. Wolkers, Physica,'s Grav. <u>17</u> 801 (1951).

(*). 8-term fit of PV data (P → 3000 atm).

Class I.

T	B	C
273.15	-10.05	1 319
298.15	- 4.46	1 063
323.15	- 0.25	1 157
348.15	+ 3.38	1 115
373.15	6.50	1 092
398.15	9.21	1 038
423.15	11.51	1 036

11. D. White, J.-H. Hu, and H.L. Johnston, J. chem. Phys. <u>21</u> 1149 (1953). Values of B determined using a constant volume gas thermometer. Class II.

T	B	B†
80	-265	-251
85	-247	-223
90	-213	-199
95	-174	-178
100	-158	-161
110	-142	-134
125	- 90	-105
150	- 56	- 71
175	- 29	- 50
200	- 34	- 35
250	- 23	- 16

†Obtained from gas density measurements (Friedman, D. White, and H.L. Johnston). These results are considered by the authors to be the more accurate.

12. W.C. Pfefferle, Jr., J.A. Goff, and J.G. Miller, J. chem. Phys. <u>23</u> 509 (1955).

T	B	C
303.15	-4.17 ± 0.06	1 485
303.15	-4.13 ± 0.08	1 485
303.15	-4.17 ± 0.04	1 495

Burnett method. Maximum pressure 55 atm.

13. A. van Itterbeek, H. Lambert, and G. Forres, Appl. scient. Res. 6A 15 (1956)(*).

Values of B calculated from velocity of sound measurements.

Class II.

T	B	T	B
70	-306	120	-113
80	-239	130	- 97
90	-193	140	- 83
100	-159	150	- 72
110	-134		

14. R.D. Gunn, M.S. Thesis, University of California (Berkeley)(1958).

Values of B given by J.A. Huff amd T.M. Reed, J. chem. Engng Data 8 306 (1963).

Class I.

T	B	T	B
277.6	- 8.5	398.2	9.05
298.2	- 4.84	427.6	11.6
310.9	- 2.0	444.3	13.1
323.2	- 0.52	460.9	14.2
348.2	+ 3.31	477.6	15.4
373.2	6.19	510.9	17.4

15. R.A.H. Pool, G. Saville, T.M. Herrington, B.D.C. Shields, and L.A.K. Staveley, Trans. Faraday Soc. 58 1692 (1962).

T	B
90	-201 ± 2

16. R.J. Witonsky and J.G. Miller, J. Am. chem. Soc. 85 282 (1963)(*).

Maximum pressure 100 atm.

Class I.

T	B
448.15	14.26
523.15	18.32
598.15	20.80
673.15	23.41
748.15	24.73

17. F.B. Canfield, T.W. Leland, and R. Kobayashi, Adv. cryogen. Engng
    8 146 (1963).
    Class I.

T	B	C
133.15	-91.95	3 100
143.14	-79.56	2 920
158.15	-63.50	2 414
183.15	-45.35	2 132
223.13	-25.17	1 636
273.15	- 9.70	1 416

Burnett method. Maximum pressure 500 atm.

18. A.E. Hoover, F.B. Canfield, R. Kobayashi, and T.W. Leland, Jr.,
    J. chem. Engng Data 9 568 (1964).
    Class I.

T	B	C
273.15	-10.56	1 573
223.13	-26.05	1 850
183.15	-45.15	2 119
158.15	-64.14	2 530
143.14	-76.59	2 914
133.15	-91.99	3 119

19. R.W. Crain, Jr. and R.E. Sonntag, Adv. cryogen. Engng 11 379 (1966).
    Class I.

T	B	C
143.15	-79.45	2 889
163.15	-59.42	2 392
203.15	-33.85	1 837
273.15	-10.26	1 517

Burnett method. Maximum pressure 500 atm.

20. D.S. Tsiklis, L.R. Linshits, and I.B. Rodkina, Zh. fiz. Khim. 40
    2823 (1966); Russ. J. phys. Chem. 40 1516 (1966).
    PVT data given: P range 100 - 580 atm, T: 320 K.
    Burnett method.

21. P.S. Ku and B.F. Dodge, J. chem. Engng Data 12 158 (1967)(*).
3-term fit of PV data.
Class I.

T	B	C
311.65	-2.73	1 385
373.15	+5.97	1 195

22. S.L. Robertson and S.E. Babb, Jr., J. chem. Phys. 50 4560 (1969).
PVT data given: P range 2000 - 10000 atm., T range 308 - 673 K.

23. K.R. Hall and F.B. Canfield, Physica,'s Grav. 47 219 (1970).
Burnett method. Maximum pressure just below saturation pressure.

T	B	C
103.15	-148.46	
113.15	-117.78	

24. L.A. Weber, J. chem. Thermodyn. 2 839 (1970).
PVT data given: P range - 110 atm, T range - 140 K.

25. D.R. Roe and G. Saville (unpublished results). See also D.R. Roe,
Ph.D. thesis, University of London (1972).
Estimated errors are ± 0.20 in B and ± 125 in C.

T	B	C
155.89	-65.95	2540
181.86	-45.85	2070
192.64	-39.60	1980
204.61	-33.60	1900
204.61	-33.50	1870
218.87	-27.30	1810
218.87	-27.30	1790
234.05	-21.60	1680
248.54	-17.00	1640
248.54	-16.90	1610
263.08	-12.95	1590
276.94	- 9.40	1500
291.41	- 6.20	1460

26. G. Pocock and C.J. Wormald, J.C.S. Faraday I 71 705 (1975).

B values calculated from measured isothermal Joule-Thomson coefficients.

Errors in B estimated at ± 1% below 130 K, and ± 1 above 130 K.

T	B	T	B
75	-302	200	- 36.4
80	-264	225	- 25.0
85	-233	250	- 16.3
90	-207	275	- 9.6
95	-187	300	- 4.2
100	-169	350	+ 3.8
110	-140	400	9.5
120	-118	450	13.6
130	-100	500	16.7
140	- 86.2	600	21.2
150	- 74.3	700	24.1
175	- 52.0		

NITROUS OXIDE N_2O

There are a number of discrepancies in the reported values of the second virial coefficient of nitrous oxide, but the measurements made since 1960 are in good agreement. The following smoothed values are based on these results.

T	B	T	B
240	-219 ± 10	330	-103 ± 10
260	-181 ± 10	360	- 85 ± 10
280	-151 ± 10	400	- 68 ± 10
300	-128 ± 10		

1. A. Leduc and P. Sacerdote, C. r. hebd. Seanc. Acad. Sci., Paris 125 297 (1897).

T	B
289.2	-140.0

2. Lord Rayleigh, Phil. Trans. R. Soc. A204 351 (1905).

T	B
284.2	-152.5

3. E. Briner, H. Biedermann, and A. Rothen, Helv. chim. Acta <u>8</u> 923
    (1925).

    PVT data given: P range 30 - 160 atm., T range 195 - 282 K.

4. W. Cawood and H.S. Patterson, J. chem. Soc. 619 (1933).
    (a) Phil. Trans. R. Soc. <u>A236</u> 77 (1937)(*).

T	B
273.15	-158.9
294.15	-136.8
294.15	-134.9(a)
309.65	-126.8

5. H.L. Johnston and H.R. Weimer, J. Am. chem. Soc. <u>56</u> 625 (1934).
    Measurements made with a constant volume gas thermometer. Gas in
    bulb 1. Maximum pressure 1.2 atm.
    Class II.

(a) T	B	T	B
198.68	-349.4	261.71	-198.8
204.89	-325.9	273.13	-183.6
213.80	-299.8	273.14	-182.5
220.02	-281.5	273.15	-182.8
230.10	-257.8	280.67	-174.8
240.07	-235.5	290.29	-163.5
250.16	-217.2	296.43	-155.3

    The experiment was repeated with a lower initial gas pressure, and
    then with gas in bulb 2. Values of B at corresponding temperatures
    agreed to within ± 3.

6. G.A. Bottomley, D.S. Massie, and R. Whytlaw-Gray, Proc. R. Soc.
    <u>A200</u> 201 (1950)(*).

T	B
295.21	-139.5

7. E.J. Couch, L.J. Hirth, and K.A. Kobe, J. chem. Engng Data <u>6</u> 229
    (1961).

    Estimated maximum error in B is ± 25 at 243.15 K decreasing to
    ± 10 at 423.15 K.

T	B	T	B
243.15	-212.0	323.15	-109.1
258.15	-186.9	348.15	- 92.0
273.15	-160.9	373.15	- 79.4
288.15	-142.3	398.15	- 68.0
303.15	-127.8	423.15	- 58.2

8. H.W. Schamp, E.A. Mason, and K. Su, Physics Fluids $\underline{5}$ 769 (1962).
   4-term fit of PV data.

T	B	C
273.15	-160.92	5 920
298.15	-133.18	5 640
323.15	-111.51	4 990
348.15	- 94.30	4 400
373.15	- 80.21	3 880
398.15	- 68.40	3 420
423.15	- 58.30	2 980

An alternative calculation of B from the same PV data leads to values that differ by less than 0.4 from the values quoted above.

9. S. Kirouac and T.K. Bose, J. chem. Phys. $\underline{59}$ 3043 (1973).
   B derived from low pressure dielectric measurments.

T	B
279.7	-146.4 ± 1.5
303.3	-119.7 ± 1.2
348.3	-101.2 ± 4.0

SODIUM Na

1. J.P. Stone, C.T. Ewing, J.R. Spann, E.W. Steinkuller, D.D. Williams, and R.R. Miller, J. chem. Engng Data $\underline{11}$ 309 (1966).
   C.T. Ewing, J.P. Stone, J.R. Spann, and R.R. Miller, J. chem. Engng Data $\underline{11}$ 468 (1966).

   PVT data given (P → 24 atm, T 1230 - 1695 K).

NEON   Ne

There is good agreement between the various sets of values given for the second virial coefficient of neon. The only exception is the value at 170.15 K given by Crommelin, Martinez, and Onnes (2) which lies below a smooth curve drawn through all the other points. The following values were taken from that curve.

T	B	T	B
60	-24.8 ± 1	150	+ 3.2 ± 1
70	-17.9 ± 1	200	7.6 ± 1
80	-12.8 ± 1	300	11.3 ± 1
100	- 6.0 ± 1	400	12.8 ± 1
125	- 0.4 ± 1	600	13.8 ± 0.5

1.   H. Kamerlingh Onnes and C.A. Crommelin, Communs phys. Lab. Univ. Leiden 147d (1915)(*).

Class I.

T	B	C
273.15	9.27	580
293.15	10.77	388

2.   C.A. Crommelin, J.P. Martinez, and H. Kamerlingh Onnes, Communs phys. Lab. Univ. Leiden 154a (1919)(*).

6-term fit of PV data (ρ series). Maximum pressure 80 atm. (The 170.15 K value appears to be in error).

Class I.

T	B	C
55.64	-32.30	900
60.08	-25.13	483
65.06	-21.60	534
73.08	-16.49	523
90.56	- 9.09	510
131.94	- 1.18	749
170.15	+ 0.25	929

3.   L. Holborn and J. Otto, Z. Phys. 33 1 (1925): 38 359 (1926)(*).

5-term fit of PV data (P series: terms in P^0, P^1, P^2, P^4, and P^6). Maximum pressure 100 atm.

Class I.

T	B	C
65.25	- 20.98	563
90.65	- 8.18	442
123.15	+ 0.10	309
173.15	6.45	199
223.15	9.12	228
273.15	10.67	256
373.15	11.86	432
473.15	13.06	335
573.15	13.77	-
673.15	13.74	-

4. A. Michels and R.O. Gibson, Annln Phys. <u>87</u> 850 (1928)(*).
    (P → 500 atm).
    Class I.

T	B		T	B
273.15	10.34		345.98	11.86
295.62	11.08		373.98	12.18
323.55	11.26			

5. G.A. Nicholson and W.G. Schneider, Can. J. Chem. <u>33</u> 589 (1955)(*).
    3-term fit of PV data (P series). Maximum pressure 80 atm.
    Class I.

T	B	C
273.15	10.90	220
323.15	12.31	180
373.15	12.75	
473.15	13.42	
573.15	13.69	
673.15	13.77	
773.15	13.73	
873.15	13.88	
973.15	13.98	

6. A. Michels, T. Wassenaar, and P. Louwerse, Physica,'s Grav. <u>26</u> 539
    (1960).
    Class I.

(a) 6-term fit of PV data (P → 2900 atm).

T	B	C
273.15	10.83	225
298.15	11.42	221
323.15	11.86	224
348.15	12.21	224
373.15	12.52	224
398.15	12.86	205
423.15	13.10	197

(b) 3-term fit of PV data (P → 80 atm).

T	B	C
273.15	10.77	246
298.15	11.39	233
323.15	11.82	234
348.15	12.19	236
373.15	12.48	238
398.15	12.82	217
423.15	13.08	208

7. R.M. Gibbons, Cryogenics 9 251 (1969).
    PVT data given for P → 200 atm.

T	B	C
44	-46.14	610.4
46	-42.17	556.2
50	-35.41	475.5
60	-24.91	462.9
70	-17.05	395.6

8. W.B. Streett, J. chem. Engng Data 16 289 (1971).
    PVT data given: T 80 - 130 K, P → 2000 atm.

OXYGEN O_2

The second virial coefficient data for oxygen are in good agreement.
Smoothed values are given below.

T	B	T	B
90	-241 ± 10	200	- 49 ± 2
100	-194 ± 7	250	- 28 ± 2
110	-161 ± 7	300	- 16 ± 1
125	-126 ± 5	350	- 7.5 ± 1
150	- 89 ± 3	400	- 1.0 ± 1
175	- 65 ± 3		

1. H.A. Kuypers and H. Kamerlingh Onnes, Archs neerl. Sci. $\underline{6}$ 277 (1923). (*).

    3-term fit of PV data.

    Class I.

T	B	C
273.15	- 21.43	1 033
293.15	- 16.76	963

2. L. Holborn and J. Otto, Z. Phys. $\underline{33}$ 1 (1925)(*).

    5-term fit of PV data (P series, terms in P^0, P^1, P^2, P^4, and P^6). Maximum pressure 100 atm.

    Class I.

T	B	C
273.15	- 22.14	1 560
323.15	- 10.81	1 068
373.15	- 3.46	906

3. G.P. Nijhoff and W.H. Keesom, Communs phys. Lab. Univ. Leiden $\underline{179b}$ (1925)(*).

    Class II.

T	B	T	B
120.60	- 134.83	157.15	- 79.77
127.77	- 121.76	159.22	- 78.24
137.87	- 104.42	163.17	- 73.72
148.21	- 90.40	170.67	- 67.51
154.58	- 82.65	193.16	- 51.23
156.15	- 81.38	233.15	- 32.98

4. G.A. Bottomley, D.S. Massie, and R. Whytlaw-Gray, Proc. R. Soc. A200 201 (1950)(*).

T	B
295.21	- 15.5

5. D. White, J.-H. Hu, and H.L. Johnston, J. chem. Phys. 21 1149 (1953). Values of B determined using a constant-volume gas thermometer. Class II.

T	B	T	B
80	- 339	150	- 94
90	- 237	175	- 62
100	- 170	200	- 42
125	- 112	250	- 35

6. A. Michels, H.W. Schamp, and W. de Graaff, Physica,'s Grav. 20 1209 (1954)(*).
Class I.
(a) 3-term fit of PV data. Maximum pressure 135 atm.

T	B	C
273.15	- 21.89	1 230
298.15	- 16.24	1 163
323.15	- 11.62	1 146

(b) 3-term fit of PV data. Maximum pressure 60 atm.

T	B	C
273.15	- 21.80	1 189
298.15	- 16.50	1 300
323.15	- 11.91	1 290

7. T.L. Cottrell, R.A. Hamilton, and R.P. Taubinger, Trans. Faraday Soc. 52 1310 (1956).

T	B
303	- 16.6 ± 5.1
333	- 7.3 ± 2.0
363	- 2.5 ± 2.3

8. R.A.H. Pool, G. Saville, T.M. Herrington, B.D.C. Shields, and L.A.K. Staveley, Trans. Faraday Soc. 58 1692 (1962).

T	B
90	- 245 ± 2

9. L.A. Weber, J. Res. natn. Bur. Stand. 74A 93 (1970).

3-term fit of PV data at densities up to 0.006709 mol cm^{-3}.

T	B	C
85	-267.78 ± 30	-21462 ± 15000
90	-240.67	-12764
95	-217.51	- 7058
100	-197.54 ± 15	- 3326 ± 4000
105	-180.20	- 904
110	-165.05	+ 644
115	-151.71	1609
120	-139.91 ± 5	2187 ± 1000
125	-129.41	2507
130	-120.02	2659
135	-111.59	2702
140	-103.98 ± 1	2677 ± 100
145	- 97.08	2611
150	- 90.81 ± 0.25	2522 ± 40
155	- 85.09	2423
160	- 79.84 ± 0.25	2320 ± 30
165	- 75.02	2219
170	- 70.58	2122
175	- 66.48	2031
180	- 62.67	1948
185	- 59.14	1871
190	- 55.85	1801
195	- 52.77	1738
200	- 49.89 ± 0.3	1680 ± 30
205	- 47.20	1628
210	- 44.66	1580
215	- 42.27	1537
220	- 40.02	1498
225	- 37.90	1461
230	- 35.89	1428

T	B	C
235	- 33.98	1397
240	- 32.17	1368
245	- 30.45	1342
250	- 28.81	1317
255	- 27.25	1294
260	- 25.77	1273
265	- 24.34	1253
270	- 22.98	1234
275	- 21.68	1217
280	- 20.44	1201
285	- 19.24	1186
290	- 18.09	1172
295	- 16.98	1160
300	- 15.92 ± 0.30	1149 ± 30

SULPHUR DIOXIDE SO_2

1. A. Leduc and P. Sacerdote, C. r. hebd. Seanc. Acad. Sci., Paris 125 297 (1897).

T	B
289.2	- 454.5

2. D. Le B. Cooper and O. Maass, Can. J. Res. 4 495 (1931)(*).
    Class II.

T	B
265.33	- 577
273.15	- 534
298.43	- 412
323.15	- 332
345.62	- 280

3. W. Cawood and H.S. Patterson, J. chem. Soc. 619 (1933)(*).

T	B
303.15	- 411.2
323.15	- 350.5

4. L. Riedel, Z. ges. Kalteind. <u>46</u> 22 (1939).

Following B values given by V.F. Baibuz, Zh. fiz. Khim. <u>32</u> 2644 (1958).

T	B	T	B
278.2	- 501	303.2	- 395
283.2	- 475	313.2	- 360
293.2	- 433	318.2	- 347
298.2	- 414		

5. T.L. Kang, L.J. Hirth, K.A. Kobe, and J.J. McKetta, J. chem. Engng Data <u>6</u> 220 (1961).

Estimated maximum error in B ± 20 at 283K decreasing to ± 7 at 473 K.

T	B	T	B
283.15	- 500.0	373.15	- 232.5
293.15	- 452.0	398.15	- 201.0
303.15	- 404.0	423.15	- 171.1
313.15	- 367.5	448.15	- 144.1
323.15	- 332.8	473.15	- 125.8
348.15	- 279.0		

6. J.H. Dymond and E.B. Smith, (1962)(†).

T	B
298.15	- 406 ± 10

7. R. Stryjek and A. Kreglewski, Bull. Acad. pol. Sci. Ser. Sci. chim. <u>13</u> 201 (1965).

Maximum pressure less than 1 atm.

Class II.

T	B
298.15	- 440
323.25	- 345
348.15	- 285
368.15	- 245

XENON   Xe

The second virial coefficients reported by different workers are
generally in very good agreement. The following values were obtained
from a smooth curve through all the data.

T	B	T	B
160	-425 ± 10	275	-156 ± 4
170	-378 ± 10	300	-133 ± 3
180	-337 ± 8	325	-109 ± 2
190	-306 ± 8	350	- 93.2 ± 2
200	-276 ± 7	400	- 69.4 ± 2
210	-254 ± 5	450	- 51.8 ± 2
225	-224 ± 5	550	- 28.0 ± 2
250	-184 ± 4	650	- 13.0 ± 2

1. J.A. Beattie, R.J. Barriault, and J.S. Brierley, J. chem. Phys. 19
    1222 (1951)(*).

    4-term fit of PV data (V series; terms in V^0, V^{-1}, V^{-2}, and V^{-4}).
    Maximum pressure 410 atm.

    Class I.

T	B	C
289.80	- 137.8	6 343
298.15	- 130.2	6 069
323.15	- 110.6	5 306
348.15	- 94.5	4 635
373.16	- 81.2	4 115
398.17	- 70.1	3 739
423.18	- 60.7	3 469
448.20	- 52.6	3 240
473.21	- 45.4	3 031
498.23	- 39.1	2 869
523.25	- 33.2	2 686
548.26	- 28.0	2 526
573.28	- 23.5	2 423

2. A. Michels, T. Wassenaar, and P. Louwerse, Physica,'s Grav. 20 99
    (1954)(*).

    Class I.

(a) 9-term fit of PV data (P → 2800 atm).

T	B	C
273.15	- 154.21	6 255
298.15	- 130.81	6 551
303.15	- 126.51	6 078
313.15	- 118.56	5 632
323.15	- 111.18	5 239
348.15	- 95.38	4 673
373.15	- 81.65	3 555
398.15	- 70.61	3 247
423.15	- 61.32	3 096

(b) 4-term fit of PV data (P → 80 atm).

T	B	C
273.15	- 154.74	6 760
298.15	- 130.27	5 661
323.15	- 110.98	5 042
348.15	- 95.10	4 470
373.15	- 81.90	3 954
398.15	- 70.62	3 461
423.15	- 61.02	2 973

3. C.G. Reeves and R. Whytlaw-Gray, Proc. R. Soc. A232 173 (1955). Class II.

T	B
273.15	- 157.1
283.15	- 147.5
293.15	- 139.2
303.15	- 129.7
313.15	- 120.2

4. E. Whalley, Y. Lupien, and W.G. Schneider, Can. J. Chem. 33 633 (1955)(*).

3-term fit to PV data (P series).

T	B	C
273.15	- 147.3 ± 1.3	-8 507 ± 1 250
323.15	- 107.8 ± 0.5	+ 395 ± 240
373.15	- 83.45 ± 0.6	4 620 ± 175

T	B	C
423.15	$-$ 62.58 ± 0.6	4 650 ± 155
473.15	$-$ 47.36 ± 0.6	4 700 ± 260
573.15	$-$ 25.07 ± 0.2	2 650 ± 100
673.15	$-$ 10.78 ± 0.4	2 330 ± 210
773.15	$-$ 0.13 ± 0.2	1 130 ± 155
873.15	$+$ 7.95 ± 0.2	
973.15	14.20 ± 0.4	

5. C.M. Greenlief and G. Constabaris, J. chem. Phys. <u>44</u> 4649 (1966).
    PVT data. Pressure less than 1 atm.

T	B
298.15	$-$ 133.0 ± 0.6

6. C.A. Pollard and G. Saville (unpublished results). See also C.A.
    Pollard, Ph.D. thesis, University of London (1971).

    Standard deviation of the curve fit for B is 2 except at 160.22 K
    where σ_B is 5.

T	B	T	B
160.22	-424.3	251.12	-181.9
165.02	-400.3	265.91	-165.2
170.12	-377.6	268.31	-164.2
175.09	-355.2	280.24	-153.4
184.98	-320.7	281.01	-151.2
195.05	-290.3	290.53	-143.3
205.20	-264.3	293.43	-139.6
220.59	-231.8	294.71	-136.8
237.40	-202.0	300.63	-134.8

7. R. Hahn, K. Schäfer, and B. Schramm, Ber. (dtsch.) Bunsenges. phys.
    Chem. <u>78</u> 287 (1974).

    B values determined assuming B (296 K) = -132. Quoted accuracy of B
    is ± 2.

T	B
201.2	-274
211.1	-251
230.8	-212
251.1	-182
273.2	-155

254

8. H.-P. Rentschler and B. Schramm, Ber. (dtsch.) Bunsenges. phys. Chem. 81 319 (1977).

Maximum error in B estimated to be ± 4.

T	B	T	B
309	-120.7	553	- 28.1
332	-104.4	625	- 15.7
418	- 61.6	713	- 7.4
483	- 41.5		

9. B. Schramm, H. Schmiedel, R. Gehrmann, and R. Bartl, Ber. (dtsch.) Bunsenges. phys. Chem. 81 316 (1977).

Maximum error in B estimated to be ± 4.

T	B	T	B
230.7	-213.1	372.2	- 81.1
262.8	-167.6	411.8	- 64.8
271.9	-150.3	433.2	- 57.0
278.2	-146.9	456.8	- 51.0
295.2	-131.0	491.3	- 40.5
332.5	-103.9		

MIXTURES

ARGON + CARBON TETRACHLORIDE   Ar + CCl_4

1. M.S. Vigdergauz and V.I. Semkin, Russ. J. phys. Chem. 45 518 (1971);
    Zh. fiz. Khim. 45 931 (1971). Values also given in M. Vigdergauz
    and V. Semkin, J. Chromat. 58 95 (1971).

    Chromatographic measurements using dinonyl phthalate.

T	B_{12}
323.2	-115 ± 15
353.2	- 67 ± 15

2. S.K. Gupta and A.D. King, Jr., Can. J. Chem. 50 660 (1972).

    B_{12} determined from solubility measurements of carbon tetrachloride
    in the compressed gas: P → 60 atm.

T	B_{12}
273.2	-155 ± 3
298.2	-124 ± 2
323.2	-100 ± 2
348.2	- 77 ± 2

ARGON + CARBON TETRAFLUORIDE   Ar + CF_4

1. E.M. Dantzler Siebert and C.M. Knobler, J. phys. Chem., Ithaca 75
    3863 (1971).

    E determined from measurements of pressure changes on mixing.

T	E	B_{11}	B_{22}	B_{12}
373.15	10 ± 1	-4	-43.1	-14

ARGON + CHLOROFORM   Ar + $CHCl_3$

1. M.S. Vigdergauz and V.I. Semkin, Russ. J. phys. Chem. 45 518 (1971);
    Zh. fiz. Khim. 45 931 (1971). Values also given in M. Vigdergauz and
    V. Semkin, J. Chromat. 58 95 (1971).

Chromatographic measurements using dinonyl phthalate.

T	B_{12}
323.2	-60 ± 15
353.2	-59 ± 15

ARGON + TRIFLUOROMETHANE   Ar + CHF_3

1. T.G. Copeland and R.H. Cole, J. chem. Phys. <u>64</u> 1747 (1976).

T	B_{12}
323.2	-25 ± 4

ARGON + METHYL BROMIDE   Ar + CH_3Br

1. R.N. Lichtenthaler and K. Schäfer, Ber. (dtsch.) Bunsenges. phys. Chem. <u>73</u> 42 (1969).
   Class II.

T	B_{12}
287.8	-73.00
296.0	-63.20
303.2	-56.10
323.1	-37.80

ARGON + METHYL CHLORIDE   Ar + CH_3Cl

1. G.A. Bottomley and T.H. Spurling, Aust. J. Chem. <u>20</u> 1789 (1967).
   Low pressure differential method. B_{22} interpolated from values given. Estimated error in B_{12} greater than ± 25.

T	B_{11}	B_{22}	B_{12}
293.82	-16.8	-420	-28.5
324.45	-10.8	-333	-13.6
347.36	- 7.4	-286	- 4.6
376.11	- 3.6	-243	+ 0.8
430.25	+ 1.9	-177	16.1

2. R.N. Lichtenthaler and K. Schäfer, Ber. (dtsch.) Bunsenges. phys. Chem <u>73</u> 42 (1969).

   Class 1.

T	B_{12}
288.2	-66.30
296.0	-57.87
303.2	-51.60
313.2	-43.60

ARGON + METHYL FLUORIDE   Ar + CH_3F

1. T.G. Copeland and R.H. Cole, J. chem. Phys. **64** 1747 (1976).

T	B_{12}
323.2	-28 ± 5

ARGON + NITROMETHANE   Ar + CH_3NO_2

1. G.A. Bottomley and T.H. Spurling, Aust. J. Chem. **16** 1 (1963).
   Differential compressibility apparatus.

T	B_{11}	B_{22}	B_{12}
323.2	-22	-2866	-72 ± 20

2. M.S. Vigdergauz and V.I. Semkin, Russ. J. phys. Chem. **45** 518 (1971);
   Zh. fiz. Khim. **45** 931 (1971). Values also given in M. Vigdergauz and
   V. Semkin, J. Chromat. **58** 95 (1971).

   Chromatographic measurements using dinonyl phthalate.

T	B_{12}
323.2	-91 ± 15
353.2	-62 ± 15

ARGON + METHANE   Ar + CH_4

1. G. Thomaes, R. Van Steenwinkel, and W. Stone, Molec. Phys. **5** 301
   (1962).

   Volume expansion relative to hydrogen of mixture with X_{Ar} = 0.6130.

T	B_{11}	B_{22}	B_{12}
108.61	-165.5	-301	-252.9
142.6	- 94	-205.6	-138.6
176.7	- 62.2	-135	- 86.7
239.8	- 31.4	- 73	- 48.1
295.0	- 16.5	- 44.5	- 26.9

2. M.A. Byrne, M.R. Jones, and L.A.K. Staveley, Trans. Faraday Soc. <u>64</u> 1747 (1968).

Class 1.

T	B_{12}	T	B_{12}
107.10	-233.1	159.88	-119.2
109.69	-222.5	174.20	- 94.0
113.68	-207.5	191.01	- 78.8
119.48	-189.4	208.79	- 66.1
127.63	-168.4	228.96	- 54.1
138.99	-143.3	249.95	- 43.9
149.27	-126.1	274.15	- 35.2

3. R.N. Lichtenthaler and K. Schäfer, Ber. (dtsch.) Bunsenges. phys. Chem. <u>73</u> 42 (1969).

Estimated absolute error in B_{12} $\pm 1 \to 2$.

T	B_{12}	T	B_{12}
288.2	-23.34	313.2	-17.96
296.0	-21.90	323.1	-15.69
303.2	-20.19		

4. K. Strein, R.N. Lichtenthaler, B. Schramm, and Kl. Schäfer, Ber. (dtsch.) Bunsenges. phys. Chem. <u>75</u> 1308 (1971).

Estimated accuracy of B_{12} is ± 1.

T	B_{12}	T	B_{12}
296.15	-21.1	413.8	- 2.0
308.0	-18.6	434.0	+ 0.1
333.5	-13.7	453.4	1.9
353.8	-10.2	473.5	3.6
374.0	- 7.0	493.0	5.3
393.9	- 4.2		

5. J. Bellm, W. Reineke, K. Schäfer, and B. Schramm, Ber. (dtsch.) Bunsenges. phys. Chem. <u>78</u> 282 (1974).

Estimated accuracy of B_{12} is ± 2.

T	B_{12}	T	B_{12}
300	-27.9	430	- 4.7
320	-23.4	460	- 1.0
340	-19.2	490	+ 1.8
370	-13.6	520	3.9
400	- 8.7	550	5.5

6. R. Hahn, K. Schäfer, and B. Schramm, Ber. (dtsch.) Bunsenges. phys. Chem. <u>78</u> 287 (1974).

B_{12} determined assuming B_{12} (296 K) = -22.

Quoted accuracy of B_{12} is ± 2.

T	B_{12}	T	B_{12}
201.2	-62	251.5	-37
211.4	-54	272.2	-29
231.5	-45		

ARGON+METHANOL   Ar + CH_3OH

1. B. Hemmaplardh and A.D. King, Jr., J. phys. Chem., Ithaca <u>76</u> 2170 (1972).

B_{12} derived from measurements of solubility of methanol in the compressed gas.

T	B_{12}	T	B_{12}
288.15	-102 ± 2	323.15	- 72 ± 2
298.15	- 87 ± 4	333.15	- 71 ± 4
310.15	- 77 ± 3		

ARGON + CARBON MONOXIDE

1. B. Schramm and R. Gehrmann (1979) (†).

Estimated error in B_{12} is ± 5.

T	B_{12}	T	B_{12}
213	-44.2	262	-21.6
223	-37.5	276	-18.1
242	-28.7		

2. B. Schramm and H. Schmiedel (1979) (†).
    Estimated error in B_{12} is ±4.

T	B_{12}	T	B_{12}
295	-13.4	425	3.8
330	- 8.1	450	6.3
365	- 3.3	475	7.9
400	+ 0.7		

ARGON + CARBON DIOXIDE Ar + CO_2

1. T.L. Cottrell, R.A. Hamilton and R.P. Taubinger, Trans. Faraday Soc.
    52 1310 (1956).
    Gas expansions at P < 1 atm, X_2 is 0.5.

T	B_{11}	B_{22}	B_{12}
303.2	-14.9	-119.2	-31.8 ± 4.6
333.2	- 9.7	- 97.1	-25.8 ± 4.2
363.2	- 5.3	- 79.6	-19.6 + 4.2

2. W.H. Abraham and C.O. Bennett, A.I. Ch. E. Jl 6 257 (1960).
    PVT data for seven mixtures at 323.15K; P range 50 - 1000 atm.

3. R.N. Lichtenthaler and K. Schäfer, Ber. (dtsch.) Bunsenges. phys.
    Chem. 73 42 (1969).
    Estimated absolute error in B_{12} ± 1 → 2.

T	B_{12}	T	B_{12}
288.2	-40.30	313.2	-31.20
296.0	-37.05	323.1	-28.30
303.2	-34.21		

4. H. Sutter and R.H. Cole, J. chem. Phys. 52 140 (1970).
    Low pressure measurements.

T	B_{11}	B_{22}	B_{12}
322.85	-15.8	-109.3	-30.1
322.85	-11.3	-103.9	-40.0

5. J.P. Strakey, C.O. Bennett and B.F. Dodge, A. I. Ch. E. Jl <u>20</u> 803 (1974).

Adiabatic Joule-Thomson coefficient measurement.

T	$T \dfrac{dB_{12}}{dT} - B_{12}$
233.2	275 ± 23
273.2	209 ± 18
313.2	153 ± 13
353.2	114 ± 10
383.2	95 ± 8

6. B. Schramm and R. Gehrmann (1979) (†).

Estimated error in B_{12} is ± 6.

T	B_{12}	T	B_{12}
213	-93.9	262	-53.8
223	-84.5	276	-48.7
242	-67.2		

7. B. Schramm and H. Schmiedel (1979) (†).

Estimated error in B_{12} is ± 5.

T	B_{12}	T	B_{12}
295	-47.7	425	-14.9
330	-37.4	450	-10.1
365	-26.2	475	- 8.0
400	-17.5		

ARGON + ETHYLENE   Ar + $CH_2:CH_2$

1. I. Masson and L.G.F. Dolley, Proc. R. Soc. <u>A103</u> 524 (1923).

PVTx data given for five mixtures; P → 125 atm and T = 298K.

2. T.K. Bose and R.H. Cole, J. chem. Phys. <u>54</u> 3829 (1971).

B_M values determined at 323 K for five mixtures with mole fraction ethylene in range 0.12 - 0.45.

ARGON + NITROETHANE   Ar + $CH_3.CH_2NO_2$

1. M.S. Vigdergauz and V.I. Semkin, Russ. J. phys. Chem. 45 518 (1971); Zh. fiz. Khim. 45 931 (1971). Values also given in M. Vigdergauz and V. Semkin, J. Chromat. 58 95 (1971).

Chromatographic measurements using dinonyl phthalate.

T	B_{12}
323.2	-125 ± 15
353.2	- 97 ± 15

ARGON + ETHANE   Ar + $CH_3.CH_3$

1. R.N. Lichtenthaler and K. Schäfer, Ber. (dtsch.) Bunsenges. phys. Chem. 73 42 (1969).

Estimated absolute error in B_{12} ± 1 → 2.

T	B_{12}	T	B_{12}
288.2	-58.73	313.2	-46.72
296.0	-54.40	323.1	-42.68
303.2	-50.97		

2. R. Hahn, K. Schäfer, and B. Schramm, Ber. (dtsch.) Bunsenges. phys. Chem. 78 287 (1974).

B_{12} determined assuming B_{12} (296 K) = -52.

Quoted accuracy of B_{12} is ± 2.

T	B_{12}	T	B_{12}
199.4	-134	273.2	- 64
211.4	-113	333	- 40
231.5	- 95	365	- 29
251.5	- 79		

3. K. Schäfer, B. Schramm, and J.S.U. Navarro, Z. phys. Chem. Frankf. Ausg. 93 203 (1974).

T	B_{12}	T	B_{12}
296.2	-54.4	432.1	-11.5
353.2	-31.6	472.3	- 4.8
392.8	-20.8	510.6	+ 2.2

ARGON + ETHANOL   Ar + $CH_3.CH_2OH$

1. M.S. Vigdergauz and V.I. Semkin, Russ. J. phys. Chem. 45 518 (1971);
    Zh. fiz. Khim. 45 931 (1971). Values also given in M. Vigdergauz
    and V. Semkin, J. Chromat. 58 95 (1971).

    Chromatographic measurements using dinonyl phthalate.

T	B_{12}
323.2	-118 ± 15
353.2	- 46 ± 15

2. S.K. Gupta, R.D. Lesslie and A.D. King, Jr., J. phys. Chem., Ithaca
    77 2011 (1973).

    B_{12} derived from measurements of solubility of ethanol in the
    compressed gas.

T	B_{12}
298.15	-73 ± 5
323.15	-63 ± 2
348.15	-50 ± 3

ARGON + ACETONE   Ar + $CH_3.CO.CH_3$

1. M.S. Vigdergauz and V.I. Semkin, Russ. J. phys. Chem. 45 518 (1971);
    Zh. fiz. Khim. 45 931 (1971). B_{12} given also in M. Vigdergauz and
    V. Semkin, J. Chromat. 58 95 (1971).

    Chromatographic measurements using dinonyl phthalate.

T	B_{12}
353.2	-58 ± 15

ARGON + PROPANE   Ar + $CH_3.CH_2.CH_3$

1. R.N. Lichtenthaler and K. Schäfer, Ber. (dtsch.) Bunsenges. phys.
    Chem. 73 42 (1969).

    Estimated absolute error in B_{12} ± 1 → 2.

T	B_{12}	T	B_{12}
288.2	-84.45	313.2	-68.67
296.0	-79.00	322.8	-62.88
303.2	-74.60		

2. R. Hahn, K. Schäfer, and B. Schramm, Ber. (dtsch.) Bunsenges. phys. Chem. **78** 287 (1974).

B_{12} determined assuming B_{12} (296 K) = -81.

Quoted accuracy of B_{12} is ± 2.

T	B_{12}	T	B_{12}
211.4	-183	273.2	- 98
231.5	-152	333	- 60
251.5	-116	365	- 45

3. K. Schäfer, B. Schramm, and J.S.U. Navarro, Z. phys. Chem. Frankf. Ausg. **93** 203 (1974).

T	B_{12}	T	B_{12}
296.2	-78.8	432.3	-24.4
353.2	-48.9	473.0	-15.5
392.9	-34.1	511.8	- 6.6

ARGON + ISOPROPANOL $Ar + (CH_3)_2.CHOH$

1. M.S. Vigdergauz and V.I. Semkin, Russ. J. phys. Chem. **45** 518 (1971); Zh. fiz. Khim. **45** 931 (1971). Values also given in M. Vigdergauz and V. Semkin, J. Chromat. **58** 95 (1971).

Chromatographic measurements using dinonyl phthalate.

T	B_{12}
323.2	-80 ± 15
353.2	-51 ± 15

ARGON + METHYL ETHYL KETONE $Ar + CH_3.CO.CH_2.CH_3$

1. M.S. Vigdergauz and V.I. Semkin, Russ. J. phys. Chem. **45** 518 (1971); Zh. fiz. Khim. **45** 931 (1971). B_{12} given also in M. Vigdergauz and V. Semkin, J. Chromat. **58** 95 (1971).

Chromatographic measurements using dinonyl phthalate.

T	B_{12}
323.2	-86 ± 15
353.2	-65 ± 15

ARGON + DIOXAN   Ar + $CH_2.CH_2.O.(CH_2)_2.O$

1. M.S. Vigdergauz and V.I. Semkin, Russ. J. phys. Chem. 45 518 (1971); Zh. fiz. Khim. 45 931 (1971). Values also given in M. Vigdergauz and V. Semkin, J. Chromat. 58 95 (1971).

Chromatographic measurements using dinonyl phthalate.

T	B_{12}
323.2	-124 ± 15
353.2	$- 86 \pm 15$

ARGON + n-BUTANE   Ar + $CH_3.(CH_2)_2.CH_3$

1. R.N. Lichtenthaler and K. Schäfer, Ber. (dtsch.) Bunsenges. phys. Chem. 73 42 (1969).

Class I.

T	B_{12}	T	B_{12}
288.2	-120.9	313.2	-100.3
296.0	-113.2	323.1	- 93.00
303.2	-107.7		

2. K. Schäfer, B. Schramm, and J.S.U. Navarro, Z. phys. Chem. Frankf. Ausg. 93 203 (1974).

T	B	T	B
296.2	-113	432.2	- 40
353.1	- 73	472.5	- 32
393.3	- 54	511.1	- 28

ARGON + 1-BUTANOL   Ar + $CH_3.(CH_2)_2.CH_2OH$

1. R. Massoudi and A.D. King, Jr., J. phys. Chem., Ithaca 77 2016 (1973).

B_{12} derived from measurements of solubility of 1-butanol in the compressed gas.

T	B_{12}
298.15	-90 ± 11

ARGON + DIETHYL ETHER   Ar + $(C_2H_5)_2O$

1. R. Massoudi and A.D. King, Jr., J. phys. Chem., Ithaca 77 2016 (1973).
    B_{12} derived from measurements of solubility of diethyl ether in the compressed gas.

T	B_{12}
298.15	-89 ± 6

ARGON + TETRAMETHYLSILANE   Ar + $(CH_3)_4Si$

1. J. Bellm, W. Reineke, K. Schäfer, and B. Schramm, Ber. (dtsch.) Bunsenges. phys. Chem. 78 282 (1974).

    Class III.

T	B_{12}	T	B_{12}
300	-147.2	430	- 77.1
320	-127.0	460	- 67.3
340	-119.0	490	- 60.3
370	-102.0	520	- 54.2
400	- 89.5	550	- 47.6

ARGON + PYRIDINE   Ar + CH : CH.CH : CH.CH : N

1. M.S. Vigdergauz and V.I. Semkin, Russ. J. phys. Chem. 45 518 (1971); Zh. fiz. Khim. 45 931 (1971). Values also given in M. Vigdergauz and V. Semkin, J. Chromat. 58 95 (1971).

    Chromatographic measurements using dinonyl phthalate.

T	B_{12}
323.2	-115 ± 15
353.2	$- 95 \pm 15$

ARGON + CYCLOPENTANE   Ar + $CH_2.(CH_2)_3.CH_2$

1. M.S. Vigdergauz and V.I. Semkin, Russ. J. phys. Chem. 45 518 (1971); Zh. fiz. Khim. 45 931 (1971). Value also given in M. Vigdergauz and V. Semkin, J. Chromat. 58 95 (1971).

B_{12} determined from pressure variation of chromatographic retention indices, using the following B_{12}(298.2K) reference values: -98 for argon + n-pentane, -124 for argon + n-hexane.
Class II.

T	B_{12}
298.2	-108

ARGON + n PENT-1-ENE   Ar + $CH_3 \cdot (CH_2)_2 \cdot CH:CH_2$

1. A.J.B. Cruikshank, B.W. Gainey and C.L. Young, Trans. Faraday Soc. 64 337 (1968).

    Gas chromatographic measurements.

T	B_{12}
308.15	- 84 ± 12

ARGON + n-PENTANE   Ar + $CH_3 \cdot (CH_2)_3 \cdot CH_3$

1. A.J.B. Cruikshank, M.L. Windsor and C.L. Young, Proc. R. Soc. A295 271 (1966).

    Gas chromatographic measurements.

T	B_{12}
298.2	- 98 ± 6

2. E.M. Dantzler, C.M. Knobler and M.L. Windsor, J. Chromat. 32 433 (1968).

    B_{12} determined from measurements of pressure change on mixing gases at constant volume and constant temperature, and given values of B_{11} and B_{22}.

T	B_{11}	B_{22}	B_{12}
298.2	-16.4	-1197	-136 ± 18
298.2	-16.4	-1172	-125 ± 23

3. M.S. Vigdergauz and V.I. Semkin, Russ. J. phys. Chem. 45 518 (1971); Zh. fiz. Khim. 45 931 (1971). Values also given by M. Vigdergauz and V. Semkin. J. Chromat. 58 95 (1971).

    Chromatographic measurements using dinonyl phthalate.

T	B_{12}
323.2	-82 ± 15
353.2	-68 ± 15

4. R. Massoudi and A.D. King, Jr., J. phys. Chem., Ithaca <u>77</u> 2016 (1973).

B_{12} derived from measurements of solubility of n-pentane in the compressed gas.

T	B_{12}
289.15	-97 ± 8

ARGON + 2-METHYL BUTANE (ISOPENTANE) $Ar + (CH_3)_2 CH.CH_2.CH_3$

1. A.J.B. Cruikshank, M.L. Windsor and C.L. Young, Proc. R. Soc. <u>A295</u> 271 (1966).

Gas chromatographic measurements.

T	B_{12}
298.2	-94 ± 6

2. A.J.B. Cruikshank, B.W. Gainey and C.L. Young, Trans. Faraday Soc. <u>64</u> 337 (1968).

Gas chromatographic measurements.

T	B_{12}
308.15	-78 ± 12

3. E.M. Dantzler, C.M. Knobler, and M.L. Windsor, J. Chromat. <u>32</u> 433 (1968).

B_{12} determined from measurements of pressure changes on mixing gases at constant temperature and constant volume, and given values of B_{11} and B_{22}.

T	B_{11}	B_{22}	B_{12}
298.2	-16.4	-1160	-178 ± 14

ARGON + NEOPENTANE $Ar + (CH_3)_4 C$

1. K. Strein, R.N. Lichtenthaler, B. Schramm, and Kl. Schäfer, Ber. (dtsch.) Bunsenges. phys. Chem. <u>75</u> 1308 (1971).

Estimated accuracy of B_{12} is $\pm 1\%$.

T	B_{12}	T	B_{12}
296.15	-126.7	413.8	- 48.5
308.0	-114.0	433.8	- 41.2
334.0	- 90.8	453.6	- 34.0
353.5	- 75.3	473.8	- 28.5
373.6	- 65.1	492.6	- 22.9
394.3	- 53.7		

2. J. Bellm, W. Reineke, K. Schäfer, and B. Schramm, Ber. (dtsch.)
    Bunsenges. phys. Chem. __78__ 282 (1974).

    Class 111.

T	B_{12}	T	B_{12}
300	-118.2	430	- 44.1
320	-100.1	460	- 35.6
340	- 89.0	490	- 27.9
370	- 72.9	520	- 20.2
400	- 56.6	550	- 14.1

3. G.L. Baughman, S.P. Westhoff, S. Dincer, D.D. Duston and A.J. Kidnay,
    J. chem. Thermodyn. __7__ 875 (1975).

B_{12} determined from measurements of the solubility of neopentane in
the compressed gas. Maximum error in B_{12} is estimated to be 10%.

T	B_{12}	T	B_{12}
199.62	-290	240.15	-172
209.59	-242	249.45	-143
219.56	-224	257.91	-111
229.53	-201		

ARGON + BENZENE   Ar + CH:(CH)$_4$:CH

1. A.J.B. Cruikshank, M.L. Windsor and C.L. Young, Proc. R. Soc. __A295__
    271 (1966).

    Gas chromatographic measurements.

T	B_{12}
298.2	-88 ± 6

2. H. Bradley, Jr., and A.D. King, Jr., J. chem. Phys. __47__ 1189 (1967).

B_{12} derived from measurements of solubility of benzene in compressed
argon: P   160 atm.

T	B_{12}
298.2	-100 ± 7

3. B.W. Gainey and C.L. Young, Trans. Faraday Soc. __64__ 349 (1968).

    Gas chromatographic measurements.

T	B_{12}
305.15	-135 ± 13
313.15	-126 ± 12
323.15	$- 90 \pm 10$

4. D.H. Everett, B.W. Gainey and C.L. Young, Trans. Faraday Soc. <u>64</u> 2667 (1968).

Gas chromatographic measurements, using squalane.

T	B_{12}
323.15	-85 ± 8

Gas chromatographic measurements, using n-octadecane.

T	B_{12}
323.15	-79 ± 8

5. C.R. Coan and A.D. King, Jr., J. Chromat. <u>44</u> 429 (1969).

B_{12} determined from measurements of solubility of benzene in the compressed gas: $P \rightarrow 65$ atm.

T	B_{12}
305.2	-122 ± 3
323.2	$- 95 \pm 3$

6. M.S. Vigdergauz and V.I. Semkin, Russ. J. phys. Chem. <u>45</u> 518 (1971); Zh. fiz. Khim. <u>45</u> 931 (1971). Values also given in M. Vigdergauz and V. Semkin, <u>J.</u> Chromat. <u>58</u> 95 (1971).

Chromatographic measurements using dinonyl phthalate.

T	B_{12}
323.2	-117 ± 15
353.2	-105 ± 15

7. M.S. Vigdergauz and V.I. Semkin, Russ. J. phys. Chem. <u>45</u> 518 (1971); Zh. fiz. Khim. <u>45</u> 931 (1971). Value also given in M. Vigdergauz and V. Semkin, J. Chromat. <u>58</u> 95 (1971).

B_{12} determined from the pressure variation of chromatographic retention indices using the following $B_{12}(298.2K)$ reference values: -98 for argon + n-pentane, -124 for argon + n-hexane.

Class II.

T	B_{12}
298.2	-112

ARGON + CYCLOHEXANE   Ar + $CH_2 \cdot (CH_2)_4 \cdot CH_2$

1. M.S. Vigdergauz and V.I. Semkin, Russ. J. phys. Chem. 45 518 (1971); Zh. fiz. Khim. 45 931 (1971). Values also given in M. Vigdergauz and V. Semkin, J. Chromat. 58 95 (1971).

   Chromatographic measurements using dinonyl phthalate.

T	B_{12}
323.2	-102 ± 15
353.2	∼ 86 ± 15

2. M.S. Vigdergauz and V.I. Semkin, Russ. J. phys. Chem. 45 518 (1971); Zh. fiz. Khim. 45 931 (1971). Value also given in M. Vigdergauz and V. Semkin, J. Chromat. 58 95 (1971).

   B_{12} determined from the pressure variation of chromatographic retention indices using the following B_{12} (298.2K) reference values: -98 for argon + n-pentane, -124 for argon + n-hexane. Class II.

T	B_{12}
298.2	-130

ARGON + n HEX-1-ENE   Ar + $CH_2:CH(CH_2)_3 \cdot CH_3$

1. A.J.B. Cruikshank, B.W. Gainey and C.L. Young, Trans. Faraday Soc. 64 337 (1968).

   Gas chromatographic measurements.

T	B_{12}
308.15	-110 ± 12

ARGON + METHYL CYCLOPENTANE   Ar + $CH(CH_3) \cdot (CH_2)_3 \cdot CH_2$

1. M.S. Vigdergauz and V.I. Semkin, Russ. J. phys. Chem. 45 518 (1971); Zh. fiz. Khim. 45 931 (1971). Value also given in M. Vigdergauz and V. Semkin, J. Chromat. 58 95 (1971).

   B_{12} determined from the pressure variation of chromatographic retention indices using the following B_{12}(298.2K) reference values: -98 for argon + n-pentane, -124 for argon + n-hexane. Class II.

272

T	B_{12}
298.2	-127

ARGON + BUTYL ACETATE   Ar + $CH_3.COO(CH_2)_3.CH_3$

1. M.S. Vigdergauz and V.I. Semkin, Russ. J. phys. Chem. 45 518 (1971); Zh. fiz. Khim. 45 931 (1971). Values also given in M. Vigdergauz and V. Semkin, J. Chromat. 58 95 (1971).

   Chromatographic measurements using dinonyl phthalate.

T	B_{12}
323.2	-138 ± 15
353.2	-119 ± 15

ARGON + n-HEXANE   Ar + $CH_3.(CH_2)_4CH_3$

1. A.J.B. Cruikshank, M.L. Windsor and C.L. Young, Proc. R. Soc. A295 271 (1966).

   Gas chromatographic measurements.

T	B_{12}	T	B_{12}
293.2	-130 ± 6	323.2	-107 ± 6
298.2	-127 ± 6	338.6	-100 ± 6
310.6	-104 ± 6		

2. E.M. Dantzler, C.M. Knobler, and M.L. Windsor, J. Chromat. 32 433 (1968).

   B_{12} determined from measurements of pressure change on mixing gases at constant temperature and constant volume, and given values of B_{11} and B_{22}.

T	B_{11}	B_{22}	B_{12}
298.2	-16.4	-1896	-149 ± 65
298.2	-16.4	-1758	- 80 ± 73
298.2	-16.4	-1958	-180 ± 50
323.2	-11.5	-1423	-108 ± 32
323.2	-11.5	-1456	-120 ± 9
323.2	-11.5	-1513	-148 ± 27

3. M.S. Vigdergauz and V.I. Semkin, Russ. J. phys. Chem. 45 518 (1971); Zh. fiz. Khim. 45 931 (1971). Values also given by M. Vigdergauz and V. Semkin, J. Chromat. 58 95 (1971).

Chromatographic measurements using dinonyl phthalate.

T	B_{12}
323.2	-106 ± 15
353.2	$- 81 \pm 15$

ARGON + 2-METHYLPENTANE   Ar + $(CH_3)_2CH.(CH_2)_2.CH_3$

1. A.J.B. Cruikshank, M.L. Windsor and C.L. Young, Proc. R. Soc. A295 271 (1966).

    Gas chromatographic measurements.

T	B_{12}
298.2	-125 ± 6

2. E.M. Dantzler, C.M. Knobler, and M.L. Windsor, J. Chromat. 32 433 (1968).

    B_{12} determined from measurements of pressure change on mixing gases at constant temperature and constant volume, and given values of B_{11} and B_{22}.

T	B_{11}	B_{22}	B_{12}
298.2	-16.4	-1680	-147 ± 49

3. M.S. Vigdergauz and V.I. Semkin, Russ. J. phys. Chem. 45 518 (1971); Zh. fiz. Khim. 45 931 (1971). Value also given in M. Vigdergauz and V. Semkin, J. Chromat. 58 95 (1971).

    B_{12} determined from the pressure variation of chromatographic retention indices using the following $B_{12}(298.2K)$ reference values: -98 for argon + n-pentane, -124 for argon + n-hexane.

    Class II.

T	B_{12}
298.2	-115

ARGON + 3 METHYL PENTANE   Ar + $C_2H_5.CH(CH_3)C_2H_5$

1. A.J.B. Cruikshank, B.W. Gainey and C.L. Young, Trans. Faraday Soc. 64 337 (1968).

    Gas chromatographic measurements.

T	B_{12}
308.15	-95 ± 12

2. M.S. Vigdergauz and V.I. Semkin, Russ. J. phys. Chem. 45 518 (1971); Zh. fiz. Khim. 45 931 (1971). Value also given in M. Vigdergauz and V. Semkin, J. Chromat. 58 95 (1971).

B_{12} determined from the pressure variation of chromatographic retention indices using the following B_{12}(298.2K) reference values: -98 for argon + n-pentane, -124 for argon + n-hexane. Class II.

T	B_{12}
298.2	-116

ARGON + 2,2-DIMETHYLBUTANE $(CH_3)_3C.C_2H_5$

1. A.J.B. Cruikshank, M.L. Windsor and C.L. Young, Proc. R. Soc. A295 271 (1966).

Gas chromatographic measurements.

T	B_{12}
298.2	-115 ± 6

2. E.M. Dantzler, C.M. Knobler, and M.L. Windsor, J. Chromat. 32 433 (1968).

B_{12} determined from measurements of pressure change on mixing gases at constant temperature and constant volume, and given values of B_{11} and B_{22}.

T	B_{11}	B_{22}	B_{12}
298.2	-16.4	-1450	-156 ± 26

3. M.S. Vigdergauz and V.I. Semkin, Russ. J. phys. Chem. 45 518 (1971); Zh. fiz. Khim. 45 931 (1971). Value also given in M. Vigdergauz and V. Semkin, J. Chromat. 58 95 (1971).

B_{12} determined from pressure variation of chromatographic retention indices, using the following B_{12}(298.2K) reference values: -98 for argon + n-pentane, -124 for argon + n-hexane. Class II.

T	B_{12}
298.2	-108

ARGON + 2,3 DIMETHYL BUTANE $Ar + (CH_3)_2.CH.CH.(CH_3)_2$

1. A.J.B. Cruikshank, B.W. Gainey and C.L. Young, Trans. Faraday Soc.

<u>64</u> 337 (1968).

Gas chromatographic measurements.

T	B_{12}
308.15	-98 ± 12

2. M.S. Vigdergauz and V.I. Semkin, Russ. J. phys. Chem. <u>45</u> 518 (1971); Zh. fiz. Khim. <u>45</u> 931 (1971). Value also given in M. Vigdergauz and V. Semkin, J. Chromat. <u>58</u> 95 (1971).

B_{12} determined from pressure variation of chromatographic retention indices, using the following B_{12}(298.2K) reference values: -98 for argon + n-pentane, -124 for argon + n-hexane.

Class II.

B	B_{12}
298.2	-112

ARGON + TOLUENE   Ar + $C_6H_5.CH_3$

1. M.S. Vigdergauz and V.I. Semkin, Russ. J. phys. Chem. <u>45</u> 518 (1971); Zh. fiz. Khim. <u>45</u> 931 (1971). Values also given by M. Vigdergauz and V. Semkin, J. Chromat. <u>58</u> 95 (1971).

Chromatographic measurements using dinonyl phthalate.

T	B_{12}
323.2	-135 ± 15
353.2	-120 ± 15

2. M. Vigdergauz and V. Semkin, J. Chromat. <u>58</u> 95 (1971).

B_{12} determined from chromatographic retention data with reference B_{12} (353.2K) values of -105 for argon + benzene, and -148 for argon + styrene.

Class III.

T	B_{12}
353.2	-123

ARGON + 1,1-DIMETHYLCYCLOPENTANE   Ar + $C(CH_3)_2.(CH_2)_3.CH_2$

1. M.S. Vigdergauz and V.I. Semkin, Russ. J. phys. Chem. <u>45</u> 518 (1971); Zh. fiz. Khim. <u>45</u> 931 (1971). Value also given in M. Vigdergauz and V. Semkin, J. Chromat. <u>58</u> 95 (1971).

B_{12} determined from the pressure variation of chromatographic retention indices using the following B_{12} (298.2K) reference values: -98 for argon + n-pentane, -124 for argon + n-hexane. Class II.

T	B_{12}
298.2	-134

ARGON + trans-1,2-DIMETHYLCYCLOPENTANE   Ar + $CH_3.CH.CH(CH_3).(CH_2)_2.CH_2$

1. M.S. Vigdergauz and V.I. Semkin, Russ. J. phys. Chem. 45 518 (1971); Zh. fiz. Khim. 45 931 (1971). Value also given in M. Vigdergauz and V. Semkin, J. Chromat. 58 95 (1971).

   B_{12} determined from the pressure variation of chromatographic retention indices using the following B_{12} (298.2K) reference values: -98 for argon + n-pentane, -124 for argon + n-hexane. Class II.

T	B_{12}
298.2	-136

ARGON + cis-1,3-DIMETHYLCYCLOPENTANE Ar + $CH_3.CH.CH_2.CH(CH_3).CH_2.CH_2$

1. M.S. Vigdergauz and V.I. Semkin, Russ. J. phys. Chem. 45 518 (1971); Zh. fiz. Khim. 45 931 (1971). Value also given in M. Vigdergauz and V. Semkin, J. Chromat. 58 95 (1971).

   B_{12} determined from the pressure variation of chromatographic retention indices using the following B_{12} (298.2K) reference values: -98 for argon + n-pentane, -124 for argon + n-hexane. Class II.

T	B_{12}
298.2	-136

ARGON + trans-1,3-DIMETHYLCYCLOPENTANE Ar + $CH_3.CH.CH_2.CH(CH_3).CH_2.CH_2$

1. M.S. Vigdergauz and V.I. Semkin, Russ. J. phys. Chem. 45 518 (1971); Zh. fiz. Khim. 45 931 (1971). Value also given in M. Vigdergauz and V. Semkin, J. Chromat. 58 95 (1971).

   B_{12} determined from the pressure variation of chromatographic retention indices using the following B_{12} (298.2K) reference values: -98 for argon + n-pentane, -124 for argon + n-hexane.

Class II.

T	B_{12}
298.2	-136

ARGON + n HEPT-1-ENE $CH_2:CH(CH_2)_4.CH_3$

1. A.J.B. Cruikshank, B.W. Gainey and C.L. Young, Trans. Faraday Soc. 64 337 (1968).

Gas chromatographic measurements.

T	B_{12}
308.15	-139 ± 12

ARGON + n-HEPTANE $Ar + CH_3.(CH_2)_5.CH_3$

1. M.S. Vigdergauz and V.I. Semkin, Russ. J. phys. Chem. 45 518 (1971); Zh. fiz. Khim. 45 931 (1971). Values also given by M. Vigdergauz and V. Semkin, J. Chromat. 58 95 (1971).

Chromatographic measurements using dinonyl phthalate.

T	B_{12}
323.2	-136 ± 15
353.2	$- 97 \pm 15$

ARGON + 2-METHYLHEXANE $Ar + (CH_3)_2CH.(CH_2)_3.CH_3$

1. M.S. Vigdergauz and V.I. Semkin, Russ. J. phys. Chem. 45 518 (1971); Zh. fiz. Khim. 45 931 (1971). Value also given in M. Vigdergauz and V. Semkin, J. Chromat. 58 95 (1971).

B_{12} determined from the pressure variation of chromatographic retention indices using the following $B_{12}(298.2K)$ reference values: -98 for argon + n-pentane, -124 for argon + n-hexane.

Class II.

T	B_{12}
298.2	-137

ARGON + 3-METHYLHEXANE $Ar + CH_3.CH_2.CH(CH_3).(CH_2)_2.CH_3$

1. M.S. Vigdergauz and V.I. Semkin, Russ. J. phys. Chem. 45 518 (1971); Zh. fiz. Khim. 45 931 (1971). Value also given in M. Vigdergauz and

V. Semkin, J. Chromat. <u>58</u> 95 (1971).

B_{12} determined from the pressure variation of chromatographic retention indices using the following B_{12}(298.2K) reference values: -98 for argon + n-pentane, -124 for argon + n-hexane.
Class II.

T	B_{12}
298.2	-138

ARGON + 2,2-DIMETHYLPENTANE Ar + $(CH_3)_3 \cdot C \cdot (CH_2)_2 \cdot CH_3$

1. M.S. Vigdergauz and V.I. Semkin, Russ. J. phys. Chem. <u>45</u> 518 (1971); Zh. fiz. Khim. <u>45</u> 931 (1971). Value also given in M. Vigdergauz and V. Semkin, J. Chromat. <u>58</u> 95 (1971).

   B_{12} determined from the pressure variation of chromatographic retention indices using the following B_{12}(298.2K) reference values: -98 for argon + n-pentane, -124 for argon + n-hexane.
   Class II.

T	B_{12}
298.2	-128

ARGON + 2,3-DIMETHYLPENTANE Ar + $(CH_3)_2 \cdot CH \cdot CH(CH_3) \cdot CH_2 \cdot CH_3$

1. M.S. Vigdergauz and V.I. Semkin, Russ. J. phys. Chem. <u>45</u> 518 (1971); Zh. fiz. Khim. <u>45</u> 931 (1971). Value also given in M. Vigdergauz and V. Semkin, J. Chromat. <u>58</u> 95 (1971).

   B_{12} determined from the pressure variation of chromatographic retention indices using the following B_{12}(298.2K) reference values: -98 for argon + n-pentane, -124 for argon + n-hexane.
   Class II.

T	B_{12}
298.2	-135

ARGON + 2,4-DIMETHYLPENTANE Ar + $(CH_3)_2 \cdot CH \cdot CH_2 \cdot CH(CH_3) \cdot CH_3$

1. M.S. Vigdergauz and V.I. Semkin, Russ. J. phys. Chem. <u>45</u> 518 (1971); Zh. fiz. Khim. <u>45</u> 931 (1971). Value also given in M. Vigdergauz and V. Semkin, J. Chromat. <u>58</u> 95 (1971).

   B_{12} determined from the pressure variation of chromatographic retention indices using the following B_{12}(298.2K) reference values:

-98 for argon + n-pentane and -124 for argon + n-hexane.

Class II.

T	B_{12}
298.2	-130

ARGON + 3,3-DIMETHYLPENTANE   Ar + $(CH_3.CH_2)_2.C.(CH_3)_2$

1. M.S. Vigdergauz and V.I. Semkin, Russ. J. phys. Chem. <u>45</u> 518 (1971); Zh. fiz. Khim. <u>45</u> 931 (1971). Value also given in M. Vigdergauz and V. Semkin, J. Chromat. <u>58</u> 95 (1971).

    B_{12} determined from the pressure variation of chromatographic retention indices using the following B_{12}(298.2K) reference values: -98 for argon + n-pentane, -124 for argon + n-hexane.

    Class II.

T	B_{12}
298.2	-132

ARGON + 2,2,3-TRIMETHYLBUTANE   Ar + $(CH_3)_3.C.CH(CH_3).CH_3$

1. M.S. Vigdergauz and V.I. Semkin, Russ. J. phys. Chem. <u>45</u> 518 (1971); Zh. fiz Khim. <u>45</u> 931 (1971). Value also given in M. Vigdergauz and V. Semkin, J. Chromat. <u>58</u> 95 (1971).

    B_{12} determined from the pressure variation of chromatographic retention indices using the following B_{12}(298.2K) reference values: -98 for argon + n-pentane, -124 for argon + n-hexane.
    Class II.

T	B_{12}
298.2	-127

ARGON + STYRENE   Ar + $C_6H_5.CH:CH_2$

1. M.S. Vigdergauz and V.I. Semkin, Russ. J. phys. Chem. <u>45</u> 518 (1971); Zh. fiz. Khim. <u>45</u> 931 (1971). Values also given by M. Vigdergauz and V. Semkin, J. Chromat. <u>58</u> 95 (1971).

    Chromatographic measurements using dinonyl phthalate.

T	B_{12}
323.2	-169 ± 15
353.2	-148 ± 15

ARGON + ETHYLBENZENE   Ar + $C_6H_5.CH_2.CH_3$

1. M. Vigdergauz and V. Semkin, J. Chromat. $\underline{58}$ 95 (1971).

    B_{12} determined from chromatographic retention data with reference
    B_{12} (353.2K) values of -105 for argon + benzene, and -148 for argon
    + styrene.

    Class III.

T	B_{12}
353.2	-138

ARGON + o-XYLENE   Ar + $C_6H_4.(CH_3)_2$

1. M. Vigdergauz and V. Semkin, J. Chromat. $\underline{58}$ 95 (1971).

    B_{12} determined from chromatographic retention data with reference
    B_{12} (353.2K) values of -105 for argon + benzene, and -148 for argon +
    styrene.

    Class III.

T	B_{12}
353.2	-146

ARGON + m-XYLENE   Ar + $C_6H_4.(CH_3)_2$

1. M. Vigdergauz and V. Semkin, J. Chromat. $\underline{58}$ 95 (1971).

    B_{12} determined from chromatographic retention data with reference
    B_{12} (353.2K) values of -105 for argon + benzene, and -148 for argon +
    styrene.

    Class III.

T	B_{12}
353.2	-141

ARGON + p-XYLENE   Ar + $C_6H_4.(CH_3)_2$

1. M. Vigdergauz and V. Semkin, J. Chromat. $\underline{58}$ 95 (1971).

    B_{12} determined from chromatographic retention data with reference
    B_{12} (353.2K) values of -105 for argon + benzene and -148 for argon +
    styrene.

    Class III.

T	B_{12}
353.2	-143

ARGON + n OCT-1-ENE $Ar + CH_2:CH(CH_2)_5CH_3$

1. A.J.B. Cruikshank, B.W. Gainey and C.L. Young, Trans. Faraday Soc. 64 337 (1968).

   Gas chromatographic measurements.

T	B_{12}
308.15	-136 ± 12

ARGON + n-OCTANE $Ar + CH_3 \cdot (CH_2)_6 \cdot CH_3$

1. M.S. Vigdergauz and V.I. Semkin, Russ. J. phys. Chem. 45 518 (1971); Zh. fiz. Khim. 45 931 (1971). Values also given by M. Vigdergauz and V. Semkin, J. Chromat. 58 95 (1971).

   Chromatographic measurements using dinonyl phthalate.

T	B_{12}
323.2	-160 ± 15
353.2	-122 ± 15

ARGON + 2,2,4-TRIMETHYL PENTANE $Ar + (CH_3)_3 \cdot C \cdot CH_2 \cdot CH(CH_3)_2$

1. M.S. Vigdergauz and V.I. Semkin, Russ. J. phys. Chem. 45 518 (1971); Zh. fiz. Khim. 45 931 (1971). Values also given in M. Vigdergauz and V. Semkin, J. Chromat. 58 95 (1971).

   Chromatographic measurements using dinonyl phthalate.

T	B_{12}
323.2	-128 ± 15
353.2	- 87 ± 15

ARGON + n-PROPYLBENZENE $Ar + C_6H_5 \cdot (CH_2)_2 \cdot CH_3$

1. M. Vigdergauz and V. Semkin, J. Chromat. 58 95 (1971).

   B_{12} determined from chromatographic retention data with reference B_{12} (353.2K) values of -105 for argon + benzene, and -148 for argon + styrene.

   Class III.

T	B_{12}
353.2	-152

ARGON + ISOPROPYLBENZENE   Ar + $C_6H_5 \cdot CH \cdot (CH_3)_2$

1. M. Vigdergauz and V. Semkin, J. Chromat. 58 95 (1971).
    B_{12} determined from chromatographic retention data with reference
    B_{12} (353.2K) values of -105 for argon + benzene, and -148 for argon +
    styrene.
    Class III.

T	B_{12}
353.2	-148

ARGON + 1-METHYL-2-ETHYLBENZENE   Ar + $C_6H_4 \cdot (CH_3)(CH_2CH_3)$

1. M. Vigdergauz and V. Semkin, J. Chromat. 58 95 (1971).
    B_{12} determined from chromatographic retention data with reference
    B_{12} (353.2K) values of -105 for argon + benzene, and -148 for argon +
    styrene.
    Class III.

T	B_{12}
353.2	-158

ARGON + 1-METHYL-3-ETHYLBENZENE   Ar + $C_6H_4 \cdot (CH_3)(CH_2CH_3)$

1. M. Vigdergauz and V. Semkin, J. Chromat. 58 95 (1971).
    B_{12} determined from chromatographic retention data with reference
    B_{12} (353.2K) values of -105 for argon + benzene, and -148 for argon +
    styrene.
    Class III.

T	B_{12}
353.2	-154

ARGON + 1,2,3-TRIMETHYLBENZENE   Ar + $C_6H_3 \cdot (CH_3)_3$

1. M. Vigdergauz and V. Semkin, J. Chromat. 58 95 (1971).

B_{12} determined from chromatographic retention data with reference
B_{12} (353.2K) values of -105 for argon + benzene, and -148 for argon
+ styrene.

Class III.

T	B_{12}
353.2	-165

ARGON + 1,2,4-TRIMETHYLBENZENE   Ar + $C_6H_3 \cdot (CH_3)_3$

1. M. Vigdergauz and V. Semkin, J. Chromat. <u>58</u> 95 (1971).

   B_{12} determined from chromatographic retention data with reference
   B_{12} (353.2K) values of -105 for argon + benzene, and -148 for argon
   + styrene.

   Class III.

T	B_{12}
353.2	-156

ARGON + 1,3,5-TRIMETHYLBENZENE   Ar + $C_6H_3 \cdot (CH_3)_3$

1. M. Vigdergauz and V. Semkin, J. Chromat. <u>58</u> 95 (1971).

   B_{12} determined from chromatographic retention data with reference
   B_{12} (353.2K) values of -105 for argon + benzene, and -148 for argon
   + styrene.

   Class III.

T	B_{12}
353.2	-160

ARGON + n-NONANE   Ar + $CH_3 \cdot (CH_2)_7 \cdot CH_3$

1. M.S. Vigdergauz and V.I. Semkin, Russ. J. phys. Chem. <u>45</u> 518 (1971);
   Zh. fiz. Khim. <u>45</u> 931 (1971). Values also given by M. Vigdergauz and
   V. Semkin, J. Chromat. <u>58</u> 95 (1971).

   Chromatographic measurements using dinonyl phthalate.

T	B_{12}
323.2	-181 ± 15
353.2	-146 ± 15

ARGON + NAPHTHALENE   Ar + $C_{10}H_8$

1. A.D. King, Jr. and W.W. Robertson, J. chem. Phys. <u>37</u> 1453 (1962).
   Naphthalene solubility measurements.

T	B_{12}
298.2	-178 ± 10
347.2	-104 ± 6

2. H. Bradley, Jr., and A.D. King, Jr., J. chem. Phys. <u>47</u> 1189 (1967).
   B_{12} determined from measurements of solubility of naphthalene in
   the compressed gas: $P \to 160$ atm.

T	B_{12}
297.2	-176 ± 7

ARGON + n-BUTYLBENZENE   Ar + $C_6H_5 \cdot (CH_2)_3 \cdot CH_3$

1. M. Vigdergauz and V. Semkin, J. Chromat. <u>58</u> 95 (1971).
   B_{12} determined from chromatographic retention data with reference
   B_{12} (353.2K) values of -105 for argon + benzene, and -148 for argon
   + styrene.
   Class III.

T	B_{12}
353.2	-168

ARGON + SEC-BUTYLBENZENE   Ar + $C_6H_5 \cdot CH(CH_3)(CH_2CH_3)$

1. M. Vigdergauz and V. Semkin, J. Chromat. <u>58</u> 95 (1971).
   B_{12} determined from chromatographic retention data with reference
   B_{12} (353.2K) values of -105 for argon + benzene, and -148 for argon
   + styrene.
   Class III.

T	B_{12}
353.2	-160

ARGON + ISOBUTYLBENZENE   Ar + $C_6H_5 \cdot CH(CH_3)_2$

1. M. Vigdergauz and V. Semkin, J. Chromat. <u>58</u> 95 (1971).

B_{12} determined from chromatographic retention data with reference B_{12} (353.2K) values of -105 for argon + benzene, and -148 for argon + styrene.

Class III.

T	B_{12}
353.2	-159

ARGON + 1-METHYL-2-PROPYLBENZENE   Ar + $C_6H_4 \cdot (CH_3)(CH_2 \cdot CH_2 \cdot CH_3)$

1. M. Vigdergauz and V. Semkin, J. Chromat. $\underline{58}$ 95 (1971).

B_{12} determined from chromatographic retention data with reference B_{12} (353.2K) values of -105 for argon + benzene, and -148 for argon + styrene.

Class III.

T	B_{12}
353.2	-171

ARGON + 1-METHYL-2-ISOPROPYLBENZENE   Ar + $C_6H_4 \cdot (CH_3)(CH.(CH_3)_2)$

1. M. Vigdergauz and V. Semkin, J. Chromat. $\underline{58}$ 95 (1971).

B_{12} determined from chromatographic retention data with reference B_{12} (353.2K) values of -105 for argon + benzene, and -148 for argon + styrene.

Class III.

T	B_{12}
353.2	-168

ARGON + 1-METHYL-3-PROPYLBENZENE   Ar + $C_6H_4 \cdot (CH_3)(CH_2 \cdot CH_2 \cdot CH_3)$

1. M. Vigdergauz and V. Semkin, J. Chromat. $\underline{58}$ 95 (1971).

B_{12} determined from chromatographic retention data with reference B_{12} (353.2K) values of -105 for argon + benzene, and -148 for argon + styrene.

Class III.

T	B_{12}
353.2	-167

ARGON + 1-METHYL-4-PROPYLBENZENE Ar + $C_6H_4 \cdot (CH_3)(CH_2 \cdot CH_2 \cdot CH_3)$

1. M. Vigdergauz and V. Semkin, J. Chromat. 58 95 (1971).

   B_{12} determined from chromatographic retention data with reference B_{12} (353.2K) values of -105 for argon + benzene, and -148 for argon + styrene.

   Class III.

T	B_{12}
353.2	-169

ARGON + 1,4-DIETHYLBENZENE Ar + $C_6H_4 \cdot (CH_2 \cdot CH_3)_2$

1. M. Vigdergauz and V. Semkin, J. Chromat. 58 95 (1971).

   B_{12} determined from chromatographic retention data with reference B_{12} (353.2K) values of -105 for argon + benzene, and -148 for argon + styrene.

   Class III.

T	B_{12}
353.2	-168

ARGON + 1,2-DIMETHYL-3-ETHYLBENZENE Ar + $C_6H_3 \cdot (CH_3)_2 \cdot (CH_2 \cdot CH_3)$

1. M. Vigdergauz and V. Semkin, J. Chromat. 58 95 (1971).

   B_{12} determined from chromatographic retention data with reference B_{12} (353.2K) values of -105 for argon + benzene, and -148 for argon + styrene.

   Class III.

T	B_{12}
353.2	-168

ARGON + 1,2-DIMETHYL-4-ETHYLBENZENE Ar + $C_6H_3 \cdot (CH_3)_2 (CH_2 \cdot CH_3)$

1. M. Vigdergauz and V. Semkin, J. Chromat. 58 95 (1971).

   B_{12} determined from chromatographic retention data with reference B_{12} (353.2K) values of -105 for argon + benzene, and -148 for argon + styrene.

   Class III.

T	B_{12}
353.2	-177

ARGON + 1,3-DIMETHYL-2-ETHYLBENZENE   Ar + $C_6H_3 \cdot (CH_3)_2(CH_2 \cdot CH_3)$

1. M. Vigdergauz and V. Semkin, J. Chromat. **58** 95 (1971).

    B_{12} determined from chromatographic retention data with reference B_{12} (353.2K) values of -105 for argon + benzene, and -148 for argon + styrene.

    Class III.

T	B_{12}
353.2	-167

ARGON + 1,3-DIMETHYL-4-ETHYLBENZENE   Ar + $C_6H_3 \cdot (CH_3)_2(CH_2 \cdot CH_3)$

1. M. Vigdergauz and V. Semkin, J. Chromat. **58** 95 (1971).

    B_{12} determined from chromatographic retention data with reference B_{12} (353.2K) values of -105 for argon + benzene, and -148 for argon + styrene.

    Class III.

T	B_{12}
353.2	-174

ARGON + 1,3-DIMETHYL-5-ETHYLBENZENE   Ar + $C_6H_3 \cdot (CH_3)_2(CH_2 \cdot CH_3)$

1. M. Vigdergauz and V. Semkin, J. Chromat. **58** 95 (1971).

    B_{12} determined from chromatographic retention data with reference B_{12} (353.2K) values of -105 for argon + benzene, and -148 for argon + styrene.

    Class III.

T	B_{12}
353.2	-173

ARGON + 1,4-DIMETHYL-2-ETHYLBENZENE   Ar + $C_6H_3 \cdot (CH_3)_2(CH_2 \cdot CH_3)$

1. M. Vigdergauz and V. Semkin, J. Chromat. **58** 95 (1971).

B_{12} determined from chromatographic retention data with reference B_{12} (353.2K) values of -105 for argon + benzene, and -148 for argon + styrene.

Class III.

T	B_{12}
353.2	-174

ARGON + 1,2,3,4-TETRAMETHYLBENZENE   Ar + $C_6H_2 \cdot (CH_3)_4$

1. M. Vigdergauz and V. Semkin, J. Chromat. <u>58</u> 95 (1971).

B_{12} determined from chromatographic retention data with reference B_{12} (353.2K) values of -105 for argon + benzene, and -148 for argon + styrene.

Class III.

T	B_{12}
353.2	-178

ARGON + 1,2,3,5-TETRAMETHYLBENZENE   Ar + $C_6H_2 \cdot (CH_3)_4$

1. M. Vigdergauz and V. Semkin, J. Chromat. <u>58</u> 95 (1971).

B_{12} determined from chromatographic retention data with reference B_{12} (353.2K) values of -105 for argon + benzene, and -148 for argon + styrene.

Class III.

T	B_{12}
353.2	-183

ARGON + 1,2,4,5-TETRAMETHYLBENZENE   Ar + $C_6H_2 \cdot (CH_3)_4$

1. M. Vigdergauz and V. Semkin, J. Chromat. <u>58</u> 95 (1971).

B_{12} determined from chromatographic retention data with reference B_{12} (353.2K) values of -105 for argon + benzene, and -148 for argon + styrene.

Class III.

T	B_{12}
353.2	-183

ARGON + ANTHRACENE Ar + C_6H_4 $(CH)_2$ C_6H_4

1. H. Bradley, Jr., and A.D. King, Jr., J. chem. Phys. <u>47</u> 1189 (1967).
    B_{12} determined from measurements of solubility of anthracene in the compressed gas: P → 160 atm.

T	B_{12}
348.2	-183 ± 8

ARGON + HYDROGEN CHLORIDE

1. B. Schramm and U. Leuchs (1979) (†).
    Estimated error in B_{12} ± 5 at 480 K, ± 7 at 200 K.

T	B_{12}	T	B_{12}
190	-99.3	330	-26.4
200	-90.7	370	-17.8
225	-70.5	400	-13.7
250	-54.6	420	-10.5
275	-42.1	450	- 7.6
295	-36.1	480	- 4.4
300	-34.4		

ARGON + SULPHUR HEXAFLUORIDE Ar + SF_6

1. J. Bellm. W. Reineke, K. Schäfer, and B. Schramm, Ber. (dtsch.) Bunsenges. phys. Chem. <u>78</u> 282 (1974).
    Estimated accuracy of B_{12} is ± 2.

T	B_{12}	T	B_{12}
300	-66.2	430	-23.4
320	-57.8	460	-16.2
340	-51.3	490	-10.8
370	-41.6	520	- 6.8
400	-31.7	550	- 3.9

2. J. Santafe, J.S. Urieta and C.G. Losa, Chem. phys. <u>18</u> 341 (1974).
    Compressibility measurements. Accuracy of B estimated to be ± 3.

T	B_{12}	T	B_{12}
273.2	-76.1	303.2	-57.4
283.2	-67.2	313.2	-53.0
293.2	-62.2	323.2	-50.0

ARGON + HYDROGEN   Ar + H_2

1. C.C. Tanner and I. Masson, Proc. R. Soc. A126 268 (1930) (*).
    PVTx data for x_{H_2} range 0.16 → 0.84; P → 125 atm.

T	B_{12}
298.2	8.15

2. C.M. Knobler, J.J.M. Beenakker, and H.F.P. Knaap, Physica, 's Grav.
    25 909 (1959).
    Low   pressure differential method. Precision ± 3.

T	B_{12}
90	-52.1

3. P. Zandbergen and J.J.M. Beenakker, Physica, 's Grav. 33 343 (1967).
    Measurements of volume changes on mixing gases at constant pressure.

B	B_{12}	B	B_{12}
170	-10.3	230	+ 0.3
180	- 7.8	240	1.5
190	- 5.7	250	2.5
200	- 4.0	260	3.5
210	- 2.4	270	4.2
220	- 0.9		

4. J. Brewer and G.W. Vaughn, J. chem. Phys. 50 2960 (1969).
   Measurement of pressure changes on mixing gases at constant volume.

T	B_{11}	B_{22}	B_{12}
148.15	-88.43	6.78	-16.82
173.15	-65.19	9.15	- 9.58
198.15	-49.13	10.79	- 4.16
223.15	-37.42	12.10	- 0.48

248.15	-28.56	13.14	2.63
273.15	-21.45	13.98	5.31
298.15	-15.75	14.64	7.80
323.15	-11.23	15.15	9.41

ARGON + WATER   Ar + H_2O

1. M. Rigby and J.M. Prausnitz, J. phys. Chem., Ithaca 72 330 (1968).
    B_{12} derived from measurements of solubility of water in the
    compressed gas.

T	B_{12}	T	B_{12}
298.15	-37 ± 6	348.15	-20 ± 4
323.15	-25 ± 5	373.15	-14 ± 3

ARGON + HELIUM   Ar + He

1. C.C. Tanner and I. Masson, Proc. R. Soc. A126 268 (1930) (*).
    PVTx data for x_{He} range 0.16 → 0.84; P → 125 atm.

T	B_{12}
298.2	18.4

2. C.M. Knobler, J.J.M. Beenakker, and H.F.P. Knaap, Physica, 's Grav.
    25 909 (1959).
    Low pressure differential method. Precision ± 3.

T	B_{12}
90	6.6

3. N.K. Kalfoglou and J.G. Miller, J. phys. Chem., Ithaca 71 1256 (1967).
    Burnett method. Maximum pressure 80 atm.

T	B_{11}	B_{22}	B_{12}
303.15	-15.10	11.52	18.77 ± 0.10
373.15	- 4.04	11.30	19.73 ± 0.08
473.15	+ 5.16	10.91	20.26 ± 0.2
573.15	10.63	10.53	20.45 ± 0.2
673.15	14.20	9.72	20.35 ± 0.3
773.15	17.00	9.64	19.75 ± 0.3

4. J. Brewer and G.W. Vaughn, J. chem. Phys. $\underline{50}$ 2960 (1969).

Measurements of pressure changes on mixing gases at constant volume.

T	B_{11}	B_{22}	B_{12}
148.15	-88.43	11.75	12.20
173.15	-65.19	11.91	13.87
198.15	-49.13	11.98	15.26
223.15	-37.42	11.93	15.94
248.15	-28.56	11.84	16.82
273.15	-21.45	11.76	17.05
298.15	-15.75	11.73	18.12
323.15	-11.23	11.74	18.41

5. K.R. Hall and F.B. Canfield, Physica. 's Grav, $\underline{47}$ 75 (1969).

Burnett method. Maximum pressure 700 atm.

T	B_{11}	B_{22}	B_{12}
223.15	-37.30	12.16	16.09
273.15	-20.90	11.94	17.58
323.15	-10.82	11.76	18.44

6. J.A. Provine and F.B. Canfield, Physica, 's Grav. $\underline{52}$ 79 (1971).

Burnett method. Maximum pressure 700 atm.

B_M values given for T 143.15, 158.15, 183.15 K.

ARGON + MERCURY Ar + Hg

1. D. Stubley and J.S. Rowlinson, Trans. Faraday Soc. $\underline{57}$ 1275 (1961).

Solubility measurements of mercury in compressed argon, maximum pressure 400 atm. Estimated accuracy of B_{12} ± 10.

T	B_{12}	T	B_{12}
457.2	-47	529.2	-19
491.2	-45	578.2	-11

ARGON + KRYPTON Ar + Kr

1. B.E.F. Fender and G.D. Halsey, Jr., J. chem. Phys. $\underline{36}$ 1881 (1962).
Pressure range 60 - 80 cm. Hg.

T	B_{12}	T	B_{12}
108.001	-236.16	117.357	-204.26
110.567	-224.95	119.002	-198.42
113.272	-216.14	120.780	-193.74
115.315	-207.99	123.616	-186.44
117.083	-206.72		

2. M.A. Byrne, M.R. Jones, and L.A.K. Staveley, Trans. Faraday Soc. <u>64</u> 1747 (1968).

Class I.

T	B_{12}	T	B_{12}
116.53	-207.8	149.47	-133.1
119.38	-198.5	163.29	-115.1
124.43	-184.8	179.02	- 98.4
129.42	-172.4	196.09	- 82.0
131.01	-168.7	219.51	- 68.0
139.53	-151.0	253.85	- 47.9

3. B. Schramm, H. Schmiedel, R. Gehrmann, and R. Bartl, Ber. (dtsch.) Bunsenges. phys. Chem. <u>81</u> 316 (1977).

Maximum error in B_{12} estimated to be ± 6.

T	B_{12}	T	B_{12}
202	-75.9	330	-20.3
213	-67.6	355	-15.5
223	-60.0	370	-13.1
233	-53.9	400	- 8.8
243	-49.0	425	- 5.8
253	-43.7	435	- 4.8
263	-40.0	465	- 1.9
278	-34.0	500	+ 0.9
295	-29.0		

4. H.-P. Rentschler and B. Schramm, Ber. (dtsch.) Bunsenges. phys. Chem. <u>81</u> 319 (1977).

Maximum error in B_{12} estimated to be ± 6.

T	B_{12}	T	B_{12}
345	-17.4	552	6.1

410	- 6.5	623	9.1
484	+ 0.1	695	11.5

ARGON + NITROGEN   Ar + N_2

1. C.M. Knobler, J.J.M. Beenakker, and H.F.P. Knaap, Physica, 's Grav.
    25 909 (1959).

    Low pressure differential method.   Precision ± 3.

T	B_{12}
90	-195

2. P. Zandbergen and J.J.M. Beenakker, Physica, 's Grav. 33 343 (1967).
    Measurements of volume changes on mixing gases at constant pressure.

T	B_{12}	T	B_{12}
170	-60.5	230	-29.0
180	-53.5	240	-25.6
190	-47.4	250	-22.5
200	-42.0	260	-19.4
210	-37.2	270	-16.8
220	-32.8		

3. J. Brewer and G.W. Vaughn, J. chem. Phys. 50 2960 (1969).
    Measurement of pressure changes on mixing gases at constant volume.

T	B_{11}	B_{22}	B_{12}
148.15	-88.43	-73.54	-81.58
173.15	-65.19	-51.86	-59.13
198.15	-49.13	-37.48	-43.96
223.15	-37.42	-26.38	-32.56
248.15	-28.56	-17.59	-23.73
273.15	-21.45	-10.57	-16.43
298.15	-15.75	- 4.86	-10.88
323.15	-11.23	- 0.26	- 6.19

4. B. Schramm and R. Gehrmann (1979) (†).
    Estimated error in B_{12} is ± 5.

T	B_{12}	T	B_{12}
213	-43.3	262	-20.7
223	-38.3	276	-17.2
242	-29.1		

5. B. Schramm and H. Schmiedel (1979) (†).

Estimated error in B_{12} is ± 4.

T	B_{12}	T	B_{12}
295	-11.5	425	8.2
330	- 4.2	450	9.7
365	+ 2.4	475	10.4
400	6.4		

ARGON + NEON   Ar + Ne

1. C.M. Knobler, J.J.M. Beenakker, and H.F.P. Knaap, Physica, 's Grav. 25 909 (1959).

Low pressure differential method. Precision ± 3.

T	B_{12}
90	-35.3

2. J. Brewer and G.W. Vaughn, J. chem. Phys. 50 2960 (1969).

Measurements of pressure changes on mixing gases at constant volume.

T	B_{11}	B_{22}	B_{12}
148.15	-88.43	4.36	-8.36
173.15	-65.19	6.45	-2.57
198.15	-49.13	7.90	1.46
223.15	-37.42	9.12	4.45
248.15	-28.56	10.14	6.86
273.15	-21.45	10.94	8.91
298.15	-15.75	11.54	10.89
323.15	-11.23	12.32	12.15

3. B. Schramm and R. Gehrmann (1979) (†).

Estimated error in B_{12} is ± 5.

T	B_{12}	T	B_{12}
213	1.8	262	7.8
223	3.8	276	8.0
242	5.6		

4. B. Schramm and H. Schmiedel (1979) (†).
    Estimated error in B_{12} is ± 5.

T	B_{12}	T	B_{12}
295	10.9	425	11.4
330	11.4	450	11.3
365	11.7	475	10.0
400	11.5		

ARGON + OXYGEN Ar + O_2

1. I. Masson and L.G.F. Dolley, Proc. R. Soc. A103 524 (1923).
    PVT data given for X_2 = 0.5; P → 125 atm and T = 298 K.

2. C.M. Knobler, J.J.M. Beenakker, and H.F.P. Knaap, Physica, 's Grav.
    25 909 (1959).
    Low pressure differential method. Precision ± 3.

T	B_{12}
90	-217

ARGON + XENON Ar + Xe

1. B. Schramm, H. Schmiedel, R. Gehrmann, and R. Bartl, Ber. (dtsch.)
    Bunsenges. phys. Chem. 81 316 (1977).
    Maximum error in B_{12} estimated to be ± 6.

T	B_{12}	T	B_{12}
203	-95.0	355	-26.2
213	-84.5	370	-23.2
233	-72.8	400	-17.5
253	-63.0	415	-15.0
269	-55.5	430	-12.7
278	-51.8	460	- 8.7
295	-43.5	500	- 4.3
330	-32.9		

2. H.-P. Rentschler and B. Schramm, Ber. (dtsch.) Bunsenges. phys. Chem. **81** 319 (1977).

    Maximum error in B_{12} estimated to be ± 6.

T	B_{12}	T	B_{12}
343	-30.5	555	+ 2.4
410	-15.6	626	4.9
482	- 5.1	695	9.6

BORON TRIFLUORIDE + NITROGEN $BF_3 + N_2$

1. G.L. Brooks and C.J.G. Raw, Trans. Faraday Soc. **54** 972 (1958).

    Boyle's Law apparatus. Estimated accuracy in B ± 20.

T	B_{11}	B_{22}	$B_{12}(*)$
303.2	-98	-4	-27

CHLOROTRIFLUOROMETHANE + TRICHLOROFLUOROMETHANE $CClF_3 + CCl_3F$

1. J. Bougard and R. Jadot, J. Chim. phys. **73** 415 (1976).

    Quoted accuracy of B_{12} is ± 4%.

T	B_{12}
302.2	-448.4
302.2	-427.5

CHLOROTRIFLUOROMETHANE + CHLORODIFLUOROMETHANE $CClF_3 + CHClF_2$

1. J. Bougard and R. Jadot, J. Chim. phys. **73** 415 (1976).

    Quoted accuracy in B_{12} is ± 4%.

T	B_{12}
300.7	-275.3
300.7	-265.9

CHLOROTRIFLUOROMETHANE + FLUOROFORM $CClF_3 + CHF_3$

1. J.V. Sinka, E. Rosenthal and R.P. Dixon, J. chem. Engng Data **15** 73 (1970).

    Constant volume cell method: $X_1 = 0.5$.

T 298 - 492 K, P 14.7 → 75 atm.

2. J. Bougard and R. Jadot, J. Chim. phys. <u>73</u> 415 (1976).
    Quoted accuracy in B_{12} is ± 4%.

T	B_{12}
300.3	-163.0
300.3	-170.4

CHLOROTRIFLUOROMETHANE + SULPHUR HEXAFLUORIDE $CClF_3$ + SF_6

1. R.D. Nelson, Jr. and R.H. Cole, J. chem. Phys. <u>54</u> 4033 (1971).

T	B_{12}
323.2	-223

DICHLORODIFLUOROMETHANE + TRICHLOROFLUOROMETHANE CCl_2F_2 + CCl_3F

1. J. Bougard and R. Jadot, J. Chim. phys. <u>73</u> 415 (1976).
    Quoted accuracy of B_{12} is ± 4%.

T	B_{12}
302.2	-446.4
302.2	-432.2

DICHLORODIFLUOROMETHANE + CHLORODIFLUOROMETHANE CCl_2F_2 + $CHClF_2$

1. J. Bougard and R. Jadot, J. Chim. phys. <u>73</u> 415 (1976).
    Quoted accuracy in B_{12} is ± 4%.

T	B_{12}
300.5	-378.5
300.5	-370.6
300.5	-362.1

DICHLORODIFLUOROMETHANE + FLUOROFORM CCl_2F_2 + CHF_3

1. J. Bougard and R. Jadot, J. Chim. phys. <u>73</u> 415 (1976).
    Quoted accuracy in B_{12} is ± 4%.

T	B_{12}
300.5	-297.7
300.5	-300.7
300.5	-293.8

DICHLORODIFLUOROMETHANE + 1,1-DIFLUOROETHANE CCl_2F_2 + $CHF_2.CH_3$

1. J.V. Sinka and K.P. Murphy. J. chem. Engng Data 12 315 (1967).
    PVT data; P range 14-58 atm, T range 320-480K.

2. M. Prasad and A.P. Kudchadker, J. chem. Engng Data 23 190 (1978).
    Burnett method. Measurements made on the minimum boiling azeotrope
    (refrigerant 500).

T	B_M	C_M
298.15	-365.5	-544540
313.15	-334.1	-457460
333.15	-299.3	-352510
353.15	-267.5	-250120
373.15	-237.5	-159990
393.15	-212.1	- 82120
413.15	-190.0	- 25020

TRICHLOROFLUOROMETHANE + FLUOROFORM CCl_3F + CHF_3

1. J. Bougard and R. Jadot, J. Chim. phys. 73 415 (1976).
    Quoted accuracy of B_{12} is ± 4%.

T	B_{12}
302.2	-247.6
302.2	-239.0

CARBON TETRACHLORIDE + CHLOROFORM CCl_4 + $CHCl_3$

1. P.G. Francis and M.L. McGlashan, Trans. Faraday Soc. 51 593 (1955).
    Boyles apparatus. B_M values given, T range 310-343 K, X_2
    approximately 0.24 and 0.5

CARBON TETRACHLORIDE + METHANE $CCl_4 + CH_4$

1. S.K. Gupta and A.D. King, Jr., Can. J. Chem. <u>50</u> 660 (1972).

    B_{12} determined from solubility measurements of carbon tetrachloride
    in the compressed gas: P → 60 atm.

T	B_{12}	T	B_{12}
273.2	-252 ± 8	323.2	-174 ± 3
298.2	-205 ± 4	348.2	-129 ± 5

CARBON TETRACHLORIDE + CARBON DIOXIDE $CCl_4 + CO_2$

1. J.M. Prausnitz and P.R. Benson, A.I. Ch. E. Jl <u>5</u> 161 (1959).
    Vapour phase solubility measurements.

T	B_{12}
323.15	-205 ± 8
348.15	-163 ± 9

2. M.S. Vigdergauz and V.I. Semkin, Russ. J. phys. Chem. <u>45</u> 518 (1971);
    Zh. fiz. Khim. <u>45</u> 931 (1971).   Value also given by M. Vigdergauz and
    V. Semkin, J. Chromat. <u>58</u> 95 (1971).

    Chromatographic measurements using dinonyl phthalate.
    Class III.

T	B_{12}
353.2	-154

3. S.K. Gupta and A.D. King, Jr., Can. J. Chem. <u>50</u> 660 (1972).

    B_{12} determined from solubility measurements of carbon tetrachloride
    in the compressed gas: P → 60 atm.

T	B_{12}
298.2	-287 ± 7
323.2	-213 ± 5
348.2	-185 ± 7

CARBON TETRACHLORIDE+ETHYLENE $CCl_4 + CH_2:CH_2$

1. S.K. Gupta and A.D. King, Jr., Can. J. Chem. <u>50</u> 660 (1972).

    B_{12} determined from solubility measurements of carbon tetrachloride in the compressed gas: $P \rightarrow 60$ atm.

T	B_{12}	T	B_{12}
273.2	-457 ± 10	323.2	-299 ± 5
298.2	-384 ± 7	348.2	-248 ± 3

CARBON TETRACHLORIDE + BENZENE $CCl_4 + C_6H_6$

1. M. Rätzsch and E. Rasenberger, Z. Chem. **8** 156 (1968).

T	B_{12}
363.2	-939 ± 10

CARBON TETRACHLORIDE + HYDROGEN $CCl_4 + H_2$

1. J.M. Prausnitz and P.R. Benson, A.I. Ch. E. Jl **5** 161 (1959).
    Vapour phase solubility measurements.

T	B_{12}
348.15	34.9 ± 5.5

2. S.K. Gupta and A.D. King, Jr., Can. J. Chem. **50** 660 (1972).
    B_{12} determined from solubility measurements of carbon
    tetrachloride in the compressed gas: $P \rightarrow 60$ atm.

T	B_{12}	T	B_{12}
262.5	-20 ± 5	323.2	$+ 9 \pm 5$
273.2	$- 8 \pm 5$	348.2	27 ± 5
298.2	$- 6 \pm 4$		

CARBON TETRACHLORIDE + HELIUM $CCl_4 + He$

1. S.K. Gupta and A.D. King, Jr., Can. J. Chem. **50** 660 (1972).
    B_{12} determined from solubility measurements of carbon
    tetrachloride in the compressed gas: $P \rightarrow 60$ atm.

T	B_{12}	T	B_{12}
262.5	37 ± 5	323.2	56 ± 2
273.2	46 ± 4	348.2	59 ± 2
298.2	61 ± 4		

CARBON TETRACHLORIDE + NITROGEN CCl_4 + N_2

1. J.M. Prausnitz and P.R. Benson, A.I. Ch. E. Jl 5 161 (1959).
   Vapour phase solubility measurements.

T	B_{12}
323.15	-58.5 ± 5.9
348.15	-49.6 ± 5.5

2. M. Vigdergauz and V. Semkin, J. Chromat. 58 95 (1971).
   Chromatographic measurements using dinonyl phthalate.
   Class III.

T	B_{12}
353.2	-24

3. S.K. Gupta and A.D. King, Jr., Can. J. Chem. 50 660 (1972).

   B_{12} determined from solubility measurements of carbon tetrachloride in the compressed gas: P → 60 atm.

T	B_{12}	T	B_{12}
273.2	-132 ± 4	323.2	- 91 ± 2
298.2	-102 ± 4	348.2	- 64 ± 2

CARBON TETRAFLUORIDE + FLUOROFORM CF_4 + CHF_3

1. H.B. Lange Jr. and F.P. Stein, J. chem. Engng Data 15 56 (1970).
   Burnett method.

T	B_{11}	B_{22}	B_{12}
203.15	-216.75 ± 0.7		
223.15	-177.71 ± 0.5		
243.15	-146.36 ± 0.2	-311.60 ± 1.0	-159.33 ± 1.8
273.15	-110.87 ± 0.2	-233.60 ± 0.4	-122.75 ± 1.4
313.15	-77.26 ± 0.2	-165.50 ± 0.2	-85.28 ± 1.4
368.15	-45.75 ± 0.2	-109.50 ± 0.2	-51.13 ± 1.0

T	C_{111}	C_{222}	C_{112}	C_{122}
203.15	9320 ± 500			
223.15	9645 ± 300			
243.15	8660 ± 100	15000 ± 1000	7200 ± 500	11800 ± 600
273.15	7070 ± 100	15700 ± 175	6400 ± 200	10250 ± 300
313.15	6030 ± 125	11630 ± 100	4950 ± 200	7100 ± 200
368.15	4890 ± 150	8410 ± 250	4700 ± 500	4200 ± 300

2. K. Juris and L.A. Wenzel, A. I. Ch. E. Symp. Ser. $\underline{70}$ 70 (1974).

Integral Joule-Thomson effect for three mixtures;

T 248-318K, P → 75 atm.

CARBON TETRAFLUORIDE + METHANE CF_4 + CH_4

1. D.R. Douslin, R.H. Harrison and R.T. Moore, J. phys. Chem., Ithaca $\underline{71}$ 3477 (1967).

PVT data given: P range 16-400 atm, X_2 0.25, 0.5 and 0.75.

T	B_{11}	B_{22}	B_{12}
273.15	-111.00	-53.35	-62.07
298.15	- 88.30	-42.82	-48.48
303.15	- 84.40	-40.91	-46.09
323.15	- 70.40	-34.23	-37.36
348.15	- 55.70	-27.06	-28.31
373.15	- 43.50	-21.00	-20.43
398.15	- 33.20	-15.87	-13.98
423.15	- 24.40	-11.40	- 8.33
448.15	- 16.80	- 7.56	- 3.21
473.15	- 10.10	- 4.16	+ 1.02
498.15	- 4.25	- 1.16	4.94
523.15	+ 1.00	+ 1.49	8.28
548.15	5.60	3.89	11.39
573.15	9.80	5.98	14.10
598.15	13.60	7.88	16.55
623.15	17.05	9.66	18.88

2. E.M. Dantzler Siebert and C.M. Knobler, J. phys. Chem., Ithaca $\underline{75}$ 3863 (1971).

E determined from measurements of pressure changes on mixing.

T	E	B_{11}	B_{22}	B_{12}
298.15	13 ± 1	-88.3	-42	-52
323.15	13 ± 2	-70.4	-33	-39
373.15	10 ± 1	-43.1	-20	-22

CARBON TETRAFLUORIDE + PERFLUOROETHANE CF_4 + C_2F_6

1. E.M. Dantzler and C.M. Knobler, J. phys. Chem., Ithaca $\underline{73}$ 1335 (1969).

E determined from measurements of pressure changes on mixing.

T	E	B_{11}	B_{22}	B_{12}
323.15	15 ± 1	-70.4	-199	-120
373.15	12 ± 1	-43.1	-121	- 70

CARBON TETRAFLUORIDE + ETHANE $CF_4 + C_2H_6$

1. E.M. Dantzler Siebert and C.M. Knobler, J. phys. Chem., Ithaca <u>75</u> 3863 (1971).

    E determined from measurements of pressure changes on mixing.

T	E	B_{11}	B_{22}	B_{12}
323.15	27 ± 1	-70.4	-157	-87

CARBON TETRAFLUORIDE + PERFLUORO   PROPANE $CF_4 + CF_3.CF_2.CF_3$

1. E.M. Dantzler and C.M. Knobler, J. phys. Chem., Ithaca <u>73</u> 1335 (1969).

    E determined from measurements of pressure changes on mixing.

T	E	B_{11}	B_{22}	B_{12}
323.15	74 ± 2	-70.4	-435	-179
373.15	54 ± 2	-43.1	-290	-113

CARBON TETRAFLUORIDE + PERFLUORO-n-BUTANE $CF_4 + CF_3.(CF_2)_2CF_3$

1. E.M. Dantzler and C.M. Knobler, J. phys. Chem., Ithaca <u>73</u> 1335 (1969).

    E determined from measurements of pressure changes on mixing.

T	E	B_{11}	B_{22}	B_{12}
323.15	183 ± 4	-70.4	-744	-224
373.15	127 ± 2	-43.1	-492	-141

CARBON TETRAFLUORIDE+n-BUTANE $CF_4 + CH_3(CH_2)_2CH_3$

1. E.M. Dantzler Siebert and C.M. Knobler, J. phys. Chem., Ithaca <u>75</u> 3863 (1971).

    E determined from measurements of pressure changes on mixing.

T	E	B_{11}	B_{22}	B_{12}
323.15	166 ± 1	-70.4	-593	-166

CARBON TETRAFLUORIDE+PERFLUORO-n-PENTANE $CF_4 + CF_3.(CF_2)_3.CF_3$

1. E.M. Dantzler and C.M. Knobler, J. phys. Chem., Ithaca <u>73</u> 1335 (1969).

    E determined from measurements of pressure changes on mixing.

T	E	B_{11}	B_{22}	B_{12}
323.15	370 ± 6	-70.4	-1188	-259
373.15	245 ± 11	-43.1	- 793	-173

CARBON TETRAFLUORIDE + NEOPENTANE CF_4 + $(CH_3)_4.C$

1. E.M. Dantzler Siebert and C.M. Knobler, J. phys. Chem., Ithaca <u>75</u> 3863 (1971).

   E determined from measurements of pressure changes on mixing.

T	E	B_{11}	B_{22}	B_{12}
323.15	168 ± 10	-70.4	-734	-234

CARBON TETRAFLUORIDE+PERFLUORO-n-HEXANE CF_4 + $CF_3.(CF_2)_4.CF_3$

1. E.M. Dantzler and C.M. Knobler, J. phys. Chem., Ithaca <u>73</u> 1335 (1969).

   E determined from measurements of pressure changes on mixing.

T	E	B_{11}	B_{22}	B_{12}
323.15	614 ± 10	-70.4	-1712	-277
373.15	390 ± 8	-43.1	-1128	-196

CARBON TETRAFLUORIDE+n-HEXANE CF_4 + $CH_3.(CH_2)_4.CH_3$

1. E.M. Dantzler Siebert and C.M. Knobler, J. phys. Chem., Ithaca <u>75</u> 3863 (1971).

   E determined from measurements of pressure changes on mixing.

T	E	B_{11}	B_{22}	B_{12}
323.15	537 ± 10	-70.4	-1513	-255

CARBON TETRAFLUORIDE + SULPHUR HEXAFLUORIDE CF_4 + SF_6

1. E.M. Dantzler Siebert and C.M. Knobler, J. phys. Chem.,Ithaca <u>75</u> 3863 (1971).

   E determined from measurements of pressure changes on mixing.

T	E	B_{11}	B_{22}	B_{12}
323.15	19 ± 1	-70.4	-232	-132

2. P.M. Sigmund, I.H. Silberberg and J.J. McKetta, J. chem. Engng Data
    <u>17</u> 168 (1972).

Burnett method: standard errors given.

T	X_1	B_M	C_M
271.61	1.0000	-112.33 ± 0.09	7620 ± 20
	0.8970	-131.29 ± 0.71	8600 ± 170
	0.7869	-150.84 ± 0.87	9640 ± 170
	0.5562	-194.23 ± 1.69	12600 ± 580
	0.3925	-235.30 ± 0.70	13540 ± 130
	0.1797	-288.47 ± 0.29	16820 ± 150
	0.0000	-339.26 ± 0.60	18970 ± 210
	0.0000	-339.19 ± 0.53	18390 ± 150
	0.0000	-339.10 ± 0.63	18570 ± 170
303.15	0.0000	-266.06 ± 0.11	19340 ± 100
308.12	1.0000	$- 81.82 \pm 0.71$	6880 ± 90
	0.9408	$- 88.91 \pm 0.23$	6730 ± 30
	0.6475	-131.65 ± 0.21	9290 ± 30
	0.3453	-184.62 ± 0.14	13370 ± 20
	0.2712	-200.02 ± 0.27	13640 ± 50
	0.0000	-255.84 ± 0.10	19150 ± 30
323.55	1.0000	$- 70.95 \pm 0.21$	5430 ± 100
	1.0000	$- 70.61 \pm 0.40$	5530 ± 100
	0.7956	$- 96.15 \pm 0.65$	6140 ± 150
	0.5889	-124.44 ± 0.63	8020 ± 140
	0.3580	-162.86 ± 0.76	11130 ± 190
	0.1896	-189.97 ± 0.57	15390 ± 220
	0.0000	-230.88 ± 0.43	17700 ± 200
	0.0000	-229.54 ± 0.55	17880 ± 110
348.10	1.0000	$- 56.18 \pm 0.69$	4060 ± 160
	1.0000	$- 57.38 \pm 0.50$	4260 ± 190
	0.7980	$- 77.22 \pm 0.87$	5210 ± 200
	0.5820	-104.04 ± 0.59	6070 ± 150
	0.3649	-135.01 ± 0.85	9600 ± 250
	0.1908	-161.10 ± 0.46	11830 ± 140
	0.0000	-193.46 ± 0.57	15330 ± 180
	0.0000	-195.50 ± 0.63	15460 ± 430
373.15	1.0000	$- 44.77 \pm 0.76$	4670 ± 210
	1.0000	$- 44.50 \pm 0.69$	4510 ± 180

T	X_1	B_M	C_M
	0.8077	$- 61.71 \pm 0.80$	4810 ± 200
	0.5896	$- 85.61 \pm 0.35$	5760 ± 200
	0.4459	-102.66 ± 0.47	6450 ± 250
	0.1880	-136.16 ± 0.74	9600 ± 290
	0.0000	-163.08 ± 0.48	12040 ± 210
	0.0000	-163.86 ± 0.59	12210 ± 300
423.15	1.0000	$- 23.39 \pm 0.62$	4390 ± 90
	0.8336	$- 37.64 \pm 1.44$	1110 ± 650
	0.5800	$- 54.21 \pm 2.44$	3620 ± 680
	0.3049	$- 80.28 \pm 0.35$	8110 ± 70
	0.0755	-105.20 ± 0.74	9540 ± 140

The above results give the following values for B_{12}, C_{12}:

T	B_{11}	B_{22}	B_{12}
271.61	-112.36 ± 0.13	-338.92 ± 0.44	-193.24 ± 1.26
308.12	$- 81.22 \pm 0.56$	-255.87 ± 0.23	-144.56 ± 0.66
323.55	$- 70.93 \pm 0.41$	-230.00 ± 0.72	-125.42 ± 1.93
348.10	$- 56.93 \pm 0.55$	-194.29 ± 0.54	-104.14 ± 1.27
373.15	$- 44.48 \pm 0.30$	-163.41 ± 0.23	$- 88.08 \pm 0.42$
423.15	$- 23.71 \pm 0.89$	-113.61 ± 1.53	$- 55.00 \pm 2.22$

T	C_{111}	C_{222}	C_{112}	C_{122}
271.61	7620 ± 40	18640 ± 190	10260 ± 780	14530 ± 690
308.12	6440 ± 500	19120 ± 360	8210 ± 1370	12810 ± 1000
323.55	5500 ± 240	17870 ± 230	4640 ± 930	12440 ± 1060
348.10	4190 ± 240	15380 ± 330	4220 ± 870	8770 ± 890
373.15	4580 ± 100	12120 ± 130	4490 ± 410	6670 ± 500

CARBON TETRAFLUORIDE + HELIUM CF_4 + He

1. N.K. Kalfoglou and J.G. Miller, J. phys. Chem., Ithaca <u>71</u> 1256 (1967).

    Burnett Method. Maximum pressure 50 atm.

T	B_{11}	B_{22}	B_{12}
303.15	-84.55	11.52	26.51 ± 0.8
373.15	-44.00	11.30	30.19 ± 0.9

473.15	- 9.74	10.91	33.20 ± 0.5
573.15	+ 9.41	10.53	37.42 ± 0.8
673.15	23.03	9.72	40.47 ± 0.4
773.15	32.86	9.64	43.13 ± 0.4

CHLORODIFLUOROMETHANE + FLUOROFORM $CHClF_2$ + CHF_3

1. J. Bougard and R. Jadot, J. Chim. phys. 73 415 (1976).
    Quoted accuracy in B_{12} is ± 4%.

T	B_{12}
300.3	-261.0
300.3	-252.8

CHLOROFORM+METHANE $CHCl_3$ + CH_4

1. S.K. Gupta, R.D. Lesslie and A.D. King, Jr., J. phys. Chem., Ithaca 77 2011 (1973).

    B_{12} derived from measurements of solubility of chloroform in the compressed gas.

T	B_{12}
298.15	-199 ± 7

CHLOROFORM + CARBON DIOXIDE $CHCl_3$ + CO_2

1. M.S. Vigdergauz and V.I. Semkin, Russ. J. phys. Chem. 45 518 (1971); Zh. fiz. Khim. 45 931 (1971). Value also given by M. Vigdergauz and V. Semkin, J. Chromat. 58 95 (1971).

    Chromatographic measurements using dinonyl phthalate.

    Class III.

T	B_{12}
353.2	-169

2. S.K. Gupta, R.D. Lesslie and A.D. King, Jr., J. phys. Chem., Ithaca 77 2011 (1973).

    B_{12} derived from measurements of solubility of chloroform in the compressed gas.

T	B_{12}
298.15	-316 ± 15

CHLOROFORM + ETHYLENE $CHCl_3 + CH_2:CH_2$

1. S.K. Gupta, R.D. Lesslie and A.D. King, Jr., J. phys. Chem., Ithaca __77__ 2011 (1973).

   B_{12} derived from measurements of solubility of chloroform in the compressed gas.

T	B_{12}
298.15	-373 ± 15

CHLOROFORM + METHYL FORMATE $CHCl_3 + HCOOCH_3$

1. J.D. Lambert, J.S. Clarke, J.F. Duke, C.L. Hicks, S.D. Lawrence, D.M. Morris and M.G.T. Shone, Proc. R. Soc. __A249__ 414 (1959).

   Boyle's Law apparatus.

T	B_{11}	B_{22}	B_{12}
323.2	-990	-770	-1511
350.2	-850	-610	- 864
368.2	-820	-580	- 723

CHLOROFORM + METHYL ACETATE $CHCl_3 + CH_3.COO\ CH_3$

1. J.D. Lambert, J.S. Clarke, J.F. Duke, C.L. Hicks, S.D. Lawrence, D.M. Morris and M.G.T. Shone, Proc. R. Soc. __A249__ 414 (1959).

   Boyle's Law apparatus.

T	B_{11}	B_{22}	B_{12}
323.2	-1025	-1240	-1615
338.2	- 880	-1080	-1475
353.2	- 760	- 960	-1330
368.2	- 700	- 850	-1240

CHLOROFORM + ETHYL ACETATE $CHCl_3 + CH_3.COOC_2H_5$

1. J.D. Lambert, J.S. Clarke, J.F. Duke, C.L. Hicks, S.D. Lawrence, D.M. Morris and M.G.T. Shone, Proc. R. Soc. __A249__ 414 (1959).

   Boyle's Law apparatus.

T	B_{11}	B_{22}	B_{12}
323.2	-945	-1830	-2150
338.2	-850	-1440	-1720

353.2	-820	-1240	-1495
368.2	-750	-1080	-1380

CHLOROFORM + ACETONE $CHCl_3$ + $(CH_3)_2$ CO

1. Sh. D. Zaalishvili and L.E. Kolysko, Russ. J. phys. Chem. <u>35</u> 1291 (1961);   Zh. fiz. Khim. <u>35</u> 2613 (1961).

   Boyle's Law apparatus.

T	B_{11}	B_{22}	B_{12}
333.2	-910	-1330	-2005
343.2	-855	-1140	-1300
383.2	-775	-1010	- 995
393.2	-710	- 845	- 930

CHLOROFORM + n PROPYL FORMATE $CHCl_3$ + $H.COOC_3H_7$

1. J.D. Lambert, J.S. Clarke, J.F. Duke, C.L. Hicks, S.D. Lawrence, D.M. Morris and M.G.T. Stone, Proc. R. Soc. <u>A249</u> 414 (1959).

   Boyle's Law apparatus.

T	B_{11}	B_{22}	B_{12}
324.2	-1600	-990	-1890
335.2	-1450	-900	-1667
353.2	-1220	-830	-1321
368.2	-1010	-564	-1133

CHLOROFORM + DIETHYL ETHER $CHCl_3$ + $(C_2H_5)_2$ O

1. J.H.P. Fox and J.D. Lambert, Proc. R. Soc. <u>A210</u> 557 (1952).

   Boyle's Law apparatus.

T	B_{11}	B_{22}	B_{12}
326.2	-990	-880	-1520
338	-875	-780	-1290
352	-755	-720	-1030
363	-630	-610	- 870
393	-500	-505	- 500

2. Sh. D. Zaalishvili and L.E. Kolysko, Russ. J. phys. Chem. <u>34</u> 1223
    (1960); Zh. fiz. Khim. <u>34</u> 2596 (1960).

    Boyle's Law apparatus.

T	B_{11}	B_{22}	$B_{12}(*)$
338.2	-880	-790	-1150

CHLOROFORM + DIETHYLAMINE $CHCl_3$ + $(C_2H_5)_2NH$

1. J.D. Lambert, J.S. Clarke, J.F. Duke, C.L. Hicks, S.D. Lawrence,
    D.M. Morris and M.G.T. Shone, Proc. R. Soc. <u>A249</u> 414 (1959).

    Boyle's Law apparatus.

T	B_{11}	B_{22}	B_{12}
323.2	-1020	-1200	-2063
336.7	- 900	-1040	-1786

CHLOROFORM + BENZENE $CHCl_3$ + C_6H_6

1. P.G. Francis and M.L. McGlashan, Trans. Faraday Soc. <u>51</u> 593 (1955).
    Boyle's apparatus. B_M values given T range 315-350 K, X_2
    approximately 0.5.

CHLOROFORM + n-HEXANE $CHCl_3$ + $CH_3(CH_2)_4CH_3$

1. J.H.P. Fox and J.D. Lambert, Proc. R. Soc. <u>A210</u> 557 (1952).
    Boyle's Law apparatus.

T	B_{11}	B_{22}	B_{12}
326.2	-990	-1380	-1190
352	-755	-1160	- 960

CHLOROFORM + NITROGEN $CHCl_3$ + N_2

1. M. Vigdergauz and V. Semkin, J. Chromat. <u>58</u> 95 (1971).
    Chromatographic measurements using dinonyl phthalate.
    Class III.

T	B_{12}
353.2	-12

2. S.K. Gupta, R.D. Lesslie and A.D. King, Jr., J. phys. Chem., Ithaca
   <u>77</u> 2011 (1973).

   B_{12} derived from measurements of solubility of chloroform in the
   compressed gas.

T	B_{12}
298.15	-100 ± 4

MONOFLUOROMONOCHLOROMETHANE + TETRAFLUORODICHLOROETHANE CH_2ClF + $C_2Cl_2F_4$

1. J.V. Sinka, J. chem. Engng Data <u>15</u> 71 (1970).
   Constant volume cell method: X_1 = 0.754.
   T 373 - 522 K, P → 72 atm.

DICHLOROMETHANE + ETHYL CHLORIDE CH_2Cl_2 + $CH_3.CH_2Cl$

1. M. Rätzsch, Z. phys. Chem. <u>238</u> 321 (1968).

T	B_{12}
303.2	-590 ± 14
313.2	-567 ± 21
333.2	-469 ± 14

DIFLUOROMETHANE + MONOCHLOROPENTAFLUOROETHANE CH_2F_2 + $CClF_2.CF_3$

1. W.H. Mears, J.V. Sinka, P.F. Malbrunot, P.A. Meunier, A.G. Dedit
   and G.M. Scatena, J. chem. Engng Data <u>13</u> 344 (1968).
   PVT data given for mixture with 73.5 mole % CH_2F_2;
   T 298-473 K, P → 100 atm.

FORMIC ACID + ACETIC ACID $H.COOH$ + $CH_3.COOH$

1. J.R. Barton and C.C. Hsu, J. chem. Engng Data <u>14</u> 184 (1969).
   PVTx measurements; T 323-398 K, P below 1 atm.

METHYL BROMIDE + METHYL CHLORIDE CH_3Br + CH_3Cl

1. R.N. Lichtenthaler and K. Schäfer, Ber. (dtsch.) Bunsenges. phys.
   Chem. <u>73</u> 42 (1969).
   Class II.

T	B_{12}
296.0	-484.1
307.7	-430.3
322.8	-381.1

METHYL BROMIDE + ETHYL CHLORIDE $CH_3Br + CH_3.CH_2Cl$

1. M. Rätzsch, Z. phys. Chem. 238 321 (1968).

T	B_{12}
313.2	-436 ± 16
333.2	-397 ± 7

METHYL BROMIDE + ETHYL BROMIDE $CH_3Br + C_2H_5Br$

1. M. Rätzsch and H.-J. Bittrich, Z. phys. Chem. 228 81 (1965).

T	B_{11}	B_{22}	B_{12}
293.1	-478	-795	-1047 ± 41
313.1	-402	-669	- 881 ± 12

METHYL BROMIDE + PROPANE $CH_3Br + CH_3.CH_2CH_3$

1. W. Kappallo, N. Lund and K. Schafer, Z. phys. Chem. Frankf. Ausg. 37 196 (1963).

     Class II.

T	B_{11}	B_{22}	B_{12}
244.0	-1040	-610	-617
273.0	- 718	-477	-474
297.0	- 567	-394	-411
321.0	- 451	-340	-356

METHYL BROMIDE + n-BUTANE $CH_3Br + CH_3.(CH_2)_2CH_3$

1. W. Kappallo, N. Lund and K. Schafer, Z. phys. Chem. Frankf. Ausg. 37 196 (1963).

T	B_{11}	B_{22}	B_{12}
244.0	-1040	-1230	-915

273.0	− 718	− 923	−661
297.0	− 567	− 758	−546
321.0	− 451	− 635	−459

METHYL BROMIDE + n-PENTANE $CH_3Br + CH_3 \cdot (CH_2)_3 \cdot CH_3$

1. M. Ratzsch and H.-J. Bittrich, Z. phys. Chem. <u>228</u> 81 (1965).

T	B_{11}	B_{22}	B_{12}
313.1	-402	-1187	-765 ± 38

METHYL CHLORIDE + CARBON DISULPHIDE $CH_3Cl + CS_2$

1. G.A. Bottomley and T.H. Spurling, Aust. J. Chem. <u>20</u> 1789 (1967). Low pressure differential method. B_{11} interpolated from values given. Estimated error in B_{12} ± 25.

T	B_{11}	B_{22}	B_{12}
325.85	-328	-659	-346
349.56	-281	-569	-318
377.43	-241	-484	-263
402.43	-210	-424	-225
430.01	-177	-375	-171

METHYL CHLORIDE + ACETONE $CH_3 \cdot Cl + (CH_3)_2CO$

1. G.A. Bottomley and T.H. Spurling, Aust. J. Chem. <u>20</u> 1789 (1967). Low pressure differential method. B_{11} and B_{22} interpolated from values given. Estimated error in B_{12} ± 25.

T	B_{11}	B_{22}	B_{12}
327.05	-326	-1365	-649
352.49	-278	-1025	-507
376.38	-243	- 815	-433
400.72	-211	- 665	-367
427.78	-180	- 520	-315

METHYL IODIDE + DIETHYL ETHER $CH_3 \cdot I + (C_2H_5)_2 O$

1. Sh. D. Zaalishvili and L.E. Kolysko, Russ. J. phys. Chem. <u>36</u> 440 (1962); Zh. fiz. Khim. <u>36</u> 846 (1962). Boyle's Law apparatus.

T	B_{11}	B_{22}	B_{12}
313.2	-706	-1046	-785
328.2	-612	- 900	-654
343.2	-517	- 768	-570
358.2	-462	- 682	-406

NITROMETHANE + CARBON DIOXIDE CH_3NO_2 + CO_2

1. M.S. Vigdergauz and V.I. Semkin, Russ. J. phys. Chem. 45 518 (1971); Zh. fiz. Khim. 45 931 (1971). Value also given by M. Vigdergauz and V. Semkin, J. Chromat. 58 95 (1971).

   Chromatographic measurements using dinonyl phthalate.

   Class III.

T	B_{12}
353.2	-188

NITROMETHANE + ACETONE $CH_3.NO_2$ + $(CH_3)_2CO$

1. G.A. Bottomley and T.H. Spurling, Aust. J. Chem. 16 1 (1963).
   Differential compressibility apparatus.

T	B_{11}	B_{22}	B_{12}
323.2	-2866	-1439	-2679 ± 50

2. I. Brown and F. Smith, Aust. J. Chem. 13 30 (1960).
   Values of B calculated by the authors from vapour pressure data.

T	B_{11}	B_{22}	B_{12}
318.15	-3110	-1660	-3830

NITROMETHANE + BENZENE $CH_3.NO_2$ + C_6H_6

1. G.A. Bottomley and T.H. Spurling, Aust. J. Chem. 16 1 (1963).
   Differential compressibility apparatus.

T	B_{11}	B_{22}	B_{12}
323.2	-2866	-1194	-1283 ± 25

NITROMETHANE + NITROGEN $CH_3NO_2 + N_2$

1. M. Vigdergauz and V. Semkin, J. Chromat. <u>58</u> 95 (1971).
    Chromatographic measurements using dinonyl phthalate.
    Class III.

T	B_{12}
353.2	-41

METHANE + METHANOL $CH_4 + CH_3.OH$

1. B. Hemmaplardh and A.D. King, Jr., J. phys. Chem., Ithaca <u>76</u> 2170
    (1972).

    B_{12} derived from measurements of solubility of methanol in the
    compressed gas.

T	B_{12}	T	B_{12}
288.15	-130 ± 6	323.15	$- 97 \pm 4$
298.15	-114 ± 7	333.15	$- 95 \pm 2$
310.15	-103 ± 4		

METHANE + CARBON MONOXIDE $CH_4 + CO$

1. D. McA. Mason and B.E. Eakin, J. chem. Engng Data <u>6</u> 499 (1961) (*).

T	B_{12}
288.70	-7.3 ± 5

METHANE + CARBON DIOXIDE $CH_4 + CO_2$

1. H.H. Reamer, R.H. Olds, B.H. Sage and W.N. Lacey, Ind. Engng Chem.
    ind. Edn <u>36</u> 88 (1944).

    Volumetric behaviour of four mixtures from 310 - 500K;   P → 670 atm.

2. D. McA. Mason and B.E. Eakin, J. chem. Engng Data <u>6</u> 499 (1961) (*).

T	B_{12}
288.7	-62.9 ± 5

METHANE + PERFLUOROETHANE $CH_4 + C_2F_6$
1. E.M. Dantzler Siebert and C.M. Knobler, J. phys. Chem., Ithaca <u>75</u>

3863 (1971).

E determined from measurements of pressure changes on mixing.

T	E	B_{11}	B_{22}	B_{12}
323.15	51 ± 1	-199	-33	-65

METHANE + ETHYLENE $CH_4 + CH_2:CH_2$

1. D. McA. Mason and B.E. Eakin, J. chem. Engng Data 6 499 (1961) (*).

T	B_{12}
288.70	-70.8 ± 5

2. H.G. McMath and W.C. Edmister, A. I. Ch. E. Jl 15 370 (1969).
   PVT data for four mixtures; P range 17 - 150 atm.
   Estimated error ± 5.

T	B_{12}
266.5	-89.8
277.6	-78.6
288.7	-71.5

3. R.C. Lee and W.C. Edmister, A. I. Ch. E. Jl 16 1047 (1970).
   Burnett method; P → 800 atm.

T	B_{12}
298.2	-61.15
323.2	-53.99
348.2	-43.75

4. J.W. Lee and G. Saville (unpublished results). See also J.W. Lee,
   Ph.D. thesis, University of London (1976).
   Estimated uncertainty in B_{12} is ± 5.

T	B_{12}	T	B_{12}
240	-118.1	270	- 90.4
245	-112.8	275	- 86.7
250	-107.8	280	- 83.2
255	-103.1	285	- 79.9
260	- 98.6	290	- 76.8
265	- 94.4		

METHANE + ETHANE CH_4 + $CH_3.CH_3$

1. R.A. Budenholtzer, B.H. Sage and W.N. Lacey, Ind. Engng Chem. ind. Edn 31 1288 (1939).

   Joule-Thomson coefficients for three mixtures from 295 - 380K.

2. A. Michels and G.W. Nederbragt, Physica, 's Grav. 6 656 (1939).

   PVT data given; P → 60 atm at 273, 298 and 323K.

3. B.H. Sage and W.N. Lacey, Ind. Engng Chem. ind. Edn 31 1497 (1939).

   Specific volumes of four mixtures from 295 - 395K; P → 320 atm.

4. R.D. Gunn, M.S. Thesis, University of California, Berkeley (1958). Values given by E.M. Dantzler, C.M. Knobler and M.L. Windsor, J. phys. Chem., Ithaca 72 676 (1968).

T	B_{12}
273.15	-111.9
298.15	-92.0
323.15	-75.6

5. D. McA. Mason and B.E. Eakin, J. chem. Engng Data 6 499 (1961) (*).

T	B_{12}
288.70	-96.2 ± 5

6. A.E. Hoover, I. Nagata, T.W. Leland, Jr., and R. Kobayashi, J. chem. Phys. 48 2633 (1968).

   Burnett method.

T	B_{12}
215.00	-192.29
240.00	-158.17
273.15	-111.86

7. E.M. Dantzler, C.M. Knobler and M.L. Windsor, J. phys. Chem., Ithaca 72 676 (1968).

   E determined from measurements of pressure changes on mixing.

T	E	B_{11}	B_{22}	B_{12}
298.15	21 ± 1	-42	-185	-93
323.15	17 ± 2	-33	-159	-79
348.15	13 ± 1	-26	-136	-68
373.15	12 ± 1	-20	-115	-56

8. K.A. Alkasab, J.M. Shah, R.J. Laverman, and R.A. Budenholtzer, Ind. Engng Chem. (Fundamentals) 10 237 (1971).

Isothermal and adiabatic Joule-Thomson coefficients given for 3 mixtures; T 223 - 273 K, P → 63 atm.

9. C.J. Wormald, E.J. Lewis, and D.J. Hutchings, J. chem. Thermodyn. 11 1 (1979).

B_{12} values derived from measured excess enthalpies.

T	B_{12}	T	B_{12}
241.1	-141 ± 7	269.2	-112 ± 6
250.6	-130 ± 6	282.2	-105 ± 4
262.4	-118 ± 6	298.2	- 93 ± 4
266.7	-116 ± 4	303.2	- 88 ± 5

METHANE + ETHANOL CH_4 + C_2H_5OH

1. S.K. Gupta, R.D. Lesslie and A.D. King, Jr., J. phys. Chem., Ithaca 77 2011 (1973).

B_{12} derived from measurements of solubility of ethanol in the compressed gas.

T	B_{12}
298.15	-122 ± 2
323.15	-107 ± 3
348.15	- 84 ± 10

METHANE + PROPANE CH_4 + $CH_3 \cdot CH_2 \cdot CH_3$

1. B.H. Sage, W.N. Lacey and J.G. Schaafsma, Ind. Engng Chem. ind. Edn 26 214 (1934).

Densities of mixtures from 295 - 395K; P range 10 - 200 atm.

2. R.A. Budenholtzer, D.F. Botkin, B.H. Sage, and W.N. Lacey, Ind. Engng Chem. ind.Edn 34 878 (1942).

Measurements of Joule-Thomson coefficients; P → 100 atm, T range 294-427K.

3. H.H. Reamer, B.H. Sage and W.N. Lacey, Ind. Engng Chem. ind. Edn 42 534 (1950).

Volumetric behaviour of four mixtures from 280 - 500K; P → 670 atm.

4. R.D. Gunn, M.S. Thesis, University of California, Berkeley (1958).
   Values of B_{12} given by E.M. Dantzler, C.M. Knobler and M.L.
   Windsor, J. phys. Chem., Ithaca <u>72</u> 676 (1968).

T	B_{12}	T	B_{12}
298.15	-136	348.15	- 93
323.15	-114	373.15	- 81

5. D. McA. Mason and B.E. Eakin, J. chem. Engng Data <u>6</u> 499 (1961) (*).

T	B_{12}
288.70	-136.2 ± 5

6. J.A. Barker and M. Linton, J. chem. Phys. <u>38</u> 1853 (1963).
   Values of B_M for an equimolar mixture given from data of H.H.
   Reamer, B.H. Sage and W.N. Lacey, Ind. Engng Chem. ind. Edn <u>42</u> 534
   (1950).

T	B_M	T	B_M
377.6	-109.5	477.6	- 59.4
410.9	- 89.6	510.9	- 47.4
444.3	- 73.6		

7. E.M. Dantzler, C.M. Knobler and M.L. Windsor, J. phys. Chem., Ithaca
   <u>72</u> 676 (1968).
   E determined from measurements of pressure changes on mixing.

T	E	B_{11}	B_{22}	B_{12}
298.15	82 ± 2	-42	-399	-139 ± 8
323.15	66 ± 2	-33	-331	-116 ± 8
348.15	58 ± 1	-26	-276	- 93 ± 11
373.15	53 ± 1	-20	-235	- 75 ± 11

8. C.J. Wormald, E.J. Lewis, and D.J. Hutchings, J. chem. Thermodyn.
   <u>11</u> 1 (1979).
   B_{12} values derived from measured excess enthalpies.

T	B_{12}	T	B_{12}
243.2	-192 ± 13	263.6	-168 ± 11
245.2	-187 ± 11	273.2	-156 ± 10
250.8	-180 ± 14	281.2	-148 ± 8

253.2	-176 ± 9	290.7	-139 ± 10
261.2	-166 ± 10	302.2	-127 ± 6

METHANE + PERFLUORO-n-BUTANE $\quad CH_4 + CF_3 \cdot (CF_2)_2 CF_3$

1. E.M. Dantzler Siebert and C.M. Knobler, J. phys. Chem., Ithaca <u>75</u> 3863 (1971).

   E determined from measurements of pressure changes on mixing.

T	E	B_{11}	B_{22}	B_{12}
323.15	260 ± 1	-33	-744	-128

METHANE + n-BUTANE $\quad CH_4 + CH_3 \cdot (CH_2)_2 \cdot CH_3$

1. B.H. Sage, R.A. Budenholtzer, and W.N. Lacey, Ind. Engng Chem. ind. Edn <u>32</u> 1262 (1940).

   Compressibility factors for 24 mixtures from 295 - 395K; P → 230 atm.

2. R.A. Budenholtzer, B.H. Sage and W.N. Lacey, Ind. Engng Chem. ind. Edn <u>32</u> 384 (1940).

   Joule-Thomson coefficients for five mixtures from 295 - 430K; P range 3 - 100 atm.

3. J.A. Beattie, W.H. Stockmayer and H.G. Ingersoll, J. chem. Phys. <u>9</u> 871 (1941).

   PVT data given: P range 30 - 350 atm, T range 348 - 573 K, X_2 is approximately 0.25, 0.50 and 0.75. B values calculated from these data given by: J.A. Beattie and W.H. Stockmayer, J. chem. Phys. <u>10</u> 473 (1942).

T	B_{22}	B_{11}	B_{12}
423.2	-328.7	-11.4	-81.6
448.2	-287.3	-7.5	-69.4
473.2	-254.2	-4.0	-60.4
498.2	-224.5	-0.9	-51.2
523.2	-198.1	$+1.9$	-42.0
548.2	-176.0	$+4.5$	-35.2
573.2	-157.4	$+6.8$	-29.2

4. H.H. Reamer, K.J. Korpi, B.H. Sage, and W.N. Lacey, Ind. Engng Chem. ind. Edn <u>39</u> 206 (1947).

   Volumetric behaviour of four mixtures from 310 - 500K; P range

27 - 670 atm.

5. R.D. Gunn, M.S. Thesis, University of California, Berkeley (1958).
   Values of B_{12} given by E.M. Dantzler, C.M. Knobler and M.L.
   Windsor, J. phys. Chem., Ithaca 72 676 (1968).

T	B_{12}	T	B_{12}
298.15	-193	348.15	-121
323.15	-158	373.15	-100

6. D. McA. Mason and B.E. Eakin, J. chem. Engng Data 6 499 (1961) (*).

T	B_{12}
288.70	-175.1 ± 8

7. E.M. Dantzler, C.M. Knobler and M.L. Windsor, J. phys.,Chem. Ithaca
   72 676 (1968).

   E determined from measurements of pressure changes on mixing.

T	E	B_{11}	B_{22}	B_{12}
298.15	204 ± 2	-42	-732	-183 ± 12
323.15	164 ± 2	-33	-594	-150 ± 12
348.15	142 ± 1	-26	-501	-122 ± 11
373.15	124 ± 1	-20	-429	-100 ± 11

8. C.J. Wormald, E.J. Lewis, and D.J. Hutchings, J. chem. Thermodyn
   11 1 (1979).

   B_{12} values derived from measured excess enthalpies.

T	B_{12}	T	B_{12}
277.0	-198 ± 21	314.2	-162 ± 25
284.2	-184 ± 17	327.8	-124 ± 12
289.0	-179 ± 18	373.4	- 97 ± 22
303.2	-161 ± 14	383.2	- 93 ± 13
306.5	-155 ± 12	394.3	- 87 ± 14

METHANE + ISOBUTANE CH_4 + $(CH_3)_3CH$

1. R.H. Olds, B.H. Sage and W.N. Lacey, Ind. Engng Chem. ind. Edn 34
   1008 (1942).

   Volumetric behaviour of four mixtures from 310 - 500K; P → 330 atm.

2. D. McA. Mason and B.E. Eakin, J. chem. Engng Data $\underline{6}$ 499 (1961) (*).

T	B_{12}
288.70	-158.7 ± 5

3. R.D. Gunn, M.S. Thesis, University of California, Berkeley (1958).
    B_{12} calculated from volumetric data given by W.K. Tang, Tech. Rep.
    WIS-OOR-13, University of Wisconsin (1956).

T	B_{12}	T	B_{12}
344.3	-125.5	444.3	- 63.4
377.6	- 96.6	477.6	- 48.9
410.9	- 81.3	510.9	- 40.4

METHANE + 1-BUTANOL $CH_4 + CH_3 \cdot (CH_2)_2 CH_2OH$

1. R. Massoudi and A.D. King, Jr., J. phys. Chem., Ithaca $\underline{77}$ 2016
    (1973).

    B_{12} derived from measurements of solubility of 1-butanol in the
    compressed gas.

T	B_{12}
298.15	-160 ± 12

METHANE + DIETHYL ETHER $CH_4 + (C_2H_5)_2 O$

1. R. Massoudi and A.D. King, Jr., J. phys. Chem., Ithaca $\underline{77}$ 2016
    (1973).

    B_{12} derived from measurements of solubility of diethyl ether in
    the compressed gas.

T	B_{12}
298.15	-161 ± 3

METHANE + TETRAMETHYL SILANE $CH_4 + Si(CH_3)_4$

1. S.D. Hamann, J.A. Lambert and R.B. Thomas, Aust. J. Chem. $\underline{8}$ 149
    (1955).

    PV measurement, maximum pressure 2 atm.

T	B_{11}	B_{22}	B_{12}
323.16	-34.6	-956	-145
333.16	-31.8	-886	-134
343.16	-29.1	-831	-124
353.16	-26.6	-784	-111
363.16	-24.2	-737	-106
383.16	-19.5	-649	- 97
403.16	-15.4	-580	- 86

2. J. Bellm. W. Reineke, K. Schäfer, and B. Schramm, Ber. (dtsch.) Bunsenges. phys. Chem. 78 282 (1974).

Class III.

T	B_{12}	T	B_{12}
300	-210.5	430	- 84.4
320	-168.0	460	- 68.2
340	-149.0	490	- 59.0
370	-123.5	520	- 50.0
400	-102.6	550	- 40.6

METHANE + PERFLUORO-n-PENTANE $CH_4 + CF_3 \cdot (CF_2)_3 \cdot CF_3$

1. E.M. Dantzler Siebert and C.M. Knobler, J. phys. Chem., Ithaca 75 3863 (1971).

    E determined from measurements of pressure changes on mixing.

T	E	B_{11}	B_{22}	B_{12}
373.15	311 ± 3	-20	-793	-95

METHANE + n-PENTANE $CH_4 + CH_3 \cdot (CH_2)_3 \cdot CH_3$

1. B.H. Sage, H.H. Reamer, R.H. Olds and W.N. Lacey, Ind. Engng Chem. ind. Edn 34 1108 (1942).

    Volumetric behaviour of six mixtures from 310 - 500K; P → 330 atm.

2. D. McA. Mason and B.E. Eakin, J. chem. Engng Data 6 499 (1961) (*).

T	B_{12}
288.70	-269.6 ± 11

3. R.L. Pecsok and M.L. Windsor, Analyt. Chem. 40 1238 (1968).

Gas chromatographic method, using squalane; P → 6 atm.

Carrier gas solubility taken into account in calculating B_{12}.

T	B_{12}
298.2	-204 ± 42
323.2	-138 ± 30

4. E.M. Dantzler, C.M. Knobler and M.L. Windsor, J. phys. Chem., Ithaca 72 676 (1968).

E determined from measurements of pressure changes on mixing.

T	E	B_{11}	B_{22}	B_{12}
298.15	397 ± 3	-42	-1195	-222 ± 13
323.15	320 ± 2	-33	- 987	-190 ± 12
348.15	270 ± 2	-26	- 810	-148 ± 12
373.15	232 ± 2	-20	- 687	-122 ± 12

5. R. Massoudi and A.D. King, Jr., J. phys. Chem., Ithaca 77 2016 (1973).

B_{12} derived from measurements of solubility of n-pentane in the compressed gas.

T	B_{12}
298.15	-170 ± 12

6. C.J. Wormald, E.J. Lewis, and D.J. Hutchings, J. chem. Thermodyn. 11 1 (1979).

B_{12} values derived from measured excess enthalpies.

T	B_{12}	T	B_{12}
318.5	-178 ± 20	373.2	-125 ± 14
333.2	-163 ± 18	383.4	-121 ± 15
343.4	-150 ± 17	393.2	-120 ± 17
353.2	-139 ± 15	403.5	-105 ± 14
363.2	-121 ± 16		

METHANE + ISOPENTANE $CH_4 + (CH_3)_2.CH.CH_2.CH_3$

1. D. McA. Mason and B.E. Eakin, J. chem. Engng Data 6 499 (1961) (*).

T	B_{12}
288.70	-220.1 ± 5

2. R.L. Pecsok and M.L. Windsor, Analyt. Chem. <u>40</u> 1238 (1968).
   Gas chromatographic method using squalane; $P \rightarrow 6$ atm.
   Carrier gas solubility taken into account in calculating B_{12}.

T	B_{12}
298.2	-199 ± 43
323.2	-123 ± 36

METHANE + NEOPENTANE $CH_4 + C.(CH_3)_4$

1. S.D. Hamann, J.A. Lambert and R.B. Thomas, Aust. J. Chem. <u>8</u> 149 (1955).
   PV measurements, maximum pressure 2 atm.

T	B_{11}	B_{22}	B_{12}
303.16	-41.6	-842	-165
323.16	-34.6	-734	-138
333.16	-31.8	-686	-132
343.16	-29.1	-643	-118
353.16	-26.6	-602	-113
363.16	-24.2	-566	-106
383.16	-19.5	-507	- 93
403.16	-15.4	-452	- 78

2. K. Strein, R.N. Lichtenthaler, B. Schramm, and Kl. Schäfer, Ber. (dtsch.) Bunsenges. phys. Chem. <u>75</u> 1308 (1971).
   Estimated accuracy of B_{12} is $\pm 1\%$.

T	B_{12}	T	B_{12}
296.15	-199.2	413.8	- 87.5
313.2	-172.0	433.8	- 75.0
334.0	-149.6	453.6	- 64.5
353.5	-131.1	473.8	- 53.2
373.6	-114.7	492.6	- 48.7
394.3	-101.3		

3. J. Bellm, W. Reineke, K. Schäfer, and B. Schramm, Ber. (dtsch.) Bunsenges. phys. Chem. <u>78</u> 282 (1974).

Class III.

T	B_{12}	T	B_{12}
300	-193.7	430	- 92.0
320	-170.8	460	- 80.2
340	-153.7	490	- 69.2
370	-130.2	520	- 59.5
400	-109.7	550	- 46.9

4. G.L. Baughman, S.P. Westhoff, S. Dincer, D.D. Duston and A.J. Kidnay, J. chem. Thermodyn. <u>7</u> 875 (1975).

B_{12} determined from measurements of the solubility of neopentane in the compressed gas. Maximum error in B_{12} is estimated to be 10%.

T	B_{12}	T	B_{12}
199.99	-434	240.00	-263
210.00	-374	249.99	-223
220.00	-331	257.90	-188
230.00	-292		

METHANE + PERFLUORO-n-HEXANE $CH_4 + CF_3 \cdot (CF_2)_4 CF_3$

1. E.M. Dantzler Siebert and C.M. Knobler, J. phys. Chem., Ithaca <u>75</u> 3863 (1971).

E determined from measurements of pressure changes on mixing.

T	E	B_{11}	B_{22}	B_{12}
373.15	484 ± 7	-20	-1128	-90

METHANE + BENZENE $CH_4 + C_6H_6$

1. D.H. Everett, B.W. Gainey and C.L. Young, Trans. Faraday Soc. <u>64</u> 2667 (1968).

   Gas chromatographic measurements, using squalane.

T	B_{12}
323.15	-155 ± 15

2. C.R. Coan and A.D. King, Jr., J. Chromat. <u>44</u> 429 (1969).

   B_{12} determined from measurements of solubility of benzene in the compressed gas: $P \to 65$ atm.

T	B_{12}
323.2	-171 ± 3

METHANE + n-HEXANE $CH_4 + CH_3 \cdot (CH_2)_4 \cdot CH_3$

1. R.L. Pecsok and M.L. Windsor, Analyt. Chem. <u>40</u> 1238 (1968).

   Gas chromatographic method, using squalane; $P \to 6$ atm.

   Carrier gas solubility taken into account in calculating B_{12}.

T	B_{12}
298.2	-292 ± 55
323.2	-280 ± 54

2. E.M. Dantzler, C.M. Knobler and M.L. Windsor, J. phys. Chem., Ithaca <u>72</u> 676 (1968).

   E determined from measurements of pressure changes on mixing.

T	E	B_{11}	B_{22}	B_{12}
298.15	720 ± 40	-42	-1919	-261
323.15	557 ± 5	-33	-1530	-225
348.15	454 ± 1	-26	-1239	-180
373.15	381 ± 2	-20	-1031	-145

3. C.J. Wormald, E.J. Lewis, and D.J. Hutchings, J. chem. Thermodyn. <u>11</u> 1 (1979).

   B_{12} values derived from measured excess enthalpies.

T	B_{12}	T	B_{12}
343.2	-180 ± 27	383.2	-134 ± 21

363.2	-144 ± 24	389.2	-124 ± 32
373.2	-139 ± 23	407.7	-108 ± 20

METHANE + 2-METHYLPENTANE $CH_4 + (CH_3)_2 \cdot CH \cdot (CH_2)_2 \cdot CH_3$

1. R.L. Pecsok and M.L. Windsor, Analyt. Chem. <u>40</u> 1238 (1968).
    Gas chromatographic method, using squalane; $P \to 6$ atm.
    Carrier gas solubility taken into account in calculating B_{12}.

T	B_{12}
298.2	-317 ± 42
323.2	-144 ± 34

METHANE + 2,2-DIMETHYLBUTANE $CH_4 + (CH_3)_3 C \cdot CH_2 \cdot CH_3$

1. R.L. Pecsok and M.L. Windsor, Analyt. Chem. <u>40</u> 1238 (1968).
    Gas chromatographic method using squalane; $P \to 6$ atm.
    Carrier gas solubility taken into account in calculating B_{12}.

T	B_{12}
298.3	-216 ± 42
323.2	-154 ± 29

METHANE + n-HEPTANE $CH_4 + CH_3 \cdot (CH_2)_5 \cdot CH_3$

1. C.J. Wormald, E.J. Lewis, and D.J. Hutchings, J. chem. Thermodyn
    <u>11</u> 1 (1979).
    B_{12} values derived from measured excess enthalpies.

T	B_{12}	T	B_{12}
389.2	-162 ± 26	408.2	-135 ± 35
390.2	-157 ± 26	413.3	-131 ± 33
405.7	-127 ± 26		

METHANE + n-OCTANE $CH_4 + CH_3 \cdot (CH_2)_6 \cdot CH_3$

1. C.J. Wormald, E.J. Lewis, and D.J. Hutchings, J. chem. Thermodyn.
    <u>11</u> 1 (1979).
    B_{12} values derived from measured excess enthalpies.

T	B_{12}
410.2	-151 ± 38
418.3	-150 ± 32

METHANE + 2,2,5-TRIMETHYLHEXANE $\quad CH_4 + (CH_3)_3C.(CH_2)_2.CH(CH_3)_2$

1. S.G. D'Avila, B.K. Kaul, and J.M. Prausnitz, J. chem. Engng Data 21 488 (1976).

   B_{12} calculated from solubility measurements of the heavy hydrocarbon in the compressed gas. Maximum uncertainty in B_{12} given as $\pm$ 10.

T	B_{12}	T	B_{12}
298.2	-243	348.2	-169
323.2	-200	373.2	-144

METHANE + NAPHTHALENE $\quad CH_4 + C_{10}H_8$

1. A.D. King, Jr. and W.W. Robertson, J. chem. Phys. 37 1453 (1962).
   Naphthalene solubility measurements.

T	B_{12}
296.2	-321 ± 10
342.2	-196 ± 6
348.2	-176 ± 6

2. G.C. Najour and A.D. King, Jr., J. chem. Phys. 45 1915 (1966).
   Naphthalene solubility measurements.

T	B_{12}	T	B_{12}
294	-363 ± 11	327	-257 ± 11
307	-314 ± 12	333	-243 ± 11
307	-324 ± 18	341	-243 ± 11

METHANE + tert-BUTYLBENZENE $\quad CH_4 + C_6H_5.C(CH_3)_3$

1. S.G. D'Avila, B.K. Kaul, and J.M. Prausnitz, J. chem. Engng Data 21 488 (1976).

   B_{12} calculated from solubility measurements of the heavy hydrocarbon in the compressed gas. Maximum uncertainty in B_{12} given as $\pm$ 10.

T	B_{12}	T	B_{12}
323.2	-224	373.2	-158
348.2	-179	398.2	-120

METHANE + n-DECANE $CH_4 + CH_3 \cdot (CH_2)_8 \cdot CH_3$

1. B.H. Sage, H.M. Lavender, and W.N. Lacey, Ind. Engng Chem. ind. Edn 32 743 (1940).

   Volumetric behaviour of six mixtures from 295 - 395K;   P → 300 atm.

2. H.H. Reamer, R.H. Olds, B.H. Sage, and W.N. Lacey, Ind. Engng Chem. ind. Edn 34 1526 (1942).

   Volumetric behaviour of five mixtures from 310 - 500K;   P → 670 atm.

3. S.G. D'Avila, B.K. Kaul, and J.M. Prausnitz, J. chem. Engng Data 21 488 (1976).

   B_{12} calculated from solubility measurements of n-decane in the compressed gas. Maximum uncertainty in B_{12} given as ± 10.

T	B_{12}	T	B_{12}
323.2	-289	373.2	-181
348.2	-232	398.2	-134

METHANE + 1-METHYL NAPHTHALENE $CH_4 + C_{10}H_7 \cdot CH_3$

1. B.K. Kaul and J.M. Prausnitz, A.I. Ch. E. Jl 24 223 (1978).

   Measurements of the heavy hydrocarbon solubility in the compressed gas.

T	B_{12}
348.2	-205 ± 10
398.2	-146 ± 10
448.2	- 91 ± 10

METHANE + BICYCLOHEXYL $CH_4 + C_{10}H_{18}$

1. B.K. Kaul and J.M. Prausnitz, A. I. Ch. E. Jl 24 223 (1978).

   Measurements of the heavy hydrocarbon solubility in the compressed gas.

T	B_{12}	T	B_{12}
323.2	-299 ± 15	403.2	-104 ± 10
363.2	-223 ± 10	443.2	$- 96 \pm 10$

METHANE + n-DODECANE $CH_4 + CH_3 \cdot (CH_2)_{10} \cdot CH_3$

1. S.G. D'Avila, B.K. Kaul, and J.M. Prausnitz, J. chem. Engng Data 21 488 (1976).

    B_{12} calculated from solubility measurements of the heavy hydro-carbon in the compressed gas. Maximum uncertainty in B_{12} given as $\pm$ 10.

T	B_{12}	T	B_{12}
348.2	-290	398.2	-200
373.2	-248	423.2	-164

METHANE + DIPHENYLMETHANE $CH_4 + (C_6H_5)_2CH_2$

1. B.K. Kaul and J.M. Prausnitz, A. I. Ch. E. Jl 24 223 (1978).

    Measurements of the heavy hydrocarbon solubility in the compressed gas.

T	B_{12}	T	B_{12}
338.2	-295 ± 15	408.2	-237 ± 10
373.2	-268 ± 12	443.2	-176 ± 10

METHANE + ANTHRACENE $CH_4 + C_6H_4(CH)_2C_6H_4$

1. G.C. Najour and A.D. King, Jr., J. chem. Phys. 52 5206 (1970).

    B_{12} determined from measurements of solubility of anthracene in the compressed gas: P → 100 atm. Probable error in B $\pm$ 12.

T	B_{12}	T	B_{12}
339	-326	382	-256
346	-324	396	-235
348	-291	407	-204
350	-316	423	-198
355	-314	445	-177
362	-290	449	-208
369	-269	458	-174
373	-292		

METHANE + PHENANTHRENE $CH_4 + C_{14}H_{10}$

1. H. Bradley, Jr., and A.D. King, Jr., J. chem. Phys. <u>52</u> 2851 (1970).
   B_{12} determined from measurements of solubility of phenanthrene in
   the compressed gas: $P \rightarrow 60$ atm. Probable error in $B \pm 13$.

T	B_{12}	T	B_{12}
313	-443	359	-331
320	-445	361	-287
327	-427	363	-298
329	-390	365	-296
334	-372	368	-290
340	-372	371	-293
348	-341	380	-270
352	-308	395	-240
356	-340	411	-203

METHANE + n-HEXADECANE $CH_4 + CH_3 \cdot (CH_2)_{14} \cdot CH_3$

1. B.K. Kaul and J.M. Prausnitz, A. I. Ch. E. Jl <u>24</u> 223 (1978).
   Measurements of the heavy hydrocarbon solubility in the compressed
   gas.

T	B_{12}	T	B_{12}
348.2	-371 ± 15	423.2	-170 ± 10
373.2	-260 ± 14	448.2	-103 ± 10
398.2	-214 ± 12		

METHANE + EICOSANE $CH_4 + CH_3 \cdot (CH_2)_{18} \cdot CH_3$

1. B.K. Kaul and J.M. Prausnitz, A. I. Ch. E. Jl <u>24</u> 223 (1978).
   Measurements of the heavy hydrocarbon solubility in the compressed
   gas.

T	B_{12}	T	B_{12}
438.2	-214 ± 10	508.2	-103 ± 10
473.2	-149 ± 10	543.2	- 56 ± 5

METHANE + SQUALANE $CH_4 + C_{30}H_{62}$

1. B.K. Kaul and J.M. Prausnitz, A. I. Ch. E. Jl <u>24</u> 223 (1978).

Measurements of the heavy hydrocarbon solubility in the compressed gas.

T	B_{12}
503.2	-48 ± 6
545.2	$+43 \pm 4$

METHANE + SULPHUR HEXAFLUORIDE $CH_4 + SF_6$

1. S.D. Hamann, J.A. Lambert and R.B. Thomas, Aust.J. Chem. **8** 149 (1955).

   PV measurements, maximum pressure 2 atm.

T	B_{11}	B_{22}	B_{12}
313.16	-37.9	-253	-85
333.16	-31.8	-223	-68
353.16	-26.6	-192	-57
373.16	-21.8	-163	-45
393.16	-17.4	-145	-33

2. J. Bellm, W. Reineke, K. Schäfer, and B. Schramm, Ber. (dtsch.) Bunsenges. phys. Chem. **78** 282 (1974).

   Class II.

T	B_{12}	T	B_{12}
300	-91.8	430	-29.2
320	-79.0	460	-18.9
340	-68.3	490	-10.9
370	-54.3	520	- 3.2
400	-41.3	550	+ 5.5

METHANE + HYDROGEN $CH_4 + H_2$

1. F.A. Freeth and T.T.H. Verschoyle, Proc. R. Soc. **A130** 453 (1931).

   Equilibrium compositions measured at 90.6K; P → 200 atm.

2. D. McA. Mason and B.E. Eakin, J. chem. Engng Data **6** 499 (1961) (*).

T	B_{12}
288.7	$+9.2 \pm 5$

3. W.H. Mueller, T.W. Leland, and R. Kobayashi, A.I. Ch. E. Jl $\underline{7}$ 267 (1961) (*).

Burnett method; P → 480 atm.

Class I.

T	B_{12}	T	B_{12}
144.3	-9.7	227.6	2.7
172.0	-4.5	255.4	4.9
199.8	-0.5	283.2	7.5

4. C.W. Solbrig and R.T. Ellington, Chem. Engng Prog. Sym.Ser. $\underline{59}$ 127 (1963).

PVT data given; T 138 - 423 K, P → 200 atm.

METHANE + WATER $CH_4 + H_2O$

1. M. Rigby and J.M. Prausnitz, J. phys. Chem., Ithaca $\underline{72}$ 330 (1968).

B_{12} derived from measurements of solubility of water in the compressed gas.

T	B_{12}	T	B_{12}
298.15	-63 ± 6	348.15	-37 ± 4
323.15	-46 ± 5	373.15	-30 ± 3

METHANE + HYDROGEN SULPHIDE $CH_4 + H_2S$

1. H.H. Reamer, B.H. Sage and W.N. Lacey, Ind. Engng Chem. ind. Edn $\underline{43}$ 976 (1951).

Volumetric behaviour of five mixtures from 280-440K; P → 670 atm.

METHANE + HELIUM $CH_4 + He$

1. H.L. Rhodes, W.E. De Vaney and P.C. Tully, J. chem. Engng Data $\underline{16}$ 19 (1971).

Phase equilibria for mixtures: T 94 - 192 K, P 66 - 255 atm.

2. W.E. De Vaney, H.L. Rhodes and P.C. Tully, J. chem. Engng Data $\underline{16}$ 158 (1971).

Phase equilibria for mixtures: T 124 - 191K, P → 66 atm.

METHANE + KRYPTON CH_4 + Kr

1. M.A. Byrne, M.R. Jones, and L.A.K. Staveley, Trans. Faraday Soc.
    <u>64</u> 1747 (1968).

    Class I.

T	B_{12}	T	B_{12}
118.65	-297.2	182.13	-134.6
121.05	-286.3	200.44	-112.1
125.38	-268.2	221.55	- 91.7
132.08	-243.9	225.96	- 86.8
141.22	-213.9	245.67	- 73.8
152.18	-185.3	270.82	- 60.6
166.44	-157.1		

METHANE + NITROGEN CH_4 + N_2

1. F.G. Keyes and H.G. Burks, J. Am. chem. Soc. <u>50</u> 1100 (1928).

    PVT data given: P range 30-330 atm, T range 273-473 K, X_2 is
    0.194, 0.204 and 0.566.

2. D. McA. Mason and B.E. Eakin, J. chem. Engng Data <u>6</u> 499 (1961) (*).

T	B_{12}
288.70	-17.8 ± 5

3. P. Bol'shakov, I.I. Gel'perin, M.G. Ostronov, and A.A. Orlova, Zh.
    fiz. Khim. <u>41</u> 689 (1967); Russ. J. phys. Chem <u>41</u> 353 (1967).

    Measurements of isothermal Joule-Thomson effect from 173 to 273K;
    P → 50 atm; mole fraction methane is 0.099.

4. D.R. Roe and G. Saville (unpublished results). See also D.R. Roe,
    Ph.D. thesis, University of London (1972).

T	B_{12}	C_{112}	C_{221}
155.88	-102.3 ± 1.0	4500 ± 1000	3100 ± 800
181.87	- 74.5 ± 0.8	3500 ± 600	2800 ± 600
192.64	- 65.2 ± 0.7	3300 ± 600	2300 ± 600
218.86	- 48.1 ± 0.6	2400 ± 500	2500 ± 500
248.53	- 33.8 ± 0.4	2200 ± 500	1900 ± 500
291.40	- 20.1 ± 0.4	2100 ± 500	1600 ± 500

METHANOL + CARBON DIOXIDE $CH_3OH + CO_2$

1. S.T. Sie, W. Van Beersum, and G.W.A. Rijnders, Sep. Sci. $\underline{1}$ 459
    (1966).

    Chromatographic method; $P \rightarrow 7$ atm.

T	B_{12}
313.2	-236

2. B. Hemmaplardh and A.D. King, Jr., J. phys. Chem., Ithaca $\underline{76}$ 2170
    (1972).

    B_{12} derived from measurements of solubility of methanol in the
    compressed gas.

T	B_{12}	T	B_{12}
288.15	-352 ± 7	323.15	-237 ± 3
298.15	-308 ± 5	333.15	-217 ± 7
310.15	-265 ± 8		

METHANOL + ETHYLENE (ETHENE) $CH_3OH + CH_2:CH_2$

1. M. Ratzsch and H. Freydank, J. chem. Thermodyn. $\underline{3}$ 861 (1971).
    Maximum error in B estimated to be ± 3%.

T	B_{12}
303.78	-872
311.46	-719
333.36	-685

2. B. Hemmaplardh and A.D. King, Jr., J. phys. Chem., Ithaca $\underline{76}$ 2170
    (1972).

    B_{12} derived from measurements of solubility of methanol in the
    compressed gas.

T	B_{12}	T	B_{12}
288.15	-279 ± 6	323.15	-192 ± 6
298.15	-247 ± 6	333.15	-179 ± 4
310.15	-213 ± 7		

METHANOL + ETHANE $CH_3OH + C_2H_6$

1. B. Hemmaplardh and A.D. King, Jr., J. phys. Chem., Ithaca $\underline{76}$

2170 (1972).

B_{12} derived from measurements of solubility of methanol in the compressed gas.

T	B_{12}	T	B_{12}
288.15	-306 ± 5	323.15	-220 ± 7
298.15	-276 ± 8	333.15	-203 ± 7
310.15	-234 ± 4		

METHANOL + n-BUTANE $CH_3OH + CH_3 \cdot (CH_2)_2 \cdot CH_3$

1. L.B. Petty and J.M. Smith, Ind. Engng Chem. ind. Edn <u>47</u> 1258 (1955).
    Volumetric behaviour of three mixtures from 320K - 410K; P → 40 atm.

METHANOL + n-PENTANE $CH_3OH + CH_3 \cdot (CH_2)_3 \cdot CH_3$

1. M. Ratzsch and H. Freydank, J. chem. Thermodyn. <u>3</u> 861 (1971).
    Maximum error in B estimated to be ± 3%.

T	B_{12}
303.69	-1019
313.00	- 950

METHANOL + BENZENE $CH_3OH + C_6H_6$

1. D.H. Knoebel and W.C. Edmister, J. chem. Engng Data <u>13</u> 312 (1968).
    Low pressure PVT measurements on several mixtures. Estimated
    standard deviation in B_{12} is ± 52.

T	B_{12}	T	B_{12}
313.15	-545	353.15	-319
333.15	-507	373.15	-281

METHANOL + MERCURY $CH_3OH + Hg$

1. H.S. Rosenberg and W.B. Kay, J. phys. Chem., Ithaca <u>78</u> 186 (1974).
    Solubility measurements at pressures up to 30 atm.

T	B_{12}
493.2	-126

513.2	-120
533.2	-112
553.2	-114
573.2	-110

METHANOL + NITROGEN $CH_3OH + N_2$

1. B. Hemmaplardh and A.D. King, Jr., J. phys. Chem., Ithaca 76 2170 (1972).

    B_{12} derived from measurements of solubility of methanol in the compressed gas.

T	B_{12}	T	B_{12}
288.15	-91 ± 3	323.15	-67 ± 3
298.15	-81 ± 4	333.15	-64 ± 2
310.15	-72 ± 2		

2. P. Neogi and A.P. Kudchadker, J.C.S. Faraday 1 73 385 (1977).

    Chromatographic measurements using di-isodecyl phthalate, P → 2.5 atm. Values of B_{12} read from graph.

T	B_{12}	T	B_{12}
303.2	-75.5 ± 17	333.2	-60.0 ± 17
318.2	-69.5 ± 17	348.2	-53.0 ± 17

METHANOL + NITROUS OXIDE $CH_3OH + N_2O$

1. B. Hemmaplardh and A.D. King, Jr., J. phys. Chem., Ithaca 76 2170 (1972).

    B_{12} derived from measurements of solubility of methanol in the compressed gas.

T	B_{12}	T	B_{12}
288.15	-253 ± 5	323.15	-176 ± 7
298.15	-231 ± 8	333.15	-161 ± 5
310.15	-194 ± 7		

CARBON MONOXIDE + CARBON DIOXIDE CO + CO_2

1. T.L. Cottrell, R.A. Hamilton and R.P. Taubinger, Trans. Faraday Soc. 52 1310 (1956).

Gas expansions at P < 1 atm, X_2 is 0.5.

T	B_{11}	B_{22}	B_{12}
303.2	-7.3	-119.2	-34.6 ± 4.6
333.2	-1.6	- 97.1	-36.6 ± 4.0
363.2	+3.0	- 79.6	-25.7 ± 6.0

CARBON MONOXIDE + ETHYLENE   CO + $CH_2:CH_2$

1. D. McA. Mason and B.E. Eakin, J. chem. Engng Data 6 499 (1961) (*).

T	B_{12}
288.70	-32.2 ± 5

CARBON MONOXIDE + PROPENE (PROPYLENE)   CO + $CH_2:CH.CH_3$

1. D. McA. Mason and B.E. Eakin, J. chem. Engng Data 6 499 (1961) (*).

T	B_{12}
288.7	-62.9 ± 5

CARBON MONOXIDE + BENZENE   CO + C_6H_6

1. J.F. Connolly, Physics Fluids 7 1023 (1964).
    Solubility measurements of benzene in the compressed gas;   P → 50 atm.

T	B_{12}
323.15	-114.4, -114.4
373.15	- 82.6, - 81.0

Compressibility measurements on five mixtures.

T	B_{12}	T	B_{12}
493.2	-35	553.2	-17
513.2	-29	573.2	-13
533.2	-23		

2. D.H. Everett, B.W. Gainey and C.L. Young, Trans. Faraday Soc. 64 2667 (1968).

    Gas chromatographic measurements, using squalane.

T	B_{12}
323.15	-113 ± 8

Gas chromatographic measurements, using dinonyl phthalate.

T	B_{12}
323.15	-122 ± 8

CARBON MONOXIDE + n-OCTANE $CO + CH_3 \cdot (CH_2)_6 \cdot CH_3$

1. J.F. Connolly, Physics Fluids $\underline{7}$ 1023 (1964).

    Solubility measurements of n-octane in the compressed gas;
    $P \rightarrow 50$ atm.

T	B_{12}
373.15	-105.8
373.15	-103.3

Compressibility measurements on five mixtures.

T	B_{12}
533.2	-17
553.2	-11
573.2	$- 6$

2. D.H. Everett, B.W. Gainey and C.L. Young, Trans. Faraday Soc.
    $\underline{64}$ 2667 (1968).

    Gas chromatographic measurements.

T	B_{12}
313.15	-153 ± 8
338.20	-123 ± 8

CARBON MONOXIDE + HYDROGEN $CO + H_2$

1. G.A. Scott, Proc. R. Soc. $\underline{A125}$ 330 (1929) (*).

    PVTx data given for x_{H_2} = 0.33, 0.52 and 0.66; $P \rightarrow 170$ atm.

T	B_{12}
298.2	11.2, 11.4, 13.9

2. D.T.A. Townend and L.A. Bhatt, Proc. R. Soc. A134 502 (1931) (*).
PVTx data for three mixtures; $P \rightarrow 600$ atm.

T	B_{12}
273.2	13.5, 11.0, 12.6
298.2	13.6, 13.5, 14.0

3. A. Van Itterbeek and W. Van Doninck, Proc. phys. Soc. B62 62 (1949).
Velocity of sound measurements.

4. Z. Dokoupil, G. Van Soest, and M.D.P. Swenker, Appl. sci. Res. A5 182 (1955).
Solid-vapour equilibrium measurements; $P \rightarrow 50$ atm for T range 32 - 70K.

5. J. Reuss and J.J.M. Beenakker, Physica, 's Grav. 22 869 (1956).
B_{12} determined from influence of pressure on vapour-solid equilibrium.

T	B_{12}	T	B_{12}
36	-256	52	-133
40	-213	56	-117
44	-179	60	-108
48	-153		

CARBON DIOXIDE + CARBON DISULPHIDE $CO_2 + CS_2$

1. A. Eucken and F. Bresler, Z. phys. Chem. 134 230 (1928) (*).

T	B_{12}
273.2	-374

CARBON DIOXIDE + ETHYLENE $CO_2 + CH_2:CH_2$

1. A.E. Edwards and W.E. Roseveare, J. Am. chem. Soc. 64 2816 (1942) (*).
Measurement of volume change on mixing gases at constant pressure of 1 atm, and 0.5 atm.

T	B_{11}	B_{22}	B_{12}
298.2	-117.6	-140.0	-125.2

2. A. Charnley, J.S. Rowlinson, J.R. Sutton and J.R. Townley, Proc. R. Soc. A230 354 (1955).

    Values of isothermal Joule-Thomson coefficient at zero pressure given, T range 273 - 318 K.

3. P.S. Ku and B.F. Dodge, J. chem. Engng Data 12 158 (1967).
    3-term of PVT data; P → 260 atm.

T	B_{12}
373.2	-93

4. A. Sass, B.F. Dodge, and R.H. Bretton, J. chem. Engng Data 12 168 (1967) (*).

    4 to 8 term fits of PVT data; P range 5 - 500 atm.

T	B_{12}
348.2	-77.1
373.2	-71.4
398.2	-64.0

CARBON DIOXIDE + 1,1-DICHLOROETHANE CO_2 + $CHCl_2.CH_3$

1. S.T. Sie, W. Van Beersum, and G.W.A. Rijnders, Sep. Sci. 1 459 (1966).

    Chromatographic method; P → 7 atm.

T	B_{12}
313.2	-215

CARBON DIOXIDE + NITROETHANE CO_2 + $C_2H_5NO_2$

1. M.S. Vigdergauz and V.I. Semkin, Russ. J. phys. Chem. 45 518 (1971); Zh. fiz. Khim. 45 931 (1971). Value also given by M. Vigdergauz and V. Semkin, J. Chromat. 58 95 (1971).

    Chromatographic measurements using dinonyl phthalate.

    Class III.

T	B_{12}
353.2	-247

CARBON DIOXIDE + ETHANE CO_2 + $CH_3.CH_3$

1. H.H. Reamer, R.H. Olds, B.H. Sage, and W.N. Lacey, Ind. Engng Chem. ind. Edn 37 688 (1945).

    Volumetric behaviour of five mixtures from 310 - 500K; P → 670 atm.

2. D. McA. Mason and B.E. Eakin, J. chem. Engng Data 6 499 (1961) (*).

T	B_{12}
288.7	-124.2 ± 5

CARBON DIOXIDE + ETHANOL CO_2 + $CH_3.CH_2OH$

1. S.T. Sie, W. Van Beersum, and G.W.A. Rijnders, Sep. Sci. 1 459 (1966).

   Chromatographic method;   P → 7 atm.

T	B_{12}
313.2	-303

2. M.S. Vigdergauz and V.I. Semkin, Russ. J. phys. Chem. 45 518 (1971); Zh. fiz. Khim. 45 931 (1971). Value also given by M. Vigdergauz and V. Semkin, J. Chromat. 58 95 (1971).

   Chromatographic measurements using dinonyl phthalate.

   Class III.

T	B_{12}
353.2	-95

3. S.K. Gupta, R.D. Lesslie and A.D. King, Jr., J. phys. Chem., Ithaca 77 2011 (1973).

   B_{12} derived from measurements of solubility of ethanol in the compressed gas.

T	B_{12}
298.15	-307 ± 6
323.15	-238 ± 3
348.15	-196 ± 3

CARBON DIOXIDE + ACETONE CO_2 + $CH_3.CO.CH_3$

1. S.T. Sie, W. Van Beersum, and G.W.A. Rijnders, Sep. Sci. 1 459 (1966).

   Chromatographic method;   P → 7 atm.

T	B_{12}
313.2	-342

2. M.S. Vigdergauz and V.I. Semkin, Russ. J. phys. Chem. 45 518 (1971); Zh. fiz. Khim. 45 931 (1971). Value also given by M. Vigdergauz and V. Semkin, J. Chromat. 58 95 (1971).

Chromatographic measurements using dinonyl phthalate.

Class III.

T	B_{12}
353.2	-129

CARBON DIOXIDE + PROPANE $CO_2 + CH_3.CH_2.CH_3$

1. H.H. Reamer, B.H. Sage and W.N. Lacey, Ind. Engng Chem. ind. Edn 43 2515 (1951).

   Volumetric behaviour of four mixtures from 280 - 500K; $P \to 670$ atm.

2. D. McA. Mason and B.E. Eakin, J. chem. Engng Data 6 499 (1961) (*).

T	B_{12}
288.7	-183.1 ± 5

3. S.T. Sie, W. Van Beersum, and G.W.A. Rijnders, Sep. Sci. 1 459 (1966).

   Chromatographic method; $P \to 7$ atm.

T	B_{12}
313.2	- 99

4. J. Bougard and R. Jadot, J. Chim. phys. 73 415 (1976).

   Quoted accuracy of B_{12} is ± 4%.

T	B_{12}
300.5	-159.31

5. R.D. Gunn, M.S. Thesis, University of California, Berkeley (1958).

   B_{12} calculated from volumetric data given by B.H. Sage and W.N. Lacey, Some properties of the lighter hydrocarbons, Amer. Petrol. Inst., New York (1955).

T	B_{12}	T	B_{12}
310.9	-154.5	444.3	- 61.2
344.3	-128.7	477.6	- 51.0
377.6	- 94.3	510.9	- 42.0

346

CARBON DIOXIDE + ISOPROPANOL CO_2 + $(CH_3)_2CHOH$

1. M.S. Vigdergauz and V.I. Semkin, Russ. J. phys. Chem. 45 518 (1971);
   Zh. fiz. Khim. 45 931 (1971). Value also given by M. Vigdergauz and
   V. Semkin, J. Chromat. 58 95 (1971).

   Chromatographic measurements using dinonyl phthalate.

   Class III.

T	B_{12}
353.2	-127

CARBON DIOXIDE + 1,3-BUTADIENE CO_2 + $CH_2:CH.CH:CH_2$

1. S.T. Sie, W. Van Beersum and G.W.A. Rijnders, Sep. Sci. 1 459 (1966).
   Chromatographic method; P → 7 atm.

T	B_{12}
313.2	-155

CARBON DIOXIDE + METHYL ETHYL KETONE CO_2 + $CH_3.CO.CH_2.CH_3$

1. M.S. Vigdergauz and V.I. Semkin, Russ. J. phys. Chem. 45 518 (1971);
   Zh. fiz. Khim. 45 931 (1971). Value also given by M. Vigdergauz and
   V. Semkin, J. Chromat. 58 95 (1971).

   Chromatographic measurements using dinonyl phthalate.

   Class III.

T	B_{12}
353.2	-184

CARBON DIOXIDE + DIOXAN CO_2 + $CH_2.CH_2.O.(CH_2)_2.O$

1. M.S. Vigdergauz and V.I. Semkin, Russ. J. phys. Chem. 45 518 (1971);
   Zh. fiz. Khim. 45 931 (1971). Value also given by M. Vigdergauz and
   V. Semkin, J. Chromat. 58 95 (1971).

   Chromatographic measurements using dinonyl phthalate.

   Class III.

T	B_{12}
353.2	-136

CARBON DIOXIDE + n-BUTANE $CO_2 + CH_3 \cdot (CH_2)_2 \cdot CH_3$

1. D. McA. Mason and B.E. Eakin, J. chem. Engng Data $\underline{6}$ 499 (1961) (*).

T	B_{12}
288.7	-232.8 ± 5

2. S.T. Sie, W. Van Beersum, and G.W.A. Rijnders, Sep. Sci. $\underline{1}$ 459 (1966).

    Chromatographic method; $P \rightarrow 7$ atm.

T	B_{12}
313.2	-153

3. R.D. Gunn, M.S. Thesis, University of California, Berkeley (1958).
    B_{12} calculated from volumetric data given by B.H. Sage and W.N. Lacey, Some properties of the lighter hydrocarbons, Amer. Petrol. Inst., New York (1955).

T	B_{12}	T	B_{12}
377.6	-130.6	444.3	-80.6
410.9	-100.3	477.6	-65.0

CARBON DIOXIDE + 1-BUTANOL $CO_2 + CH_3 \cdot (CH_2)_2 CH_2OH$

1. R. Massoudi and A.D. King, Jr., J. phys. Chem., Ithaca $\underline{77}$ 2016 (1973).
    B_{12} derived from measurements of solubility of 1-butanol in the compressed gas.

T	B_{12}
298.15	-414 ± 14

CARBON DIOXIDE + DIETHYL ETHER $CO_2 + (C_2H_5)O$

1. R. Massoudi and A.D. King, Jr., J. phys. Chem., Ithaca $\underline{77}$ 2016 (1973).
    B_{12} derived from measurements of solubility of diethyl ether in the compressed gas.

T	B_{12}
298.15	-491 ± 23

CARBON DIOXIDE + PYRIDINE CO_2 + $\underline{CH:CH.CH:CH.CH:N}$

1. M.S. Vigdergauz and V.I. Semkin, Russ. J. phys. Chem. $\underline{45}$ 518 (1971);
    Zh. fiz. Khim. $\underline{45}$ 931 (1971). Value also given by M. Vigdergauz and
    V. Semkin, J. Chromat. $\underline{58}$ 95 (1971).

    Chromatographic measurements using dinonyl phthalate.

    Class III.

T	B_{12}
353.2	-207

CARBON DIOXIDE + CYCLOPENTANE CO_2 + $\underline{CH_2 \cdot (CH_2)_3 \cdot CH_2}$

1. D.H. Desty, A. Goldup, G.R. Luckhurst, and W.T. Swanton, Gas
    Chromatography, p.76, table 3, Butterworths, London, 1962. (Ed.
    M. van Swaay).

    Gas chromatographic measurements using squalane; P → 5 atm.

    Maximum error in B_{12} estimated to be ± 30.

T	B_{12}
298.2	-197

CARBON DIOXIDE + n-PENTANE CO_2 + $CH_3 \cdot (CH_2)_3 \cdot CH_3$

1. D.H. Desty, A. Goldup, G.R. Luckhurst, and W.T. Swanton, Gas
    Chromatography, p.76, table 3, Butterworths, London, 1962. (Ed.
    M. van Swaay).

    Gas chromatographic measurements using squalane; P → 5 atm.

    Maximum error in B_{12} estimated to be ± 30.

T	B_{12}
298.2	-173

2. S.T. Sie, W. Van Beersum, and G.W.A. Rijnders, Sep. Sci. $\underline{1}$ 459
    (1966).

    Chromatographic method; P → 7 atm.

T	B_{12}
313.2	-198

3. M.S. Vigdergauz and V.I. Semkin, Russ. J. phys. Chem. $\underline{45}$ 518 (1971);
    Zh. fiz. Khim. $\underline{45}$ 931 (1971). Value also given by M. Vigdergauz and
    V. Semkin, J. Chromat. $\underline{58}$ 95 (1971).

    Chromatographic measurements using dinonyl phthalate.

Class III.

T	B_{12}
353.2	-76

4. R. Massoudi and A.D. King, Jr., J. phys. Chem., Ithaca <u>77</u> 2016 (1973).

B_{12} derived from measurements of solubility of n pentane in the compressed gas.

T	B_{12}
298.15	-273 ± 23

CARBON DIOXIDE + ISOPENTANE $CO_2 + (CH_3)_2CH.CH_2.CH_3$

1. D.H. Desty, A. Goldup, G.R. Luckhurst, and W.T. Swanton, Gas Chromatography, p.76, table 3, Butterworths, London, 1962. (Ed. M. van Swaay).

Gas chromatographic measurements using squalane; P → 5 atm.

Maximum error in B_{12} estimated to be ± 30.

T	B_{12}
298.2	-163

CARBON DIOXIDE + BENZENE $CO_2 + C_6H_6$

1. D.H. Desty, A. Goldup, G.R. Luckhurst, and W.T. Swanton, Gas Chromatography, p.76, table 3, Butterworths, London, 1962. (Ed. M. van Swaay).

Gas chromatographic measurements using squalane; P → 5 atm.

Maximum error in B_{12} estimated to be ± 30.

T	B_{12}
298.2	-251

2. S.T. Sie, W. Van Beersum, and G.W.A. Rijnders, Sep. Sci. <u>1</u> 459 (1966).

Chromatographic method; P → 7 atm.

T	B_{12}
313.2	-288

3. M.S. Vigdergauz and V.I. Semkin, Russ. J. phys. Chem. 45 518 (1971); Zh. fiz. Khim. 45 931 (1971). Value also given by M. Vigdergauz and V. Semkin, J. Chromat. 58 95 (1971).

Chromatographic measurements using dinonyl phthalate.

Class III.

T	B_{12}
353.2	-216

4. A.J.B. Cruickshank, B.W. Gainey, C.P. Hicks, T.M. Letcher, R.W. Moody and C.L. Young, Trans. Faraday Soc. 65 1014 (1969).

Gas chromatographic measurements.

T	B_{12}
323.2	-250 ± 15

CARBON DIOXIDE + CYCLOHEXANE $CO_2 + CH_2 \cdot (CH_2)_4 \cdot CH_2$

1. D.H. Desty, A. Goldup, G.R. Luckhurst, and W.T. Swanton, Gas Chromatography, p.76, table 3, Butterworths, London, 1962. (Ed. M. van Swaay).

Gas chromatographic measurements using squalane; $P \to 5$ atm.

Maximum error in B_{12} estimated to be ± 30.

T	B_{12}
298.2	-236

2. M.S. Vigdergauz and V.I. Semkin, Russ. J. phys. Chem. 45 518 (1971); Zh. fiz. Khim. 45 931 (1971). Value also given by M. Vigdergauz and V. Semkin, J. Chromat. 58 95 (1971).

Chromatographic measurements using dinonyl phthalate.

Class III.

T	B_{12}
353.2	-163

CARBON DIOXIDE + METHYLCYCLOPENTANE $CO_2 + CH(CH_3) \cdot (CH_2)_3 \cdot CH_2$

1. D.H. Desty, A. Goldup, G.R. Luckhurst, and W.T. Swanton, Gas Chromatography, p.76, table 3, Butterworths, London, 1962. (Ed. M. van Swaay).

Gas chromatographic measurements using squalane; $P \to 5$ atm.

Maximum error in B_{12} estimated to be ± 30.

T	B_{12}
298.2	-234

CARBON DIOXIDE + BUTYL ACETATE $CO_2 + CH_3.COO.(CH_2)_3.CH_3$

1. M.S. Vigdergauz and V.I. Semkin, Russ. J. phys. Chem. 45 518 (1971);
    Zh. fiz. Khim. 45 931 (1971). Value also given by M. Vigdergauz and
    V. Semkin, J. Chromat. 58 95 (1971).

    Chromatographic measurements using dinonyl phthalate.

    Class III.

T	B_{12}
353.2	-260

CARBON DIOXIDE + n-HEXANE $CO_2 + CH_3.(CH_2)_4.CH_3$

1. D.H. Desty, A. Goldup, G.R. Luckhurst, and W.T. Swanton, Gas
    Chromatography, p.76, table 3, Butterworths, London, 1962. (Ed.
    M. van Swaay).

    Gas chromatographic measurements using squalane; P → 5 atm.

    Maximum error in B_{12} estimated to be ± 30.

T	B_{12}
298.2	-233

2. M.S. Vigdergauz and V.I. Semkin, Russ. J. phys. Chem. 45 518 (1971);
    Zh. fiz. Khim. 45 931 (1971). Value also given by M. Vigdergauz and
    V. Semkin, J. Chromat. 58 95 (1971).

    Chromatographic measurements using dinonyl phthalate.

    Class III.

T	B_{12}
353.2	-147

CARBON DIOXIDE + 2- METHYLPENTANE $CO_2 + (CH_3)_2.CH.(CH_2)_2.CH_3$

1. D.H. Desty, A. Goldup, G.R. Luckhurst, and W.T. Swanton, Gas
    Chromatography, p.76, table 3, Butterworths, London, 1962. (Ed.
    M. van Swaay).

    Gas chromatographic measurements using squalane; P → 5 atm.

    Maximum error in B_{12} estimated to be ± 30.

T	B_{12}
298.2	-206

CARBON DIOXIDE + 3-METHYLPENTANE CO_2 + $CH_3.CH_2.CH(CH_3).CH_2.CH_3$

1. D.H. Desty, A. Goldup, G.R. Luckhurst, and W.T. Swanton, Gas Chromatography, p.76, table 3, Butterworths, London, 1962. (Ed. M. van Swaay).

    Gas chromatographic measurements using squalane; P → 5 atm.

    Maximum error in B_{12} estimated to be ± 30.

T	B_{12}
298.2	-211

CARBON DIOXIDE + 2,2-DIMETHYLBUTANE CO_2 + $(CH_3)_3.C.CH_2.CH_3$

1. D.H. Desty, A. Goldup, G.R. Luckhurst, and W.T. Swanton, Gas Chromatography, p.76, table 3, Butterworths, London, 1962. (Ed. M. van Swaay).

    Gas chromatographic measurements using squalane; P → 5 atm.

    Maximum error in B_{12} estimated to be ± 30.

T	B_{12}
298.2	-168

CARBON DIOXIDE + 2,3-DIMETHYLBUTANE CO_2 + $(CH_3)_2.CH.CH.(CH_3)_2$

1. D.H. Desty, A. Goldup, G.R. Luckhurst, and W.T. Swanton, Gas Chromatography, p.76, table 3, Butterworths, London, 1962. (Ed. M. van Swaay).

    Gas chromatographic measurements using squalane; P → 5 atm.

    Maximum error in B_{12} estimated to be ± 30.

T	B_{12}
298.2	-203

CARBON DIOXIDE + TOLUENE CO_2 + $C_6H_5.CH_3$

1. J.M. Prausnitz and P.R. Benson, A. I. Ch. E. Jl 5 161 (1959).
    Vapour phase solubility measurements.

T	B_{12}
323.15	-254 ± 8
348.15	-215 ± 9

2. M.S. Vigdergauz and V.I. Semkin, Russ. J. phys. Chem. <u>45</u> 518 (1971); Zh. fiz. Khim. <u>45</u> 931 (1971). Value also given by M. Vigdergauz and V. Semkin, J. Chromat. <u>58</u> 95 (1971).

   Chromatographic measurements using dinonyl phthalate.

   Class III.

T	B_{12}
353.2	-235

3. M. Vigdergauz and V. Semkin, J. Chromat. <u>58</u> 95 (1971).

   B_{12} determined from chromatographic retention data with reference B_{12} (353.2K) values of -216 for carbon dioxide + benzene, -300 for carbon dioxide + styrene.

   Class III.

T	B_{12}
353.2	-248

CARBON DIOXIDE + n-HEPTANE $CO_2 + CH_3 \cdot (CH_2)_5 \cdot CH_3$

1. D.H. Desty, A. Goldup, G.R. Luckhurst, and W.T. Swanton, Gas Chromatography, p.76, table 3, Butterworths, London, 1962. (Ed. M. van Swaay).

   Gas chromatographic measurements using squalane; $P \to 5$ atm.

   Maximum error in B_{12} estimated to be $\pm$ 30.

T	B_{12}
298.2	-294

2. M.S. Vigdergauz and V.I. Semkin, Russ. J. phys. Chem. <u>45</u> 518 (1971); Zh. fiz. Khim. <u>45</u> 931 (1971). Value also given by M. Vigdergauz and V. Semkin, J. Chromat. <u>58</u> 95 (1971).

   Chromatographic measurements using dinonyl phthalate.

   Class III.

T	B_{12}
353.2	-177

354

CARBON DIOXIDE + 2-METHYLHEXANE CO_2 + $(CH_3)_2.CH.(CH_2)_3.CH_3$

1. D.H. Desty, A. Goldup, G.R. Luckhurst, and W.T. Swanton, Gas
    Chromatography, p.76, table 3, Butterworths, London, 1962. (Ed.
    M. van Swaay).

    Gas chromatographic measurements using squalane; P → 5 atm.

    Maximum error in B_{12} estimated to be ± 30.

T	B_{12}
298.2	-272

CARBON DIOXIDE + 3-METHYLHEXANE CO_2 + $CH_3.CH_2.CH(CH_3).(CH_2)_2.CH_3$

1. D.H. Desty, A. Goldup, G.R. Luckhurst, and W.T. Swanton, Gas
    Chromatography, p.76, table 3, Butterworths, London, 1962. (Ed.
    M. van Swaay).

    Gas chromatographic measurements using squalane; P → 5 atm.

    Maximum error in B_{12} estimated to be ± 30.

T	B_{12}
298.2	-276

CARBON DIOXIDE + 3-ETHYLPENTANE CO_2 + $(CH_3.CH_2)_3CH$

1. D.H. Desty, A. Goldup, G.R. Luckhurst, and W.T. Swanton, Gas
    Chromatography, p.76, table 3, Butterworths, London, 1962. (Ed.
    M. van Swaay).

    Gas chromatographic measurements using squalane; P → 5 atm.

    Maximum error in B_{12} estimated to be ± 30.

T	B_{12}
298.2	-292

CARBON DIOXIDE + 2,2-DIMETHYLPENTANE CO_2 + $(CH_3)_3.C.(CH_2)_2.CH_3$

1. D.H. Desty, A. Goldup, G.R. Luckhurst, and W.T. Swanton, Gas
    Chromatography, p.76, table 3, Butterworths, London, 1962. (Ed.
    M. van Swaay).

    Gas chromatographic measurements using squalane; P → 5 atm.

    Maximum error in B_{12} estimated to be ± 30.

T	B_{12}
298.2	-245

CARBON DIOXIDE + 2,3-DIMETHYLPENTANE CO_2 + $(CH_3)_2.CH.CH(CH_3).CH_2.CH_3$

1. D.H. Desty, A. Goldup, G.R. Luckhurst, and W.T. Swanton, Gas
    Chromatography, p.76, table 3, Butterworths, London, 1962. (Ed.
    M. van Swaay).

    Gas chromatographic measurements using squalane; $P \to 5$ atm.

    Maximum error in B_{12} estimated to be ± 30.

T	B_{12}
298.2	-264

CARBON DIOXIDE + 2,4-DIMETHYLPENTANE CO_2 + $(CH_3)_2.CH.CH_2.CH.(CH_3)_2$

1. D.H. Desty, A. Goldup, G.R. Luckhurst, and W.T. Swanton, Gas
    Chromatography, p.76, table 3, Butterworths, London, 1962. (Ed.
    M. van Swaay).

    Gas chromatographic measurements using squalane; $P \to 5$ atm.

    Maximum error in B_{12} estimated to be ± 30.

T	B_{12}
298.2	-249

CARBON DIOXIDE + 3,3-DIMETHYLPENTANE CO_2 + $CH_3.CH_2.C(CH_3)_2.CH_2.CH_3$

1. D.H. Desty, A. Goldup, G.R. Luckhurst, and W.T. Swanton, Gas
    Chromatography, p.76, table 3, Butterworths, London, 1962. (Ed.
    M. van Swaay).

    Gas chromatographic measurements using squalane; $P \to 5$ atm.

    Maximum error in B_{12} estimated to be ± 30.

T	B_{12}
298.2	-242

CARBON DIOXIDE + 2,2,3-TRIMETHYLBUTANE CO_2 + $(CH_3)_3.C.CH.(CH_3)_2$

1. D.H. Desty, A. Goldup, G.R. Luckhurst, and W.T. Swanton, Gas
    Chromatography, p.76, table 3, Butterworths, London, 1962. (Ed.
    M. van Swaay).

    Gas chromatographic measurements using squalane; $P \to 5$ atm.

    Maximum error in B_{12} estimated to be ± 30.

T	B_{12}
298.2	-222

CARBON DIOXIDE + STYRENE CO_2 + $C_6H_5.CH:CH_2$

1. M.S. Vigdergauz and V.I. Semkin, Russ. J. phys. Chem. 45 518 (1971);
    Zh. fiz. Khim. 45 931 (1971). Value also given by M. Vigdergauz and
    V. Semkin, J. Chromat. 58 95 (1971).

    Chromatographic measurements using dinonyl phthalate.

    Class III.

T	B_{12}
353.2	-300

CARBON DIOXIDE + ETHYLBENZENE CO_2 + $C_6H_5.C_2H_5$

1. M. Vigdergauz and V. Semkin, J. Chromat. 58 95 (1971).

    B_{12} determined from chromatographic retention data with reference
    B_{12} (353.2K) values of -216 for carbon dioxide + benzene, -300 for
    carbon dioxide + styrene.

    Class III.

T	B_{12}
353.2	-271

CARBON DIOXIDE + o-XYLENE CO_2 + $C_6H_4.(CH_3)_2$

1. M. Vigdergauz and V. Semkin, J. Chromat. 58 95 (1971).

    B_{12} determined from chromatographic retention data with reference
    B_{12} (353.2K) values of -216 for carbon dioxide + benzene, -300 for
    carbon dioxide + styrene.

    Class III.

T	B_{12}
353.2	-289

CARBON DIOXIDE + m-XYLENE CO_2 + $C_6H_4.(CH_3)_2$

1. M. Vigdergauz and V. Semkin, J. Chromat. 58 95 (1971).

    B_{12} determined from chromatographic retention data with reference
    B_{12} (353.2K) values of -216 for carbon dioxide + benzene, -300 for
    carbon dioxide + styrene.

    Class III.

T	B_{12}
353.2	-282

CARBON DIOXIDE + p-XYLENE $CO_2 + C_6H_4 \cdot (CH_3)_2$

1. M. Vigdergauz and V. Semkin, J. Chromat. $\underline{58}$ 95 (1971).

   B_{12} determined from chromatographic retention data with reference B_{12} (353.2K) values of -216 for carbon dioxide + benzene, -300 for carbon dioxide + styrene.

   Class III.

T	B_{12}
353.2	-284

CARBON DIOXIDE + n-OCTANE $CO_2 + CH_3 \cdot (CH_2)_6 \cdot CH_3$

1. M.S. Vigdergauz and V.I. Semkin, Russ. J. phys. Chem. 45 518 (1971); Zh. fiz. Khim. $\underline{45}$ 931 (1971). Value also given by M. Vigdergauz and V. Semkin, J. Chromat. $\underline{58}$ 95 (1971).

   Chromatographic measurements using dinonyl phthalate.

   Class III.

T	B_{12}
353.2	-227

CARBON DIOXIDE + ISOOCTANE (2,2,4-TRIMETHYLPENTANE) $CO_2 + (CH_3)_3 \cdot C.CH_2 \cdot CH.(CH_3)_2$

1. J.M. Prausnitz and P.R. Benson, A. I. Ch. E. Jl $\underline{5}$ 161 (1959).

   Vapour phase solubility measurements.

T	B_{12}
323.15	-303 ± 8
348.15	-252 ± 9

2. M.S. Vigdergauz and V.I. Semkin, Russ. J. phys. Chem. $\underline{45}$ 518 (1971); Zh. fiz. Khim. $\underline{45}$ 931 (1971). Value also given by M. Vigdergauz and V. Semkin, J. Chromat. $\underline{58}$ 95 (1971).

   Chromatographic measurements using dinonyl phthalate.

   Class III.

T	B_{12}
353.2	-147

CARBON DIOXIDE + n-PROPYLBENZENE CO_2 + $C_6H_5 \cdot CH_2 \cdot CH_2 \cdot CH_3$

1. M. Vigdergauz and V. Semkin, J. Chromat. <u>58</u> 95 (1971).

    B_{12} determined from chromatographic retention data with reference B_{12} (353.2K) values of -216 for carbon dioxide + benzene, -300 for carbon dioxide + styrene.

    Class III.

T	B_{12}
353.2	-292

CARBON DIOXIDE + ISOPROPYLBENZENE CO_2 + $C_6H_5 \cdot CH \cdot (CH_3)_2$

1. M. Vigdergauz and V. Semkin, J. Chromat. <u>58</u> 95 (1971).

    B_{12} determined from chromatographic retention data with reference B_{12} (353.2K) values of -216 for carbon dioxide + benzene, -300 for carbon dioxide + styrene.

    Class III.

T	B_{12}
353.2	-286

CARBON DIOXIDE + 1-METHYL-2-ETHYLBENZENE CO_2 + $C_6H_4 \cdot (CH_3)(CH_2 \cdot CH_3)$

1. M. Vigdergauz and V. Semkin, J. Chromat. <u>58</u> 95 (1971).

    B_{12} determined from chromatographic retention data with reference B_{12} (353.2K) values of -216 for carbon dioxide + benzene, -300 for carbon dioxide + styrene.

    Class III.

T	B_{12}
353.2	-310

CARBON DIOXIDE + 1-METHYL-3-ETHYLBENZENE CO_2 + $C_6H_4 \cdot (CH_3)(CH_2 \cdot CH_3)$

1. M. Vigdergauz and V. Semkin, J. Chromat. <u>58</u> 95 (1971).

B_{12} determined from chromatographic retention data with reference B_{12} (353.2K) values of -216 for carbon dioxide + benzene, -300 for carbon dioxide + styrene.

Class III.

T	B_{12}
353.2	-309

CARBON DIOXIDE + 1,2,3-TRIMETHYLBENZENE $CO_2 + C_6H_3 \cdot (CH_3)_3$

1. M. Vigdergauz and V. Semkin, J. Chromat. <u>58</u> 95 (1971).

B_{12} determined from chromatographic retention data with reference B_{12} (353.2K) values of -216 for carbon dioxide + benzene, -300 for carbon dioxide + styrene.

Class III.

T	B_{12}
353.2	-335

CARBON DIOXIDE + 1,2,4 TRIMETHYLBENZENE $CO_2 + C_6H_3 \cdot (CH_3)_3$

1. M. Vigdergauz and V. Semkin, J. Chromat. <u>58</u> 95 (1971).

B_{12} determined from chromatographic retention data with reference B_{12} (353.2K) values of -216 for carbon dioxide + benzene, -300 for carbon dioxide + styrene.

Class III.

T	B_{12}
353.2	-340

CARBON DIOXIDE + 1,3,5-TRIMETHYLBENZENE $CO_2 + C_6H_3 \cdot (CH_3)_3$

1. M. Vigdergauz and V. Semkin, J. Chromat. <u>58</u> 95 (1971).

B_{12} determined from chromatographic retention data with reference B_{12} (353.2K) values of -216 for carbon dioxide + benzene, -300 for carbon dioxide + styrene.

Class III.

T	B_{12}
353.2	-325

CARBON DIOXIDE + n-NONANE $CO_2 + CH_3.(CH_2)_7.CH_3$

1. M.S. Vigdergauz and V.I. Semkin, Russ. J. phys. Chem. 45 518 (1971);
    Zh. fiz. Khim. 45 931 (1971). Value also given by M. Vigdergauz and
    V. Semkin, J. Chromat. 58 95 (1971).

    Chromatographic measurements using dinonyl phthalate.

    Class III.

T	B_{12}
353.2	-249

CARBON DIOXIDE +NAPHTHALENE $CO_2 + C_{10}H_8$

1. G.C. Najour and A.D. King, Jr., J. chem. Phys. 45 1915 (1966).

    B_{12} determined from measurements of solubility of naphthalene
    in the compressed gas: P → 130 atm.

T	B_{12}	T	B_{12}
297	-573 ± 12	332	-361 ± 13
299	-552 ± 11	333	-345 ± 13
309	-501 ± 16	337	-346 ± 13
323.5	-401 ± 15	346	-311 ± 13
328	-389 ± 13		

CARBON DIOXIDE + n-BUTYLBENZENE $CO_2 + C_6H_5.CH_2.CH_2.CH_2.CH_3$

1. M. Vigdergauz and V. Semkin, J. Chromat. 58 95 (1971).

    B_{12} determined from chromatographic retention data with reference
    B_{12} (353.2K) values of -216 for carbon dioxide + benzene, -300 for
    carbon dioxide + styrene.

    Class III.

T	B_{12}
353.2	-345

CARBON DIOXIDE + SEC-BUTYLBENZENE $CO_2 + C_6H_5.CH.(CH_3)(CH_2.CH_3)$

1. M. Vigdergauz and V. Semkin, J. Chromat. 58 95 (1971).

    B_{12} determined from chromatographic retention data with reference
    B_{12} (353.2K) values of -216 for carbon dioxide + benzene, -300 for
    carbon dioxide + styrene.

Class III.

T	B_{12}
353.2	-312

CARBON DIOXIDE + 1-METHYL-2-PROPYLBENZENE CO_2 + $C_6H_4 \cdot (CH_3)(CH_2 \cdot CH_2 \cdot CH_3)$

1. M. Vigdergauz and V. Semkin, J. Chromat. **58** 95 (1971).

B_{12} determined from chromatographic retention data with reference B_{12} (353.2K) values of -216 for carbon dioxide + benzene, -300 for carbon dioxide + styrene.

Class III.

T	B_{12}
353.2	-327

CARBON DIOXIDE + 1-METHYL-2-ISOPROPYLBENZENE CO_2 + $C_6H_4 \cdot (CH_3)(CH \cdot (CH_3)_2)$

1. M. Vigdergauz and V. Semkin, J. Chromat. **58** 95 (1971).

B_{12} determined from chromatographic retention data with reference B_{12} (353.2K) values of -216 for carbon dioxide + benzene, -300 for carbon dioxide + styrene.

Class III.

T	B_{12}
353.2	-329

CARBON DIOXIDE + 1-METHYL-3-PROPYLBENZENE CO_2 + $C_6H_4 \cdot (CH_3)(CH_2 \cdot CH_2 \cdot CH_3)$

1. M. Vigdergauz and V. Semkin, J. Chromat. **58** 95 (1971).

B_{12} determined from chromatographic retention data with reference B_{12} (353.2K) values of -216 for carbon dioxide + benzene, -300 for carbon dioxide + styrene.

Class III.

T	B_{12}
353.2	-328

CARBON DIOXIDE + 1-METHYL-4-PROPYLBENZENE CO_2 + $C_6H_4 \cdot (CH_3)(CH_2 \cdot CH_2 \cdot CH_3)$

1. M. Vigdergauz and V. Semkin, J. Chromat. <u>58</u> 95 (1971).

   B_{12} determined from chromatographic retention data with reference B_{12} (353.2K) values of -216 for carbon dioxide + benzene, -300 for carbon dioxide + styrene.

   Class III.

T	B_{12}
353.2	-318

CARBON DIOXIDE + 1,4-DIETHYLBENZENE CO_2 + $C_6H_4 \cdot (CH_2 \cdot CH_3)_2$

1. M. Vigdergauz and V. Semkin, J. Chromat. <u>58</u> 95 (1971).

   B_{12} determined from chromatographic retention data with reference B_{12} (353.2K) values of -216 for carbon dioxide + benzene, -300 for carbon dioxide + styrene.

   Class III.

T	B_{12}
353.2	-342

CARBON DIOXIDE + 1,2-DIMETHYL-3-ETHYLBENZENE CO_2 + $C_6H_3 \cdot (CH_3)_2 (CH_2 \cdot CH_3)$

1. M. Vigdergauz and V. Semkin, J. Chromat. <u>58</u> 95 (1971).

   B_{12} determined from chromatographic retention data with reference B_{12} (353.2K) values of -216 for carbon dioxide + benzene, -300 for carbon dioxide + styrene.

   Class III.

T	B_{12}
353.2	-349

CARBON DIOXIDE + 1,2-DIMETHYL-4-ETHYLBENZENE CO_2 + $C_6H_3 \cdot (CH_3)_2 (CH_2 \cdot CH_3)$

1. M. Vigdergauz and V. Semkin, J. Chromat. <u>58</u> 95 (1971).

   B_{12} determined from chromatographic retention data with reference B_{12} (353.2K) values of -216 for carbon dioxide + benzene, -300 for carbon dioxide + styrene.

   Class III.

T	B_{12}
353.2	-354

CARBON DIOXIDE + 1,3-DIMETHYL-2-ETHYLBENZENE $CO_2 + C_6H_3 \cdot (CH_3)_2 (CH_2 \cdot CH_3)$

1. M. Vigdergauz and V. Semkin, J. Chromat. 58 95 (1971).

   B_{12} determined from chromatographic retention data with reference B_{12} (353.2K) values of -216 for carbon dioxide + benzene, -300 for carbon dioxide + styrene.

   Class III.

T	B_{12}
353.2	-352

CARBON DIOXIDE + 1,3-DIMETHYL-4-ETHYLBENZENE $CO_2 + C_6H_3 \cdot (CH_3)_2 (CH_2 \cdot CH_3)$

1. M. Vigdergauz and V. Semkin, J. Chromat. 58 95 (1971).

   B_{12} determined from chromatographic retention data with reference B_{12} (353.2K) values of -216 for carbon dioxide + benzene, -300 for carbon dioxide + styrene.

   Class III.

T	B_{12}
353.2	-350

CARBON DIOXIDE + 1,3-DIMETHYL-5-ETHYLBENZENE $CO_2 + C_6H_3 \cdot (CH_3)_2 (CH_2 \cdot CH_3)$

1. M. Vigdergauz and V. Semkin, J. Chromat. 58 95 (1971).

   B_{12} determined from chromatographic retention data with reference B_{12} (353.2K) values of -216 for carbon dioxide + benzene, -300 for carbon dioxide + styrene.

   Class III.

T	B_{12}
353.2	-348

CARBON DIOXIDE + 1,4-DIMETHYL-2-ETHYLBENZENE $CO_2 + C_6H_3 \cdot (CH_3)_2 (CH_2 \cdot CH_3)$

1. M. Vigdergauz and V. Semkin, J. Chromat. 58 95 (1971).

   B_{12} determined from chromatographic retention data with reference B_{12} (353.2K) values of -216 for carbon dioxide + benzene, -300 for carbon dioxide + styrene.

   Class III.

T	B_{12}
353.2	-348

CARBON DIOXIDE + 1,2,3,4-TETRAMETHYLBENZENE $CO_2 + C_6H_2 \cdot (CH_3)_4$

1. M. Vigdergauz and V. Semkin, J. Chromat. $\underline{58}$ 95 (1971).

    B_{12} determined from chromatographic retention data with reference
    B_{12} (353.2K) values of -216 for carbon dioxide + benzene, -300 for
    carbon dioxide + styrene.

    Class III.

T	B_{12}
353.2	-386

CARBON DIOXIDE + 1,2,3,5-TETRAMETHYLBENZENE $CO_2 + C_6H_2 \cdot (CH_3)_4$

1. M. Vigdergauz and V. Semkin, J. Chromat. $\underline{58}$ 95 (1971).

    B_{12} determined from chromatographic retention data with reference
    B_{12} (353.2K) values of -216 for carbon dioxide + benzene, -300 for
    carbon dioxide + styrene.

    Class III.

T	B_{12}
353.2	-380

CARBON DIOXIDE + 1,2,4,5-TETRAMETHYLBENZENE $CO_2 + C_6H_2 \cdot (CH_3)_4$

1. M. Vigdergauz and V. Semkin, J. Chromat. $\underline{58}$ 95 (1971).

    B_{12} determined from chromatographic retention data with reference
    B_{12} (353.2K) values of -216 for carbon dioxide + benzene, -300 for
    carbon dioxide + styrene.

    Class III.

T	B_{12}
353.2	-376

CARBON DIOXIDE + n-DECANE $CO_2 + CH_3 \cdot (CH_2)_8 \cdot CH_3$

1. J.M. Prausnitz and P.R. Benson, A. I. Ch. E. Jl $\underline{5}$ 161 (1959).
    Vapour phase solubility measurements.

T	B_{12}
323.15	-417 ± 7
348.15	-321 ± 8

CARBON DIOXIDE + ANTHRACENE CO_2 + $C_6H_4(CH)_2C_6H_4$

1. G.C. Najour and A.D. King, Jr., J. chem. Phys. <u>52</u> 5206 (1970).

    B_{12} determined from measurements of solubility of anthracene in the compressed gas: $P \rightarrow 100$ atm. Probable error in $B \pm 12$.

T	B_{12}	T	B_{12}
338	-540	399	-361
348	-520	407	-352
350	-498	419	-337
355	-466	423	-309
365	-443	449	-273
378	-395		

CARBON DIOXIDE + PHENANTHRENE CO_2 + $C_{14}H_{10}$

1. H. Bradley, Jr., and A.D. King, Jr., J. chem. Phys. <u>52</u> 2851 (1970).

    B_{12} determined from measurements of solubility of phenanthrene in the compressed gas: $P \rightarrow 60$ atm. Probable error in $B \pm 24$.

T	B_{12}	T	B_{12}
312	-709	348	-512
316	-686	350	-502
319	-675	355	-448
321	-694	356	-454
328	-610	362	-439
330	-625	366	-452
334	-628	368	-405
336	-598	381	-406
338	-563	398	-358
343	-562	414	-306
346	-546		

CARBON DIOXIDE + HYDROGEN CO_2 + H_2

1. I. Kritschewsky and V. Markov, Acta phys.-chim. URSS <u>12</u> 59 (1940).

    PVTx data given for three mixtures; $P \rightarrow 500$ atm for T range 273 - 473 K.

2. A.E. Edwards and W.E. Roseveare, J. Am. chem. Soc. <u>64</u> 2816 (1942)(*).

Measurement of volume changes on mixing gases at constant pressure of 1 atm, and 0.5 atm.

T	B_{11}	B_{22}	B_{12}
298.2	-117.6	14.8	-32.9

3. T.L. Cottrell, R.A. Hamilton and R.P. Taubinger, Trans. Faraday Soc. 52 1310 (1956).

Gas expansions at $P \rightarrow 1$ atm, X_2 is 0.5.

T	B_{11}	B_{22}	B_{12}
303.2	-119.2	13.8	-1.0 ± 3.8
333.2	- 97.1	13.9	+0.4 ± 4.4
363.2	- 79.6	14.4	0.0 ± 3.2

CARBON DIOXIDE + WATER $CO_2 + H_2O$

1. F. Pollitzer and E. Strebel, Z. phys. Chem. 110 768 (1924).

T	B_{12}
323	-198
343.2	-209

2. C.R. Coan and A.D. King, Jr., J. Am. chem. Soc., 93 1857 (1971).
B_{12} derived from measurements of solubility of water in the compressed gas. $P \rightarrow 50$ atm.

T	B_{12}	T	B_{12}
298.15	-214 ± 6	348.15	-107 ± 4
323.15	-151 ± 6	373.15	- 89 ± 2

CARBON DIOXIDE + HELIUM $CO_2 + He$

1. A.E. Edwards and W.E. Roseveare, J. Am. chem. Soc. 64 2816 (1942) (*).

Measurement of volume change on mixing gases at constant pressure of 1 atm, and 0.5 atm.

T	B_{11}	B_{22}	B_{12}
298.2	-117.6	11.8	-36.1

2. W.C. Pfefferle, Jr., J. A. Goff and J.G. Miller, J. chem. Phys. 23 509 (1955) (*).

Burnett method: maximum pressure 120 atm.

T	B_{11}	B_{22}	B_{12}
303.2	-117.7	11.84	18.93

3. T.L. Cottrell and R.A. Hamilton, Trans. Faraday Soc. <u>52</u> 156 (1956).
   Gas expansions at $P \rightarrow 1$ atm, X_2 is 0.5.

T	B_{11}	B_{22}	B_{12}
303.2	-119.3	11.7	21.6 ± 3.4
333.2	- 95.5	11.5	24.6 ± 3.0
363.2	- 77.8	11.4	22.6 ± 6.8

4. R.C. Harper, Jr. and J.G. Miller, J. chem. Phys. <u>27</u> 36 (1957).
   Burnett method: maximum pressure 120 atm.

T	B_{11}	B_{22}	B_{12}
303.2	-121.9	11.81	22.4 ± 1.2

5. D.S. Tsiklis, L.R. Linshits and I.B. Rodkina, Russ. J. phys. Chem.
   <u>48</u> 906 (1974); Zh. fiz. Khim. <u>48</u> 1541 (1974).
   Burnett method.

T	B_{12}
373.2	28.1

6. D.S. Tsiklis, L.R. Linshits and I.B. Rodkina, Russ. J. phys. Chem.
   <u>48</u> 908 (1974); Zh. fiz. Khim. <u>48</u> 1544 (1974).
   Burnett method.

T	B_{12}
423.2	28.2

7. L.R. Linshits, I.B. Rodkina and D.S. Tsiklis, Russ. J. phys. Chem.
   <u>49</u> 1258 (1975); Zh. fiz. Khim. 4<u>9</u> 2141 (1975).
   Burnett method.

T	B_{12}
323.2	28.8

CARBON DIOXIDE + NITROGEN CO_2 + N_2

1. I. Kritschewsky and V. Markov, Acta phys.-chim. URSS <u>12</u> 59 (1940).

PVTx data given for three mixtures; $P \rightarrow 500$ atm for T range 273 - 473K.

2. A.E. Markham and K.A. Kobe, J. chem. Phys. 9 438 (1941).

   Percentage volume change measured on mixing gases at 298 K and 1 atm. pressure.

3. A.E. Edwards and W.E. Roseveare, J. Am. chem. Soc. 64 2816 (1942) (*).

   Measurement of volume change on mixing gases at constant pressure of 1 atm, and 0.5 atm.

T	B_{11}	B_{22}	B_{12}
298.2	-117.6	-4.5	-47.5

4. R.E.D. Haney and H. Bliss, Ind. Engng Chem. ind. Edn 36 985 (1944).

   PVT data given. $P \rightarrow 500$ atm, T range 298 - 398 K.

5. R.A. Gorski and J.G. Miller, J. Am. chem. Soc. 75 550 (1953).

   Measurement of volume change on mixing gases at constant pressure, below 1 atm.

T	B_{11}	B_{22}	B_{12}
303.15	-120.7	-4.34	-40.6 ± 0.2

6. W.C. Pfefferle, Jr., J.A. Goff and J.G. Miller, J. chem. Phys. 23 509 (1955) (*).

   Burnett method: maximum pressure 120 atm.

T	B_{11}	B_{22}	B_{12}
303.2	-117.7	-4.17	-40.49

7. T.L. Cottrell, R.A. Hamilton and R.P. Taubinger, Trans. Faraday Soc. 52 1310 (1956).

   Gas expansions at $P \rightarrow 1$ atm, X_2 is 0.5.

T	B_{11}	B_{22}	B_{12}
303.2	-119.2	-4.0	-41.4 ± 6.2
333.2	- 97.1	+1.2	-36.0 ± 4.0
363.2	- 79.6	5.4	-28.5 ± 3.6

8. D. McA. Mason and B.E. Eakin, J. chem. Engng Data $\underline{6}$ 499 (1961) (*).

T	B_{12}
288.7	-42.9 ± 5

9. N.P. Yakimenko, G.M. Glukh, and M.B. Iomtev, Russ. J. phys. Chem. $\underline{51}$ 928 (1977); Zh. fiz. Khim. $\underline{51}$ 1566 (1977).

B_{12} determined from measurements of the solubility of carbon dioxide in the compressed gas at pressures $\rightarrow$ 35 atm. Accuracy in B_{12} estimated to be better than $\pm$ 13.

T	B_{12}	T	B_{12}
110.0	-397	120.0	-278
115.0	-327	125.0	-256

10. R.D. Gunn, M.S. Thesis, University of California, Berkeley (1958).

B_{12} calculated from volumetric data given by W.K. Tang, Tech. Rep. WIS-OOR-13, University of Wisconsin (1956).

T	B_{12}	T	B_{12}
298.15	-44.1	373.15	-21.5
323.15	-33.6	398.15	-17.7
348.15	-27.4		

CARBON DIOXIDE + NITROUS OXIDE $CO_2 + N_2O$

1. A.E. Markham and K.A. Kobe, J. chem. Phys. $\underline{9}$ 438 (1941).

Percentage volume change measured on mixing gases at 298 K and 1 atm. pressure.

2. A. Charnley, J.S. Rowlinson, J.R. Sutton and J.R. Townley, Proc. R. Soc. $\underline{A230}$ 354 (1955).

Values of isothermal Joule-Thomson coefficient at zero pressure given, T range 273 -318 K.

CARBON DIOXIDE + OXYGEN $CO_2 + O_2$

1. A.E. Markham and K.A. Kobe, J. chem. Phys. $\underline{9}$ 438 (1941).

Percentage volume change measured on mixing gases at 298 K and 1 atm. pressure.

2. A.E. Edwards and W.E. Roseveare, J. Am. chem. Soc. 64 2816 (1942) (*).

Measurement of volume change on mixing gases at constant pressure of 1 atm. and 0.5 atm.

T	B_{11}	B_{22}	B_{12}
298.2	-117.6	-20.6	-56.4

3. R.A. Gorski and J.G. Miller, J. Am. chem. Soc. 75 550 (1953).

Measurement of volume change on mixing gases at constant pressure, below 1 atm.

T	B_{11}	B_{22}	B_{12}
303.15	-120.7	-15.96	-41.5 ± 0.2

4. T.L. Cottrell, R.A. Hamilton and R.P. Taubinger, Trans. Faraday Soc. 52 1310 (1956).

Gas expansions at $P \to 1$ atm, X_2 is 0.5.

T	B_{11}	B_{22}	B_{12}
303.2	-119.2	-14.7	-36.8 ± 5.2
333.2	- 97.1	- 9.3	-28.4 ± 5.6
363.2	- 79.6	- 4.9	-25.6 ± 4.4

CARBON DISULPHIDE + ACETONE CS_2 + $(CH_3)_2CO$

1. G.A. Bottomley and T.H. Spurling, Aust.J. Chem. 20 1789 (1967).

Low pressure differential method. B_{22} interpolated from values given. Estimated error in B_{12} ± 25.

T	B_{11}	B_{22}	B_{12}
324.84	-664	-1405	-600
349.85	-568	-1050	-493
378.84	-479	- 795	-406
407.38	-415	- 630	-363
432.09	-373	- 500	-326

CARBON DISULPHIDE + HYDROGEN CS_2 + H_2

1. A. Eucken and F. Bresler, Z. phys. Chem. 134 230 (1928) (*).

T	B_{12}
273.2	-70

CARBON DISULPHIDE + NITROGEN $CS_2 + N_2$

1. A. Eucken and F. Bresler, Z. phys. Chem. 134 230 (1928) (*).

T	B_{12}
273.2	-174

PERFLUOROETHANE + ETHANE $C_2F_6 + C_2H_6$

1. E.M. Dantzler Siebert and C.M. Knobler, J. phys. Chem., Ithaca 75 3863 (1971).

    E determined from measurements of pressure changes on mixing.

T	E	B_{11}	B_{22}	B_{12}
323.15	33 ± 1	-199	-157	-145
373.15	28 ± 1	-121	-115	- 90

PERFLUOROETHANE + PERFLUORO   PROPANE $C_2F_6 + CF_3 \cdot CF_2 \cdot CF_3$

1. E.M. Dantzler and C.M. Knobler, J. phys. Chem., Ithaca 73 1335 (1969).

    E determined from measurements of pressure changes on mixing.

T	E	B_{11}	B_{22}	B_{12}
323.15	18 ± 1	-199	-435	-299
373.15	14 ± 1	-121	-290	-192

PERFLUOROETHANE + PERFLUORO-n-BUTANE $C_2F_6 + CF_3 \cdot (CF_2)_2 \cdot CF_3$

1. E.M. Dantzler and C.M. Knobler, J. phys. Chem., Ithaca 73 1335 (1969).

    E determined from measurements of pressure changes on mixing.

T	E	B_{11}	B_{22}	B_{12}
323.15	84 ± 2	-199	-744	-388
373.15	52 ± 2	-121	-492	-254

PERFLUOROETHANE+n-BUTANE C_2F_6 + $CH_3(CH_2)_2CH_3$

1. E.M. Dantzler Siebert and C.M. Knobler, J. phys. Chem., Ithaca 75 3863 (1971).

   E determined from measurements of pressure changes on mixing.

T	E	B_{11}	B_{22}	B_{12}
323.15	114 ± 1	-199	-593	-282

PERFLUOROETHANE+PERFLUORO-n-PENTANE C_2F_6 + $CF_3 \cdot (CF_2)_3 \cdot CF_3$

1. E.M. Dantzler and C.M. Knobler, J. phys. Chem., Ithaca 73 1335 (1969).

   E determined from measurements of pressure changes on mixing.

T	E	B_{11}	B_{22}	B_{12}
323.15	212 ± 4	-199	-1188	-482
373.15	143 ± 5	-121	-793	-314

PERFLUOROETHANE+n-PENTANE C_2F_6 + $CH_3 \cdot (CH_2)_3 \cdot CH_3$

1. E.M. Dantzler Siebert and C.M. Knobler, J. phys. Chem., Ithaca 75 3863 (1971).

   E determined from measurements of pressure changes on mixing.

T	E	B_{11}	B_{22}	B_{12}
323.15	233 ± 2	-199	-974	-353

PERFLUOROETHANE+PERFLUORO-n-HEXANE C_2F_6 + $CF_3 \cdot (CF_2)_4 CF_3$

1. E.M. Dantzler and C.M. Knobler, J. phys. Chem., Ithaca 73 1335 (1969).

   E determined from measurements of pressure changes on mixing.

T	E	B_{11}	B_{22}	B_{12}
323.15	420 ± 12	-199	-1712	-536
373.15	272 ± 8	-121	-1128	-354

ACETYLENE + AMMONIA   CH :CH + NH_3

1. S.M. Khodeva, Russ. J. phys. Chem. 38 693 (1964);   Zh. fiz. Khim. 38 1276 (1964).

PVT data given: P range 10 - 90 atm, T range 323 - 423 K,
X_2 0.2, 0.4, 0.6, 0.8

2. H.Y. Cheh, J.P. O'Connell and J.M. Prausnitz, Can. J. Chem. 44
429 (1966).

Values of B_{12} calculated from volumetric data of C.M. Khodeva,
Zh. fiz. Khim. 38 1276 (1964).

T	B_{12}	T	B_{12}
323.2	-240	398.2	-155
348.2	-204	423.2	-134
373.2	-177		

ACETONITRILE + ACETALDEHYDE $CH_3.CN + CH_3.CHO$

1. J.M. Prausnitz and W.B. Carter, A. I. Ch. E. Jl 6 611 (1960).
PVT data; P → 35 cm Hg.

T	B_{12}	T	B_{12}
313.15	-8710 ± 150	353.15	-3450 ± 150
333.15	-6170 ± 150	373.15	-2390 ± 150

ACETONITRILE + CYCLOHEXANE $CH_3.CN + CH_2.(CH_2)_4CH_2$

1. J.D. Lambert, S.J. Murphy and A.P. Sanday, Proc. R. Soc. A226
394 (1954).

Boyle's Law apparatus.

T	B_{11}	B_{22}	$B_{12}(^*)$
326	-3500	-1300	-600
349	-2480	-1000	-460

ETHYLENE+ETHANOL $CH_2:CH_2 + C_2H_5OH$

1. S.K. Gupta, R.D. Lesslie and A.D. King, Jr., J. phys. Chem., Ithaca
77 2011 (1973).

B_{12} derived from measurements of solubility of ethanol in the
compressed gas.

T	B_{12}
298.15	-243 ± 10

ETHYLENE + BENZENE $CH_2:CH_2 + C_6H_6$

1. C.R. Coan and A.D. King, Jr., J. Chromat. <u>44</u> 429 (1969).
    B_{12} determined from measurements of solubility of benzene in the compressed gas: $P \rightarrow 46$ atm.

T	B_{12}
323.2	-282 ± 5

ETHYLENE + NAPHTHALENE $CH_2:CH_2 + C_{10}H_8$

1. A.D. King, Jr. and W.W. Robertson, J. chem. Phys. <u>37</u> 1453 (1962).
    Naphthalene solubility measurements.

T	B_{12}
296.2	-666 ± 25
337.2	-491 ± 10

2. G.C. Najour and A.D. King, Jr., J. chem. Phys. <u>45</u> 1915 (1966).
    Naphthalene solubility measurements.

T	B_{12}	T	B_{12}
296.5	-693 ± 6	326	-553 ± 12
298	-681 ± 10	333	-520 ± 12
308	-629 ± 9	342.5	-481 ± 12
312.5	-577 ± 23		

ETHYLENE + 1-METHYL NAPHTHALENE $CH_2:CH_2 + C_{10}H_7.CH_3$

1. B.K. Kaul and J.M. Prausnitz, A. I. Ch. E. Jl <u>24</u> 223 (1978).
    Measurements of the heavy hydrocarbon solubility in the compressed gas.

T	B_{12}
348.2	-571 ± 20
398.2	-381 ± 16
448.2	-235 ± 15

ETHYLENE + BICYCLOHEXYL $CH_2:CH_2 + C_{10}H_{18}$

1. B.K. Kaul and J.M. Prausnitz, A. I. Ch. E. Jl <u>24</u> 223 (1978).

Measurements of the heavy hydrocarbon solubility in the compressed gas.

T	B_{12}
323.2	-607 ± 25
383.2	-432 ± 25
443.2	-257 ± 12

ETHYLENE + DIPHENYLMETHANE $CH_2:CH_2 + (C_6H_5)_2CH_2$

1. B.K. Kaul and J.M. Prausnitz, A. I. Ch. E. Jl 24 223 (1978).

    Measurements of the heavy hydrocarbon solubility in the compressed gas.

T	B_{12}
338.2	-645 ± 27
393.2	-416 ± 22
448.2	-293 ± 10

ETHYLENE + ANTHRACENE $CH_2:CH_2 + C_6H_4(CH)_2C_6H_4$

1. G.C. Najour and A.D. King, Jr., J. chem. Phys. 52 5206 (1970).

    B_{12} determined from measurements of solubility of anthracene in the compressed gas: $P \to 100$ atm. Probable error in $B \pm 12$.

T	B_{12}	T	B_{12}
338	-686	423	-386
348	-657	445	-356
373	-552	453	-352
398	-448		

ETHYLENE + PHENANTHRENE $CH_2:CH_2 + C_{14}H_{10}$

1. H. Bradley, Jr., and A.D. King, Jr., J. chem. Phys. 52 2851 (1970).

    B_{12} determined from measurements of solubility of phenanthrene in the compressed gas: $P \to 60$ atm. Probable error in $B \pm 24$.

T	B_{12}	T	B_{12}
310	-884	349	-623
315	-817	350	-639
317	-827	357	-607
321	-855	358	-593
323	-786	361	-605
326	-751	367	-548
329	-814	380	-528
334	-734	395	-462
345	-674	415	-385
347	-659		

ETHYLENE + n-HEXADECANE $CH_2:CH_2$ + $CH_3 \cdot (CH_2)_{14} \cdot CH_3$

1. B.K. Kaul and J.M. Prausnitz, A.I. Ch. E. Jl 24 223 (1978).
    Measurements of the heavy hydrocarbon solubility in the compressed gas.

T	B_{12}
348.2	-735 ± 32
398.2	-524 ± 21
448.2	-357 ± 20

ETHYLENE + HYDROGEN $CH_2:CH_2$ + H_2

1. A.E. Edwards and W.E. Roseveare, J. Am. chem. Soc. 64 2816 (1942) (*).
    Measurement of volume change on mixing gases at constant pressure of 1 atm., and 0.5 atm.

T	B_{11}	B_{22}	B_{12}
298.2	-140.0	14.8	-39.6

2. D. McA. Mason and B.E. Eakin, J. chem. Engng Data 6 499 (1961) (*).

T	B_{12}
288.7	$+14.7 \pm 5$

ETHYLENE (ETHENE) + AMMONIA $CH_2:CH_2 + NH_3$

1. M. Ratzsch and H. Freydank, J. chem. Thermodyn. $\underline{3}$ 861 (1971).
    Maximum error in B estimated to be ± 3%.

T	B_{12}
302.84	-221
313.84	-195

ETHYLENE + HELIUM $CH_2:CH_2 + He$

1. L.R. Linshits, I.B. Rodkina and D.S. Tsiklis, Russ. J. phys. Chem.
    $\underline{51}$ 1381 (1977); Zh. fiz. Khim. $\underline{51}$ 2357 (1977).

    Burnett method.

T	B_{12}	C_{112}	C_{122}
323.2	30.2	391	2100
373.2	28.7	483	1875

ETHYLENE + NITROGEN $CH_2:CH_2 + N_2$

1. A.E. Edwards and W.E. Roseveare, J. Am. chem. Soc. $\underline{64}$ 2816 (1942)
    (*).

    Measurement of volume change on mixing gases at constant pressure
    of 1 atm, and 0.5 atm.

T	B_{11}	B_{22}	B_{12}
298.2	-140.0	-4.5	-54.7

2. W.P. Hagenbach and E.W. Comings, Ind. Engng Chem. ind. Edn $\underline{45}$
    606 (1953).

    PVTx data given for five mixtures; P → 655 atm at 323K.

ETHYLENE + NITROUS OXIDE $CH_2:CH_2 + N_2O$

1. A. Charnley, J.S. Rowlinson, J.R. Sutton and J.R. Townley, Proc.
    R. Soc. $\underline{A230}$ 354 (1955).

    Values of isothermal Joule-Thomson coefficient at zero pressure
    given, T range 273 - 318 K.

ETHYLENE + OXYGEN $CH_2:CH_2 + O_2$
1. I. Masson and L.G.F. Dolley, Proc. R. Soc. $\underline{A103}$ 524 (1923).

PVTx data given for three mixtures; $P \to 125$ atm and $T = 298$ K.

ETHYL BROMIDE + DIETHYL ETHER $C_2H_5Br + (C_2H_5)_2O$

1.   M. Ratzsch and H.-J. Bittrich, Z. phys. Chem. 228 81 (1965).

T	B_{11}	B_{22}	B_{12}
293.1	-795	-1386	-1356 ± 38
313.1	-669	-1107	-1160 ± 78

ETHYL BROMIDE + n PENTANE $C_2H_5Br + CH_3(CH_2)_3CH_3$

1.   M. Ratzsch and H.-J. Bittrich, Z. phys. Chem. 228 81 (1965).

T	B_{11}	B_{22}	B_{12}
313.1	-669	-1187	-945 ± 25

ETHYL CHLORIDE + n-PROPYL CHLORIDE $CH_3 \cdot CH_2Cl + CH_3 \cdot CH_2 \cdot CH_2Cl$

1.   M. Rätzsch, Z. phys. Chem. 238 321 (1968).

T	B_{12}
303.2	-803 ± 26
313.2	-774 ± 16
333.2	-683 ± 12

ETHYL CHLORIDE + t-BUTYL CHLORIDE $CH_3 \cdot CH_2Cl + (CH_3)_3CCl$

1.   M. Rätzsch, Z. phys. Chem. 238 321 (1968).

T	B_{12}
313.2	-712 ± 10
333.2	-591 ± 13

NITROETHANE + NITROGEN $CH_3 \cdot CH_2NO_2 + N_2$

1.   M. Vigdergauz and V. Semkin, J. Chromat. 58 95 (1971).
     Chromatographic measurements using dinonyl phthalate.
     Class III.

T	B_{12}
353.2	-71

ETHANE + ETHANOL C_2H_6 + C_2H_5OH

1. S.K. Gupta, R.D. Lesslie and A.D. King, Jr., J. phys. Chem., Ithaca 77 2011 (1973).

   B_{12} derived from measurements of solubility of ethanol in the compressed gas.

T	B_{12}
298.15	-276 ± 13
323.15	-200 ± 7
348.15	-190 ± 7

ETHANE + PROPENE $CH_3.CH_3$ + $CH_2:CH.CH_3$

1. H.W. Prengle and H. Marchman, Ind. Engng Chem. ind. Edn 42 2371 (1950).

   Compressibilities of four mixtures from 373 - 523K; P range 10 - 220 atm.

2. R.A. McKay, H.H. Reamer, B.H. Sage, and W.N. Lacey, Ind. Engng Chem. ind. Edn 43 2112 (1951).

   Volumetric behaviour of three mixtures from 265 - 480K; P → 670 atm.

3. R.D. Gunn, M.S. Thesis, University of California, Berkeley (1958).

   B_{12} calculated from volumetric data given by B.H. Sage and W.N. Lacey, Some properties of the lighter hydrocarbons, Amer. Petrol. Inst., New York (1955).

T	B_{12}	T	B_{12}
377.6	-146.3	444.3	-102.7
410.9	-120.7	477.6	- 82.3

ETHANE + PROPANE $CH_3.CH_3$ + $CH_3.CH_2.CH_3$

1. D. McA. Mason and B.E. Eakin, J. chem. Engng Data 6 499 (1961) (*).

T	B_{12}
288.70	-284.1 ± 5

2. E.M. Dantzler, C.M. Knobler and M.L. Windsor, J. phys. Chem.,
    Ithaca 72 676 (1968).

    E determined from measurements of pressure changes on mixing.

T	E	B_{11}	B_{22}	B_{12}
298.15	17 ± 1	-185	-398	-274
323.15	14 ± 1	-159	-330	-230
348.15	12 ± 2	-134	-276	-193
373.15	10 ± 1	-115	-235	-165

ETHANE+PERFLUORO-n-BUTANE $CH_3 \cdot CH_3 + CF_3 \cdot (CF_2)_2 CF_3$

1. E.M. Dantzler Siebert and C.M. Knobler, J. phys. Chem., Ithaca
    75 3863 (1971).

    E determined from measurements of pressure changes on mixing.

T	E	B_{11}	B_{22}	B_{12}
323.15	178 ± 1	-157	-744	-273

ETHANE + n-BUTANE $CH_3 \cdot CH_3 + CH_3 \cdot (CH_2)_2 \cdot CH_3$

1. D. McA. Mason and B.E. Eakin, J. chem. Engng Data 6 499 (1961) (*).

T	B_{12}
288.70	-374.1 ± 5

2. E.M. Dantzler, C.M. Knobler and M.L. Windsor, J. phys. Chem.,
    Ithaca 72 676 (1968).

    E determined from measurements of pressure changes on mixing.

T	E	B_{11}	B_{22}	B_{12}
298.15	92 ± 1	-185	-722	-362
323.15	72 ± 2	-159	-599	-307
348.15	60 ± 1	-136	-501	-258
373.15	54 ± 2	-115	-422	-215

3. C.J. Wormald, E.J. Lewis, and D.J. Hutchings, J. chem. Thermodyn.
    11 1 (1979).

    B_{12} values derived from measured excess enthalpies.

T	B_{12}
304.5	-332 ± 10
333.2	-333 ± 9
363.2	-233 ± 9

ETHANE+1-BUTANOL $CH_3 \cdot CH_3 + CH_3(CH_2)_2CH_2OH$

1.   R. Massoudi and A.D. King, Jr., J. phys. Chem., Ithaca 77 2016 (1973).

B_{12} derived from measurements of solubility of 1-butanol in the compressed gas.

T	B_{12}
298.15	-375 ± 35

ETHANE+DIETHYL ETHER $C_2H_6 + (C_2H_5)_2O$

1.   R. Massoudi and A.D. King, Jr., J. phys. Chem., Ithaca 77 2016 (1973).

B_{12} derived from measurements of solubility of diethyl ether in the compressed gas.

T	B_{12}
298.15	-388 ± 30

ETHANE + n-PENTANE $CH_3 \cdot CH_3 + CH_3 \cdot (CH_2)_3 \cdot CH_3$

1.   R.L. Pecsok and M.L. Windsor, Analyt. Chem. 40 1238 (1968).
     Gas chromatographic method using squalane;   P → 6 atm.
     Carrier gas solubility taken into account in calculating B_{12}.

T	B_{12}
298.2	-414 ± 171

2.   E.M. Dantzler, C.M. Knobler and M.L. Windsor, J. phys. Chem., Ithaca 72 676 (1968).

E determined from measurements of pressure changes on mixing.

T	E	B_{11}	B_{22}	B_{12}
298.15	242 ± 3	-185	-1195	-448
323.15	192 ± 2	-159	-980	-377

| 348.15 | 158 ± 2 | -136 | -810 | -315 |
| 373.15 | 129 ± 2 | -115 | -687 | -272 |

3. R. Massoudi and A.D. King, Jr., J. phys. Chem., Ithaca <u>77</u> 2016 (1973).

B_{12} derived from measurements of solubility of n pentane in the compressed gas.

T	B_{12}
298.15	-386 ± 30

ETHANE+n-HEXANE $CH_3 . CH_3 + CH_3 . (CH_2)_4 CH_3$

1. E.M. Dantzler, C.M. Knobler and M.L. Windsor, J. phys. Chem., Ithaca <u>72</u> 676 (1968).

E determined from measurements of pressure changes on mixing.

T	E	B_{11}	B_{22}	B_{12}
298.15	510 ± 30	-185	-1915	-540
323.15	386 ± 5	-159	-1530	-458
348.15	312 ± 2	-136	-1235	-373
373.15	254 ± 2	-115	-1031	-319

2. C.J. Wormald, E.J. Lewis, and D.J. Hutchings, J. chem. Thermodyn. <u>11</u> 1 (1979).

T	B_{12}
372.2	-330 ± 18
383.2	-310 ± 15
403.2	-274 ± 13

ETHANE + n-HEPTANE $CH_3 . CH_3 + CH_3 . (CH_2)_5 . CH_3$

1. P.C. Wu and P. Ehrlich, A. I. Ch. E. Jl <u>19</u> 533 (1973).

Molar volumes given at 353 K and 74.47 atm for $X_{heptane} \rightarrow 0.169$.

ETHANE + n-OCTANE $CH_3 . CH_3 + CH_3 (CH_2)_6 CH_3$

1. C.J. Wormald, E.J. Lewis, and D.J. Hutchings, J. chem. Thermodyn. <u>11</u> 1 (1979).

B_{12} values derived from measured excess enthalpies.

T	B_{12}
403.2	-354 ± 64
413.2	-331 ± 51

ETHANE + NAPHTHALENE $CH_3 \cdot CH_3 + C_{10}H_8$

1. A.D. King, Jr., J. chem. Phys. <u>49</u> 4083 (1968).

   B_{12} determined from measurements of solubility of naphthalene in the compressed gas: $P \rightarrow 60$ atm.

T	B_{12}	T	B_{12}
299	-724 ± 10	323	-579 ± 10
310	-699 ± 10	330	-562 ± 10
309.5	-642 ± 18	340	-528 ± 10
318	-608 ± 8		

ETHANE + 1-METHYL NAPHTHALENE $CH_3 \cdot CH_3 + C_{10}H_7 \cdot CH_3$

1. B.K. Kaul and J.M. Prausnitz, A. I. Ch. E. Jl <u>24</u> 223 (1978).

   Measurements of the heavy hydrocarbon solubility in the compressed gas.

T	B_{12}
348.2	-672 ± 28
398.2	-462 ± 23
448.2	-322 ± 15

ETHANE + BICYCLOHEXYL $CH_3 \cdot CH_3 + C_{10}H_{18}$

1. B.K. Kaul and J.M. Prausnitz, A. I. Ch. E. Jl <u>24</u> 223 (1978).

   Measurements of the heavy hydrocarbon solubility in the compressed gas.

T	B_{12}
323.2	-777 ± 32
383.2	-545 ± 27
443.2	-326 ± 15

ETHANE + DIPHENYLMETHANE $CH_3.CH_3 + (C_6H_5)_2CH_2$

1. B.K. Kaul and J.M. Prausnitz, A. I. Ch. E. Jl $\underline{24}$ 223 (1978).
   Measurements of the heavy hydrocarbon solubility in the compressed gas.

T	B_{12}
338.2	-734 ± 35
393.2	-481 ± 25
448.2	-343 ± 20

ETHANE+ANTHRACENE $CH_3.CH_3 + C_6H_4(CH)_2C_6H_4$

1. G.C. Najour and A.D. King, Jr., J. chem. Phys. $\underline{52}$ 5206 (1970).
   B_{12} determined from measurements of solubility of anthracene in the compressed gas: P → 100 atm. Probable error in B ± 12.

T	B_{12}	T	B_{12}
336	-781	398	-526
348	-719	423	-452
373	-619	448	-419

ETHANE + n-HEXADECANE $CH_3.CH_3 + CH_3.(CH_2)_{14}.CH_3$

1. B.K. Kaul and J.M. Prausnitz, A. I. Ch. E. Jl $\underline{24}$ 223 (1978).
   Measurements of the heavy hydrocarbon solubility in the compressed gas.

T	B_{12}
348.2	-928 ± 40
398.2	-645 ± 30
448.2	-434 ± 21

ETHANE + EICOSANE $CH_3.CH_3 + CH_3.(CH_2)_{18}.CH_3$

1. B.K. Kaul and J.M. Prausnitz, A. I. Ch. E. Jl $\underline{24}$ 223 (1978).
   Measurements of the heavy hydrocarbon solubility in the compressed gas.

T	B_{12}	T	B_{12}
438.2	-517 ± 25	508.2	-336 ± 15
473.2	-402 ± 20	543.2	-256 ± 10

ETHANE + SQUALANE $CH_3.CH_3 + C_{30}H_{62}$

1. B.K. Kaul and J.M. Prausnitz, A. I. Ch. E. Jl $\underline{24}$ 223 (1978).
    Measurements of the heavy hydrocarbon solubility in the compressed
    gas.

T	B_{12}
503.2	-293 $\pm$ 20
545.2	-165 $\pm$ 10

ETHANE + HYDROGEN $CH_3.CH_3 + H_2$

1. D. McA. Mason and B.E. Eakin, J. chem. Engng Data $\underline{6}$ 499 (1961) (*).

T	B_{12}
288.7	+11.1 $\pm$ 5

2. C.W. Solbrig and R.T. Ellington, Chem. Engng Prog. Symp. Ser. $\underline{59}$
    127 (1963).
    PVT data given;   T 293 - 423 K,   P $\rightarrow$ 200 atm.

ETHANE+WATER $CH_3.CH_3 + H_2O$

1. C.R. Coan and A.D. King, Jr., J. Am. chem. Soc. $\underline{93}$ 1857 (1971).
    B_{12} derived from measurements of solubility of water in the
    compressed gas.   P $\rightarrow$ 36 atm.

T	B_{12}	T	B_{12}
298.15	-125 $\pm$ 6	348.15	- 78 $\pm$ 5
323.15	- 94 $\pm$ 3	373.15	- 72 $\pm$ 2

ETHANE + HYDROGEN SULPHIDE $CH_3.CH_3 + H_2S$

1. F. Khoury and D.B. Robinson, J. chem. Phys. $\underline{55}$ 834 (1971).

T	B_{11}	B_{22}	B_{12}
323.15	-152.49	-165.94	-128.5
348.15	-129.81	-139.53	-110.4
373.15	-111.72	-118.74	- 95.2
398.15	- 96.70	-101.95	- 79.3

ETHANE + AMMONIA $CH_3.CH_3 + NH_3$

1. N.E. Khazanova, E.E. Sominskaya, and A.V. Zakharova, Russ. J. phys. Chem. <u>47</u> 467 (1973); Zh. fiz. Khim. <u>47</u> 823 (1973).

   PVTx data given: x_{NH_3} in range 0.03-0.24, P range 43-70 atm,

   T range 295-308K.

ETHANE + NITROGEN $CH_3.CH_3 + N_2$

1. H.H. Reamer, F.T. Selleck, B.H. Sage and W.N. Lacey, Ind. Engng Chem. ind. Edn <u>44</u> 198 (1952).

   Volumetric behaviour of three mixtures from 280K - 500K; P → 670 atm.

2. R.D. Gunn, M.S. thesis, University of California, Berkeley (1958).

   Values of B_{12} given by D.W. Calvin and T.M. Reed III, J. chem. Phys. <u>54</u> 3733 (1971).

T	B_{12}	T	B_{12}
277.6	-65.4	444.3	-3.8
310.9	-38.6	510.9	+5.9
377.6	-20.1		

3. D. McA. Mason and B.E. Eakin, J. chem. Engng Data <u>6</u> 499 (1961) (*).

T	B_{12}
288.70	-58.1 ± 5

4. A.L. Stockett and L.A. Wenzel, A. I. Ch. E. Jl <u>10</u> 557 (1964).

   Measurement of Joule-Thomson coefficients; P → 170 atm, T range 173 - 298 K.

ETHANOL + BENZENE $CH_3.CH_2OH + C_6H_6$

1. D.H. Knoebel and W.C. Edmister, J. chem. Engng Data <u>13</u> 312 (1968).

   Low pressure PVT measurements on several mixtures.

   Estimated standard deviation in B_{12} is ± 52.

T	B_{12}
333.15	-567
353.15	-421
373.15	-490

ETHANOL+HYDROGEN $CH_3.CH_2OH + H_2$

1. S.K. Gupta, R.D. Lesslie and A.D. King, Jr., J. phys. Chem., Ithaca
    77 2011 (1973).

    B_{12} derived from measurements of solubility of ethanol in the
    compressed gas.

T	B_{12}
298.15	5 ± 5
323.15	6 ± 4
348.15	5 ± 2

ETHANOL+HELIUM $CH_3.CH_2OH + He$

1. S.K. Gupta, R.D. Lesslie and A.D. King, Jr., J. phys. Chem., Ithaca
    77 2011 (1973).

    B_{12} derived from measurements of solubility of ethanol in the
    compressed gas.

T	B_{12}
298.15	32 ± 6
323.15	41 ± 2
348.15	42 ± 6

ETHANOL + NITROGEN $CH_3.CH_2OH + N_2$

1. M. Vigdergauz and V. Semkin, J. Chromat. 58 95 (1971).
    Chromatographic measurements using dinonyl phthalate.
    Class III.

T	B_{12}
353.2	-10

2. P. Neogi and A.P. Kudchadker, J.C.S. Faraday 1 73 385 (1977).
    Chromatographic measurements using di-isodecyl phthalate,
    $P \rightarrow 2.5$ atm. Values of B_{12} read from graph.

T	B_{12}	T	B_{12}
303.2	-94.0 ± 17	333.2	-70.0 ± 17
318.2	-83.5 ± 17	343.2	-64.5 ± 17

ETHANOL+NITROUS OXIDE $CH_3.CH_2.OH + N_2O$

1. S.K. Gupta, R.D. Lesslie and A.D. King, Jr., J. phys. Chem., Ithaca 77 2011 (1973).

    B_{12} derived from measurements of solubility of ethanol in the compressed gas.

T	B_{12}
298.15	-274 ± 5
323.15	-215 ± 5
348.15	-177 ± 5

DIMETHYL ETHER + 1-HYDROPERFLUOROPROPANE $(CH_3)_2O + CF_3.CF_2.CF_2.H$

1. T.B. Tripp and R.D. Dunlap, J. phys. Chem., Ithaca 66 635 (1962).

    Boyle apparatus. Values of B calculated by the authors from a linear equation in (a) pressure and (b) concentration.

T		B_{11}	B_{22}	B_{12}
283.15	a)	-542	-788	-1280
	b)	-531	-759	-1224
303.03	a)	-466	-674	- 951
	b)	-457	-653	- 920
323.20	a)	-411	-577	- 676
	b)	-405	-561	- 662

(B_{12} calculated from B_M derived using a quadratic equation has values -1362, -1080 and -800 at these temperatures.)

PERFLUORO-n-PROPANE +   PROPANE $CF_3.CF_2.CF_3 + CH_3.CH_2.CH_3$

1. E.M. Dantzler Siebert and C.M. Knobler, J. phys. Chem., Ithaca 75 3863 (1971).

    E determined from measurements of pressure changes on mixing.

T	E	B_{11}	B_{22}	B_{12}
323.15	70 ± 1	-435	-327	-311
373.15	54 ± 1	-290	-241	-212

PROPENE (PROPYLENE) + PROPANE $CH_2:CH.CH_3 + CH_3.CH_2.CH_3$

1. D. McA. Mason and B.E. Eakin, J. chem. Engng Data <u>6</u> 499 (1961) (*).

T	B_{12}
288.70	-407.1 ± 5

PROPENE + 1-BUTENE $CH_2:CH.CH_3 + CH_2:CH.CH_2.CH_3$

1. G.H. Goff, P.S. Farrington, and B.H. Sage, Ind. Engng Chem. ind.
    Edn <u>42</u> 735 (1950).

    Volumetric behaviour of four mixtures from 280 - 410K;   P → 670 atm.

PROPENE (PROPYLENE) + n-BUTANE $CH_2:CH.CH_3 + CH_3.(CH_2)_2.CH_3$

1. D. McA. Mason and B.E. Eakin, J. chem. Engng Data <u>6</u> 499 (1961) (*).

T	B_{12}
288.70	-529.6 ± 5

PROPENE (PROPYLENE) + ISOBUTANE $CH_2:CH.CH_3 + (CH_3)_3CH$

1. D. McA. Mason and B.E. Eakin, J. chem. Engng Data <u>6</u> 499 (1961) (*).

T	B_{12}
288.70	-510.5 ± 5

PROPENE + HEPT-1-ENE $CH_2:CH.CH_3 + CH_2:CH.(CH_2)_4CH_3$

1. M.L. McGlashan and C.J. Wormald, Trans. Faraday Soc. <u>80</u> 646 (1964)
    (*).

    Values of B_M given for X_2 0.4995.

T	B_{11}	B_{22}	B_{12}
324.3	-288	-1990	-827
328.7	-280	-1935	-750
333.2	-271	-1870	-693
338.5	-262	-1805	-666
344.8	-252	-1730	-621
348.8	-245	-1680	-603
355.5	-236	-1595	-586

363.2	-225	-1500	-551
374.3	-211	-1375	-525
384.6	-199	-1275	-495
394.6	-188	-1190	-453
403.3	-178	-1125	-416
411.5	-171	-1070	-391

PROPENE (PROPYLENE) + HYDROGEN $CH_2:CH.CH_3$ + H_2

1. D. McA. Mason and B.E. Eakin, J. chem. Engng Data 6 499 (1961) (*).

T	B_{12}
288.7	+14.5 ± 5

PROPENE + HELIUM $CH_2:CH.CH_3$ + He

1. W. Warowny and J. Stecki, J. chem. Engng Data 23 212 (1978).
    Burnett method, P → 70 atm.

T	B_{12}
393.20	40.53, 43.87, 40.94, 44.22
407.48	37.10, 39.79
407.50	33.87, 35.76
407.40	38.98, 40.09
423.03	40.98, 34.92
423.02	33.19, 36.60
423.01	31.69, 39.60

ACETONE + n BUTANE $(CH_3)_2CO$ + $CH_3(CH_2)_2CH_3$

1. W. Kappallo, N. Lund and K. Schafer, Z. phys. Chem. Frankf. Ausg.
    37 196 (1963).
    Class II.

T	B_{11}	B_{22}	B_{12}
282.3	-2733	-862	-805
297.0	-2268	-758	-656
312.0	-1876	-674	-569
321.0	-1680	-635	-504

ACETONE + DIETHYL ETHER $(CH_3)_2CO + (C_2H_5)_2O$

1. Sh. D. Zaalishvili and L.E. Kolysko, Russ. J. phys. Chem. <u>34</u> 1223
   (1960); Zh. fiz. Khim. <u>34</u> 2596 (1960).

   Boyle's Law apparatus.

T	B_{11}	B_{22}	B_{12}
323.2	-1535	-960	-820
333.2	-1370	-850	-726
343.2	-1200	-760	-632
353.2	-1035	-730	-540

ACETONE + BENZENE $(CH_3)_2CO + C_6H_6$

1. Sh. D. Zaalishvili and Z.S. Belousova, Russ. J. phys. Chem. <u>38</u> 269
   (1964); Zh. fiz. Khim. <u>38</u> 503 (1964).

   Boyle's Law apparatus.

T	B_{11}	B_{22}	B_{12}
353.2	-1024	-992	-910
363.2	- 874	-897	-807
373.2	- 740	-841	-735
383.2	- 620	-811	-710

2. D.H. Knoebel and W.C. Edmister, J. chem. Engng Data <u>13</u> 312 (1968).

   Low pressure PVT measurements on several mixtures.

   Estimated standard deviation in B_{12} is ± 52.

T	B_{12}	T	B_{12}
313.15	-1372	353.15	- 698
333.15	- 930	373.15	- 593

ACETONE + CYCLOHEXANE $(CH_3)_2\ CO + CH_2 \cdot (CH_2)_4\ CH_2$

1. J.D. Lambert, S.J. Murphy and A.P. Sanday, Proc. R. Soc. <u>A226</u> 394
   (1954).

   Boyle's Law apparatus.

T	B_{11}	B_{22}	B_{12}(*)
326	-1550	-1300	-860
349	-1240	-1110	-230

ACETONE+MERCURY $(CH_3)_2CO + Hg$

1. H.S. Rosenberg and W.B. Kay, J. phys. Chem., Ithaca $\underline{78}$ 186 (1974).
   Solubility measurements at pressures up to 30 atm.

T	B_{12}	T	B_{12}
493.2	-156	553.2	-136
513.2	-154	573.2	-123
533.2	-146		

PROPANE + n-BUTANE $CH_3 \cdot CH_2 \cdot CH_3 + CH_3 \cdot (CH_2)_2 \cdot CH_3$

1. D. McA. Mason and B.E. Eakin, J. chem. Engng Data $\underline{6}$ 499 (1961) (*).

T	B_{12}
288.70	-577.1 ± 5

2. E.M. Dantzler, C.M. Knobler and M.L. Windsor, J. phys. Chem., Ithaca $\underline{72}$ 676 (1968).
   E determined from measurements of pressure changes on mixing.

T	E	B_{11}	B_{22}	B_{12}
298.15	26 ± 1	-399	-732	-540
323.15	18 ± 1	-331	-599	-447
348.15	15 ± 1	-276	-501	-373
373.15	12 ± 1	-235	-422	-316

PROPANE + n-PENTANE $CH_3 \cdot CH_2 \cdot CH_3 + CH_3 \cdot (CH_2)_3 \cdot CH_3$

1. B.H. Sage and W.N. Lacey, Ind. Engng Chem. ind. Edn $\underline{32}$ 992 (1940).
   PVTx data given; P → 40 atm, T range 344-444K.

2. E.M. Dantzler, C.M. Knobler and M.L. Windsor, J. phys. Chem., Ithaca $\underline{72}$ 676 (1968).
   E determined from measurements of pressure changes on mixing.

T	E	B_{11}	B_{22}	B_{12}
298.15	126 ± 3	-399	-1195	-671
323.15	97 ± 2	-331	- 980	-558
348.15	76 ± 2	-276	- 809	-466
373.15	60 ± 2	-235	- 684	-399

PROPANE + ISOPENTANE $CH_3.CH_2.CH_3 + (CH_3)_2.CH.CH_2.CH_3$

1. W.E. Vaughan and F.C. Collins, Ind. Engng Chem. ind. Edn **34** 885 (1942).

   PVTx data given for five mixtures; P → 80 atm, T range 273-573K.

PROPANE+n-HEXANE $CH_3.CH_2.CH_3 + CH_3.(CH_2)_4CH_3$

1. E.M. Dantzler, C.M. Knobler and M.L. Windsor, J. phys. Chem., Ithaca **72** 676 (1968).

   E determined from measurements of pressure changes on mixing.

T	E	B_{11}	B_{22}	B_{12}
298.15	330 ± 35	-399	-1905	-822
323.15	248 ± 5	-331	-1530	-682
348.15	196 ± 2	-276	-1240	-562
373.15	154 ± 2	-235	-1030	-478

2. C.J. Wormald, E.J. Lewis, and D.J. Hutchings, J. chem. Thermodyn. **11** 1 (1979).

   B_{12} values derived from measured excess enthalpies.

T	B_{12}
344.2	-574 ± 19
363.2	-519 ± 16
396.9	-418 ± 14

PROPANE + n-HEPTANE $CH_3.CH_2.CH_3 + CH_3.(CH_2)_5.CH_3$

1. M.L. McGlashan and D.J.B. Potter, Proc. R. Soc. **A267** 478 (1962) (*).

   Low pressure differential piezometer; X = 0.5.

T	B_{12}	T	B_{12}
338.6	-666	373.2	-533
348.5	-625	383.5	-504
354.9	-601	393.4	-463
363.5	-577	403.3	-450
365.5	-554	414.2	-408

2. C.J. Wormald, E.J. Lewis, and D.J. Hutchings, J. chem. Thermodyn. **11** 1 (1979).

B_{12} values derived from measured excess enthalpies.

T	B_{12}
383.2	-537 ± 21
403.2	-490 ± 20
413.2	-448 ± 21

PROPANE + n-OCTANE $CH_3.CH_2.CH_3$ + $CH_3.(CH_2)_6.CH_3$

1. M.L. McGlashan and D.J.B. Potter, Proc. R. Soc. <u>A267</u> 478 (1962) (*).
    Low pressure differential piezometer;   X = 0.5.

T	B_{12}	T	B_{12}
353.2	-656	377.4	-574
362.6	-615	382.9	-560
363.0	-570	393.6	-498
367.6	-633	404.0	-448
373.0	-626	413.8	-434

2. C.J. Wormald, E.J. Lewis, and D.J. Hutchings, J. chem. Thermodyn.
    <u>11</u> 1 (1979).

    B_{12} values derived from measured excess enthalpies.

T	B_{12}
403.2	-541 ± 28
410.2	-512 ± 27
413.2	-508 ± 27

n-PROPANE + HYDROGEN CHLORIDE $CH_3.CH_2.CH_3$ + HCl

1. G. Glockler, D.L. Fuller, and C.P. Roe, J. chem. Phys. <u>1</u> 709 (1933)
    (*).

    pVT data given;   P → 100 atm.

    Class II.

T	B_{12}
368.7	-85
389.5	-66

PROPANE + HYDROGEN $CH_3.CH_2.CH_3 + H_2$

1. D. McA. Mason and B.E. Eakin, J. chem. Engng Data <u>6</u> 499 (1961) (*).

T	B_{12}
288.7	+5.7 ± 5

PROPANE + HELIUM $CH_3.CH_2.CH_3 + He$

1. W. Warowny and J. Stecki, J. chem. Engng Data <u>23</u> 212 (1978).
    Burnett method, P → 70 atm.

T	B_{12}
393.18	34.98, 37.53, 43.5
393.19	41.11, 42.86
407.49	39.48, 36.77, 40.53, 37.83
407.40	38.72, 43.81, 38.96, 44.08
423.00	40.75, 44.70, 40.87, 44.82
423.00	39.18, 42.63

PROPANE + MERCURY $CH_3.CH_2.CH_3 + Hg$

1. W.B. Jepson, M.J. Richardson and J.S. Rowlinson, Trans. Faraday Soc. <u>53</u> 1586 (1957).

    Mercury solubility measurements, propane pressure 0.01 - 32 atm.
    B_{12} values read from graph.

T	B_{12}
457.2	-125
491.2	-105
529.2	- 82

n-PROPANE + NITROGEN $CH_3.CH_2.CH_3 + N_2$

1. G.M. Watson, A.B. Stevens, R.B. Evans, and D. Hodges, Ind. Engng Chem. ind. Edn <u>46</u> 362 (1954).

    Compressibility factors for three mixtures at 399.31K and 422.03K;
    P range 5 - 400 atm.

2. D. McA. Mason and B.E. Eakin, J. chem. Engng Data <u>6</u> 499 (1961) (*).

T	B_{12}
288.70	-82.6 ± 5

n-PROPANOL + NITROGEN $CH_3.CH_2.CH_2OH + N_2$

1. P. Neogi and A.P. Kudchadker, J.C.S. Faraday 1 **73** 385 (1977).

    Chromatographic measurements using di-isodecyl phthalate, $P \rightarrow 2.5$ atm. Values of B_{12} read from graph.

T	B_{12}	T	B_{12}
303.2	-110.0 ± 17	333.2	-84.5 ± 17
318.2	-98.5 ± 17	343.2	-80.5 ± 17

PERFLUORO-n-BUTANE + n-BUTANE $CF_3.(CF_2)_2\ CF_3 + CH_3.(CH_2)_2\ CH_3$

1. T.B. Tripp and R.D. Dunlap, J. phys. Chem., Ithaca **66** 635 (1962).

    Boyle apparatus. Values of B calculated by the authors from a linear equation in (a) pressure and (b) concentration.

T		B_{11}	B_{22}	B_{12}
283.15	a)	-1164	-881	-792
	b)	-1098	-846	-766
303.03	a)	-942	-745	-678
	b)	-900	-715	-666
323.20	a)	-800	-641	-593
	b)	-770	-619	-604

(B_{12} calculated from B_M derived using a quadratic equation has values -1018, -686 and -636 at these temperatures.)

2. E.M. Dantzler Siebert and C.M. Knobler, J. phys. Chem., Ithaca **75** 3863 (1971).

    E determined from measurements of pressure changes on mixing.

T	E	B_{11}	B_{22}	B_{12}
298.15	158 ± 2	-891	-711	-643
323.15	130 ± 2	-744	-593	-538
373.15	90 ± 1	-492	-429	-370

PERFLUORO-n-BUTANE+PERFLUORO-n-PENTANE $CF_3 \cdot (CF_2)_2 \, CF_3 + CF_3(CF_2)_3 \cdot CF_3$

1. E.M. Dantzler and C.M. Knobler, J. phys. Chem., Ithaca $\underline{73}$ 1335 (1969).

    E determined from measurements of pressure changes on mixing.

T	E	B_{11}	B_{22}	B_{12}
323.15	27 ± 2	-744	-1188	-939

PERFLUORO-n-BUTANE+PERFLUORO-n-HEXANE $CF_3(CF_2)_2CF_3 + CF_3(CF_2)_4 \, CF_3$

1. E.M. Dantzler and C.M. Knobler, J. phys. Chem., Ithaca $\underline{73}$ 1335 (1969).

    E determined from measurements of pressure changes on mixing.

T	E	B_{11}	B_{22}	B_{12}
373.15	77 ± 4	-492	-1128	-733

PERFLUORO-n-BUTANE + n-HEXANE $CF_3 \cdot (CF_2)_2CF_3 + CH_3(CH_2)_4CH_3$

1. E.M. Dantzler Siebert and C.M. Knobler, J. phys. Chem., Ithaca $\underline{75}$ 3863 (1971).

    E determined from measurements of pressure changes on mixing.

T	E	B_{11}	B_{22}	B_{12}
323.15	298 ± 6	-744	-1513	-830

METHYLETHYL KETONE + NITROGEN $CH_3 \cdot CO \cdot CH_2 \cdot CH_3 + N_2$

1. M. Vigdergauz and V. Semkin, J. Chromat. $\underline{58}$ 95 (1971).
    Chromatographic measurements using dinonyl phthalate.
    Class III.

T	B_{12}
353.2	-36

TETRAHYDROFURAN+WATER $\underline{O \cdot CH_2CH_2CH_2CH_2} + H_2O$

1. C. Treiner, J.F. Bocquet and M. Chemla, J. Chim. phys. $\underline{70}$ 72 (1973).
    B_{12} determined from mass, composition and pressure measurement on
    the vapour mixture in a cell of known volume. At temperatures below
    304.5K, B_{22} was calculated from the Berthelot equation which, at

398

304.58 gave B_{22} in close agreement with experiment.

T	B_{11}	B_{22}	B_{12}
298.15	-1163	-1090	-8800
298.25	-1160	-1087	-8670
299.05	-1150	-1082	-7800
299.25	-1147	-1080	-7350
299.75	-1137	-1076	-6850
300.65	-1120	-1070	-6070
301.95	-1094	-1060	-5060
302.55	-1084	-1056	-4580
303.65	-1062	-1048	-4160
304.45	-1050	-1042	-4010
304.58	-1056	-1092	-3670

DIOXAN + NITROGEN $CH_2.CH_2.O.(CH_2)_2.O$ + N_2

1. M. Vigdergauz and V. Semkin, J. Chromat. 58 95 (1971).
    Chromatographic measurements using dinonyl phthalate.
    Class III.

T	B_{12}
353.2	-67

n-BUTANE + i-BUTANE $CH_3(CH_2)_2CH_3$ + $CH_3.CH.(CH_3)CH_3$

1. J.F. Connolly, J. phys. Chem., Ithaca, 66 1082 (1962) 3 term fit
    to PVT data : P range 4 atm. - S.V.P., X_2 is 0.493.

T	B_{11}	B_{22}	B_{12}
344.26	-517.0	-457.2	-483.7
360.93	-464.7	-412.7	-437.9
377.59	-418.6	-374.0	-396.9
394.26	-381.3	-341.1	-361.7
406.87	-356.1	-318.3	-338.1
410.93	-348.6	-311.5	-329.5
444.26	-289.8	-259.6	-273.1

399

n-BUTANE + n-PENTANE $CH_3 \cdot (CH_2)_2 \cdot CH_3 + CH_3 \cdot (CH_2)_3 \cdot CH_3$

1. F.W. Jessen and J.H. Lightfoot, Ind. Engng Chem. ind. Edn <u>30</u> 312 (1938) (*).

T	B_{12}
303.15	-1136

2. E.M. Dantzler, C.M. Knobler and M.L. Windsor, J. phys. Chem., Ithaca <u>72</u> 676 (1968).

    E determined from measurements of pressure changes on mixing.

T	E	B_{11}	B_{22}	B_{12}
298.15	35 ± 3	-732	-1195	-928
323.15	26 ± 2	-599	- 980	-764
348.15	20 ± 2	-501	- 810	-635
373.15	14 ± 2	-422	- 687	-540

n-BUTANE + n-HEXANE $CH_3 (CH_2)_2 CH_3 + CH_3 \cdot (CH_2)_4 CH_3$

1. E.M. Dantzler, C.M. Knobler and M.L. Windsor, J. phys. Chem., Ithaca <u>72</u> 676 (1968).

    E determined from measurements of pressure changes on mixing.

T	E	B_{11}	B_{22}	B_{12}
298.15	210 ± 35	-732	-1910	-1111
323.15	122 ± 5	-599	-1530	- 943
348.15	95 ± 2	-501	-1240	- 775
373.15	76 ± 2	-422	-1031	- 650

2. C.J. Wormald, E.J. Lewis, and D.J. Hutchings, J. chem. Thermodyn <u>11</u> 1 (1979).

    B_{12} values derived from measured excess enthalpies.

T	B_{12}	T	B_{12}
363.2	-694 ± 17	383.6	-610 ± 17
373.2	-651 ± 17	393.2	-579 ± 17

n-BUTANE + n-OCTANE $CH_3 \cdot (CH_2)_2 \cdot CH_3 + CH_3 \cdot (CH_2)_6 \cdot CH_3$

1. C.J. Wormald, E.J. Lewis, and D.J. Hutchings, J. chem. Thermodyn. <u>11</u> 1 (1979).

B_{12} values derived from measured excess enthalpies.

T	B_{12}
403.2	-735 ± 26
410.5	-714 ± 25
413.2	-700 ± 24

n-BUTANE + HYDROGEN $CH_3 \cdot (CH_2)_2 \cdot CH_3 + H_2$

1. D. McA. Mason and B.E. Eakin, J. chem. Engng Data 6 499 (1961) (*).

T	B_{12}
288.7	+9.2 ± 5

n-BUTANE + HELIUM $CH_3 \cdot (CH_2)_2 \cdot CH_3 + He$

1. A.E. Jones and W.B. Kay, A. I. Ch. E. Jl 13 720 (1967).

   Results obtained for two mixtures with slightly different composition. Precision ± 5.

T	B_{12}	T	B_{12}
373.3	45.0, 50.8	448.2	48.5, 56.2
398.2	45.5, 55.4	473.3	47.4, 46.1
423.2	51.9, 53.6	498.3	53.0, 54.3

n-BUTANE + MERCURY $CH_3 \cdot (CH_2)_2 \, CH_3 + Hg$

1. W.B. Jepson, M.J. Richardson and J.S. Rowlinson, Trans. Faraday Soc. 53 1586 (1957).

   Mercury solubility measurements in n-butane, pressure 10 - 31 atm. B_{12} values read from graph.

T	B_{12}
457.2	-195
491.2	-175
529.2	-157

2. M.J. Richardson and J.S. Rowlinson, Trans. Faraday Soc. 55 1333 (1959).

   Measurements of the solubility of mercury in the compressed gas; P → 400 atm. at 488, 529 and 573K.

n-BUTANE + NITROGEN $CH_3 \cdot (CH_2)_2 \cdot CH_3 + N_2$

1.   D. McA. Mason and B.E. Eakin, J. chem. Engng Data <u>6</u> 499 (1961) (*).

T	B_{12}
288.70	-87.1 ± 5

2.   A.J.B. Cruikshank, B.W. Gainey and C.L. Young, Trans. Faraday Soc. <u>64</u> 337 (1968).
     Gas chromatographic measurements.

T	B_{12}
308.2	-84 ± 9

3.   C.P. Hicks and C.L. Young, Trans. Faraday Soc. <u>64</u> 2675 (1968).
     Gas chromatographic measurements.

T	B_{12}
303.2	-78 ± 12
323.2	-69 ± 10

4.   C.L. Young, Trans. Faraday Soc. <u>64</u> 1537 (1968).
     Gas chromatographic measurements.

T	B_{12}
333.2	-77 ± 12

5.   R.D. Gunn, M.S. Thesis, University of California, Berkeley (1958).
     B_{12} calculated from volumetric data given by R.B. Evans and G.M.
     Watson, Ind. Engng Chem Data Ser. <u>1</u> 67 (1956).

T	B_{12}	T	B_{12}
427.6	-17.7	460.9	- 8.2
444.3	-12.7	477.6	- 2.3

1-BUTANOL+NITROGEN $CH_3 \cdot (CH_2)_2 CH_2OH + N_2$

1.   R. Massoudi and A.D. King, Jr., J. phys. Chem., Ithaca <u>77</u> 2016
     (1973).
     B_{12} derived from measurements of solubility of 1-butanol in the
     compressed gas.

T	B_{12}
298.15	-72 ± 10

2. P. Neogi and A.P. Kudchadker, J.C.S. Faraday 1 73 385 (1977).

   Chromatographic measurements using di-isodecyl phthalate, $P \to 2.5$ atm. Values of B_{12} read from graph.

T	B_{12}	T	B_{12}
303.2	-124.0 ± 17	333.2	$- 97.5 \pm 17$
318.2	-110.5 ± 17	343.2	$- 89.5 \pm 17$

DIETHYL ETHER + BENZENE $CH_3.CH_2.O.CH_2.CH_3 + C_6H_6$

1. D.H. Knoebel and W.C. Edmister, J. chem. Engng Data 13 312 (1968).

   Low pressure PVT measurements on several mixtures.

   Estimated standard deviation in B_{12} is ± 52.

T	B_{12}
333.15	-785
353.15	-769
373.15	-620

DIETHYL ETHER + n-HEXANE $(C_2H_5)_2O + CH_3.(CH_2)_4.CH_3$

1. J.H.P. Fox and J.D. Lambert, Proc. R. Soc. A210 557 (1952).

   Boyle's Law apparatus.

T	B_{11}	B_{22}	B_{12}
326.2	-880	-1380	-1130
352	-720	-1160	$- 940$

DIETHYL ETHER+NITROGEN $(C_2H_5)_2O + N_2$

1. R. Massoudi and A.D. King, Jr., J. phys. Chem., Ithaca 77 2016 (1973).

   B_{12} derived from measurements of solubility of diethyl ether in the compressed gas.

T	B_{12}
298.15	-67 ± 9

2. M. Rätzsch and H. Freydank, J. chem. Thermodyn. 3 861 (1971).

   Maximum error in B estimated to be $\pm 3\%$.

T	B_{12}
311.71	-1259
332.66	-1064

DIETHYLETHER + SULPHUR DIOXIDE $CH_3.CH_2.O.CH_2.CH_3 + SO_2$

1. R. Stryjek and A. Kreglewski, Bull. Acad. pol. Sci. Ser. Sci. chim.
    13 201 (1965).

T	B_{12}	T	B_{12}
298.15	-1283	348.15	- 702
323.25	- 907	368.15	- 471

DIETHYLAMINE + CYCLOHEXANE $(C_2H_5)_2NH + CH_2.(CH_2)_4.CH_2$

1. J.D. Lambert, S.J. Murphy and A.P. Sanday, Proc. R. Soc. A226 394
    (1954).

    Boyle's Law apparatus.

T	B_{11}	B_{22}	B_{12}
349	-1045	-1160	-990

TETRAMETHYLSILANE + NEOPENTANE $(CH_3)_4Si + (CH_3)_4C$

1. J. Bellm, W. Reineke, K. Schäfer, and B. Schramm, Ber. (dtsch.)
    Bunsenges. phys. Chem. 78 282 (1974).

    Class III.

T	B_{12}	T	B_{12}
300	-1155	430	- 627
320	-1020	460	- 573
340	- 920	490	- 530
370	- 793	520	- 494
400	- 700	550	- 462

TETRAMETHYLSILANE + SULPHUR HEXAFLUORIDE $(CH_3)_4Si + SF_6$

1. J. Bellm, W. Reineke, K. Schäfer, and B. Schramm, Ber. (dtsch.)
    Bunsenges. phys. Chem. 78 282 (1974).

    Class III.

T	B_{12}	T	B_{12}
300	-438.0	430	-195.0
320	-380.0	460	-165.0
340	-336.0	490	-142.7
370	-279.0	520	-125.0
400	-235.0	550	-105.5

PERFLUORO-n-PENTANE + n-PENTANE $\quad CF_3 \cdot (CF_2)_3 \cdot CF_3 + CH_3 \cdot (CH_2)_3 \cdot CH_3$

1. M.D.G. Garner and J.C. McCoubrey, Trans. Faraday Soc. 55 1524 (1959).
    Boyle's Law apparatus.

T	B_{11}	B_{22}	B_{12}
307.9	-1344	-1066	-969 ± 19
337.9	-1058	- 856	-693 ± 16
383.3	- 719	- 609	-625 ± 6

2. E.M. Dantzler Siebert and C.M. Knobler, J. phys. Chem., Ithaca 75
    3863 (1971).

    E determined from measurements of pressure changes on mixing.

T	E	B_{11}	B_{22}	B_{12}
323.15	228 ± 10	-1188	-974	-853
373.15	152 ± 6	- 793	-687	-588

PYRIDINE + NITROGEN $\quad CH{:}CH.CH{:}CH.CH{:}N + N_2$

1. M. Vigdergauz and V. Semkin, J. Chromat. 58 95 (1971).
    Chromatographic measurements using dinonyl phthalate.
    Class III.

T	B_{12}
353.2	-42

CIS-PENTA-1,3-DIENE + NITROGEN $\quad CH_2{:}CH.CH{:}CH.CH_3 + N_2$

1. T.M. Letcher and F. Marsicano, J. chem. Thermodyn. 6 501 (1974).
    Chromatographic measurements using octadecane.

T	B_{12}
308.15	-105 ± 20

CYCLOPENTANE + HYDROGEN $\quad CH_2 \cdot (CH_2)_3 \cdot CH_2 + H_2$

1. D.H. Desty, A. Goldup, G.R. Luckhurst, and W.T. Swanton, Gas
    Chromatography, p.76, table 3, Butterworths, London, 1962. (Ed.
    M. van Swaay).

    Gas chromatographic measurements using squalane; $P \rightarrow 5$ atm.

Maximum error in B_{12} estimated to be ± 30.

T	B_{12}
298.2	+ 5

CYCLOPENTANE + NITROGEN $CH_2 \cdot (CH_2)_3 \cdot CH_2 + N_2$

1. D.H. Desty, A. Goldup, G.R. Luckhurst, and W.T. Swanton, Gas Chromatography, p.76, table 3, Butterworths, London, 1962. (Ed. M. van Swaay).

   Gas chromatographic measurements using squalane; $P \rightarrow 5$ atm.

   Maximum error in B_{12} estimated to be ± 30.

T	B_{12}
298.2	-91

CYCLOPENTANE + OXYGEN $CH_2 \cdot (CH_2)_3 \cdot CH_2 + O_2$

1. D.H. Desty, A. Goldup, G.R. Luckhurst, and W.T. Swanton, Gas Chromatography, p.76, table 3, Butterworths, London, 1962. (Ed. M. van Swaay).

   Gas chromatographic measurements using squalane; $P \rightarrow 5$ atm.

   Maximum error in B_{12} estimated to be ± 30.

T	B_{12}
298.2	-152

n-PENT-1-ENE + HYDROGEN $CH_2 : CH \cdot (CH_2)_2 CH_3 + H_2$

1. A.J.B. Cruikshank, B.W. Gainey and C.L. Young, Trans. Faraday Soc. 64 337 (1968).

   Gas chromatographic measurements.

T	B_{12}
308.15	- 19 ± 9

n-PENT-1-ENE + NITROGEN $CH_2 : CH(CH_2)_2 CH_3 + N_2$

1. A.J.B. Cruikshank, B.W. Gainey and C.L. Young, Trans. Faraday Soc. 64 337 (1968).

   Gas chromatographic measurements.

T	B_{12}
308.15	$- 64 \pm 9$

TRANS-PENT-2-ENE + NITROGEN $CH_3 . CH:CH.CH_2 . CH_3 + N_2$

1. T.M. Letcher and F. Marsicano, J. chem. Thermodyn. $\underline{6}$ 501 (1974).
    Chromatographic measurements using octadecane.

T	B_{12}
308.15	$- 95 \pm 20$

n-PENTANE + PERFLUORO-n-HEXANE $CH_3 . (CH_2)_3 CH_3 + CF_3 (CF_2)_4 CF_3$

1. M.D.G. Garner and J.C. McCoubrey, Trans. Faraday Soc. $\underline{55}$ 1524 (1959).
    Boyle's Law apparatus.

T	B_{11}	B_{22}	B_{12}
308.0	-1066	-1970	-1184 ± 54
338.1	- 856	-1554	$- 895 \pm 15$
384.2	- 609	-1054	$- 643 \pm 41$

n-PENTANE + FLUOROBENZENE $CH_3 . (CH_2)_3 . CH_3 + C_6H_5F$

1. Z.S. Belousova, T.D. Sulimova, and V.M. Prokhorov, Russ. J. phys.
    Chem. $\underline{47}$ 235 (1973);   Zh. fiz. Khim. $\underline{47}$ 422 (1973).
    Constant volume piezometer.
    Class III.

T	B_{12}	T	B_{12}
443.2	-525	473.2	-436
453.2	-480	483.2	-416
463.2	-461	493.2	-361

n-PENTANE + BENZENE $CH_3 . (CH_2)_3 . CH_3 + C_6H_6$

1. Sh. D. Zaalishvili, Z.S. Belousova and V.P. Verkhova, Russ. J. phys.
    Chem. $\underline{45}$ 894 (1971); Zh. fiz. Khim. $\underline{45}$ 1576 (1971).
    B_M for mixtures with mole fraction n-pentane, 0.25, 0.49, 0.74:
    density range 0.2 - 0.4 mole/ℓ .

T	B_{11}	B_{22}	B_{12}
433.2	-502	-632	-487
443.2	-486	-602	-456
453.2	-458	-573	-434
463.2	-437	-545	-404
473.2	-421	-520	-380

n-PENTANE+n-HEXANE $CH_3 \cdot (CH_2)_3 \cdot CH_3 + CH_3 \cdot (CH_2)_4 CH_3$

1. E.M. Dantzler, C.M. Knobler and M.L. Windsor, J. phys. Chem., Ithaca
    72 676 (1968).

    E determined from measurements of pressure changes on mixing.

T	E	B_{11}	B_{22}	B_{12}
323.15	26 ± 6	-980	-1530	-1229
348.15	28 ± 2	-810	-1240	- 997
373.15	20 ± 2	-687	-1030	- 838

2. C.J. Wormald, E.J. Lewis, and D.J. Hutchings, J. chem. Thermodyn.
    11 1 (1979).

    B_{12} values derived from measured excess enthalpies.

T	B_{12}	T	B_{12}
343.2	-1023 ± 21	383.2	- 783 ± 18
363.2	- 893 ± 19	403.2	- 689 ± 15

n-PENTANE + n-HEPTANE $CH_3 \cdot (CH_2)_3 \cdot CH_3 + CH_3 \cdot (CH_2)_5 CH_3$

1. C.J. Wormald, E.J. Lewis, and D.J. Hutchings, J. chem. Thermodyn.
    11 1 (1979).

    B_{12} values derived from measured excess enthalpies.

T	B_{12}
393.2	-871 ± 20
403.2	-823 ± 20
413.2	-774 ± 19

n-PENTANE + n-OCTANE $CH_3 \cdot (CH_2)_3 \cdot CH_3 + CH_3 \cdot (CH_2)_6 \cdot CH_3$

1. C.J. Wormald, E.J. Lewis, and D.J. Hutchings, J. chem. Thermodyn.
    11 1 (1979).

B_{12} values derived from measured excess enthalpies.

T	B_{12}
403.2	-953 ± 25
413.2	-895 ± 23

n-PENTANE + HYDROGEN $CH_3 \cdot (CH_2)_3 \cdot CH_3 + H_2$

1. D. McA. Mason and B.E. Eakin, J. chem. Engng Data <u>6</u> 499 (1961) (*).

T	B_{12}
288.7	$+9.7 \pm 5$

2. D.H. Desty, A. Goldup, G.R. Luckhurst, and W.T. Swanton, Gas Chromatography, p.76, table 3, Butterworths, London, 1962. (Ed. M. van Swaay).

    Gas chromatographic measurements using squalane; P → 5 atm.
    Maximum error in B_{12} estimated to be ± 30.

T	B_{12}
298.2	+ 7

3. D.H. Everett, Trans. Faraday Soc. <u>61</u> 1637 (1965).
    Chromatographic method using squalane; P → 12 atm.

T	B_{12}
298.2	$+3 \pm 3$

4. A.J.B. Cruikshank, M.L. Windsor and C.L. Young, Proc. R. Soc. <u>A295</u> 271 (1966).

    Gas chromatographic measurements.

T	B_{12}
298.2	$+2 \pm 6$

n-PENTANE + HYDROGEN SULPHIDE $CH_3 \cdot (CH_2)_3 \cdot CH_3 + H_2S$

1. H.H. Reamer, B.H. Sage and W.N. Lacey, Ind. Engng Chem. ind. Edn <u>45</u> 1805 (1953).

    Volumetric behaviour of four mixtures from 280K to 500K; P → 670 atm.

n-PENTANE + NITROGEN $CH_3.(CH_2)_3.CH_3 + N_2$

1. D.H. Desty, A. Goldup, G.R. Luckhurst, and W.T. Swanton, Gas
    Chromatography, p.76, table 3, Butterworths, London, 1962. (Ed.
    M. van Swaay).

    Gas chromatographic measurements using squalane; $P \to 5$ atm.

    Maximum error in B_{12} estimated to be ± 30.

T	B_{12}
298.2	-105

2. D.H. Everett, Trans. Faraday Soc. **61** 1637 (1965).

    Chromatographic method using squalane; $P \to 12$ atm.

T	B_{12}
298.2	-76 ± 3

3. A.J.B. Cruikshank, M.L. Windsor and C.L. Young, Proc. R. Soc. **A295**
    271 (1966).

    Gas chromatographic measurements.

T	B_{12}
298.2	-103 ± 6

4. B.W. Gainey, Ph.D. Thesis, University of Bristol (1967).

    Chromatographic method.

T	B_{12}
313.2	-85 ± 6

5. A.J.B. Cruikshank, B.W. Gainey and C.L. Young, Trans. Faraday Soc.
    64 337 (1968).

    Gas chromatographic measurements.

T	B_{12}
308.15	-85 ± 13

6. C.P. Hicks and C.L. Young, Trans. Faraday Soc. **64** 2675 (1968).

    Gas chromatographic measurements.

T	B_{12}
313.2	-85 ± 10
333.2	-80 ± 10

7. C.L. Young, Trans. Faraday Soc. $\underline{64}$ 1537 (1968).
   Gas chromatographic measurements.

T	B_{12}
328.2	-66 ± 15
333.2	-72 ± 10
338.2	-78 ± 15

8. B.W. Gainey and R.L. Pecsok, J. phys. Chem., Ithaca $\underline{74}$ 2548 (1970).
   Chromatographic method.

T	B_{12}
313.2	-86 ± 6

9. M. Vigdergauz and V. Semkin, J. Chromat. $\underline{58}$ 95 (1971).
   Chromatographic measurements using dinonyl phthalate.
   Class III.

T	B_{12}
353.2	-60

10. R. Massoudi and A.D. King, Jr., J. phys. Chem., Ithaca $\underline{77}$ 2016 (1973).
    B_{12} derived from measurements of solubility of n-pentane in the
    compressed gas.

T	B_{12}
298.15	-77 ± 7

11. T.M. Letcher and F. Marsicano, J. chem. Thermodyn. $\underline{6}$ 501 (1974).
    Chromatographic measurements using octadecane.

T	B_{12}
308.15	-119 ± 20

    Chromatographic measurements using octadec-1-ene.

T	B_{12}
308.15	-76 ± 20

12. Y.-K. Leung and B.E. Eichinger, J. phys. Chem., Ithaca $\underline{78}$ 60 (1974).
    Gas chromatographic measurements using polyisobutylene. Nitrogen
    pressure up to 9 atm.

T	B_{12}
298.16	-90 ± 10

n-PENTANE + OXYGEN $CH_3.(CH_2)_3.CH_3 + O_2$

1. D.H. Desty, A. Goldup, G.R. Luckhurst, and W.T. Swanton, Gas Chromatography, p.76, table 3, Butterworths, London, 1962. (Ed. M. van Swaay).

    Gas chromatographic measurements using squalane; $P \rightarrow 5$ atm.

    Maximum error in B_{12} estimated to be $\pm$ 30.

T	B_{12}
298.2	-152

ISOPENTANE + HYDROGEN $(CH_3)_2.CH.CH_2.CH_3 + H_2$

1. D.H. Desty, A. Goldup, G.R. Luckhurst, and W.T. Swanton, Gas Chromatography, p.76, table 3, Butterworths, London, 1962. (Ed. M. van Swaay).

    Gas chromatographic measurements using squalane; $P \rightarrow 5$ atm.

    Maximum error in B_{12} estimated to be $\pm$ 30.

T	B_{12}
298.2	$+26$

ISOPENTANE + NITROGEN $(CH_3)_2.CH.CH_2.CH_3 + N_2$

1. D.H. Desty, A. Goldup, G.R. Luckhurst, and W.T. Swanton, Gas Chromatography, p.76, table 3, Butterworths, London, 1962. (Ed. M. van Swaay).

    Gas chromatographic measurements using squalane; $P \rightarrow 5$ atm.

    Maximum error in B_{12} estimated to be $\pm$ 30.

T	B_{12}
298.2	-105

2. C.L. Young, Trans. Faraday Soc. <u>64</u> 1537 (1968).

    Gas chromatographic measurements.

T	B_{12}
328.2	-64 ± 15
333.2	-76 ± 10
338.2	-81 ± 15

3. B.W. Gainey and R.L. Pecsok, J. phys. Chem., Ithaca <u>74</u> 2548 (1970).
Chromatographic method.

T	B_{12}
313.2	-92 ± 6

NEOPENTANE + SULPHUR HEXAFLUORIDE $(CH_3)_4C + SF_6$

1. J. Bellm, W. Reineke, K. Schäfer, and B. Schramm, Ber. (dtsch.)
Bunsenges. phys. Chem. <u>78</u> 282 (1974).

Class III.

T	B_{12}	T	B_{12}
300	-409.4	430	-186.5
320	-355.4	460	-160.1
340	-316.3	490	-141.0
370	-257.0	520	-119.1
400	-221.8	550	- 96.5

ISOPENTANE + OXYGEN $(CH_3)_2.CH.CH_2.CH_3 + O_2$

1. D.H. Desty, A. Goldup, G.R. Luckhurst, and W.T. Swanton, Gas
Chromatography, p.76, table 3, Butterworths, London, 1962. (Ed.
M. van Swaay).

Gas chromatographic measurements using squalane; P → 5 atm.

Maximum error in B_{12} estimated to be ± 30.

T	B_{12}
298.2	-125

NEOPENTANE + HELIUM $(CH_3)_4C + He$

1. G.L. Baughman, S.P. Westhoff, S. Dincer, D.D. Duston and A.J. Kidnay,
J. chem. Thermodyn. <u>7</u> 875 (1975).

B_{12} determined from measurements of the solubility of neopentane in
the compressed gas. Maximum error in B_{12} is estimated to be 10%.

T	B_{12}	T	B_{12}
199.99	69	240.00	60
210.00	76	249.58	60
220.00	59	257.86	59
230.00	60		

NEOPENTANE + NITROGEN $(CH_3)_4C + N_2$

1. G.L. Baughman, S.P. Westhoff, S. Dincer, D.D. Duston and A.J. Kidnay, J. chem. Thermodyn. $\underline{7}$ 875 (1975).

   B_{12} determined from measurements of the solubility of neopentane in the compressed gas. Maximum error in B_{12} is estimated to be 10%.

   (a) Graphical method.

T	B_{12}	T	B_{12}
199.62	-265	239.48	-155
209.59	-226	249.47	-137
219.56	-192	257.86	-114
229.53	-174		

   (b) non-linear regression analysis.

T	B_{12}	T	B_{12}
199.62	-241	239.48	-153
209.59	-203	249.47	-130
219.56	-180	257.86	-101
229.53	-162		

   (c) analysis described in text.

T	B_{12}	T	B_{12}
199.62	-236	239.48	-146
209.59	-210	249.47	-126
219.56	-185	257.86	-106
229.53	-170		

HEXAFLUOROBENZENE+BENZENE $C_6F_6 + C_6H_6$

1. E.M. Dantzler and C.M. Knobler, J. phys. Chem., Ithaca $\underline{73}$ 1602 (1969).

   E determined from measurements of pressure changes on mixing.

T	E	B_{11}	B_{22}	B_{12}
323.15	-1029 ± 20	-1962	-1207	-2613
348.15	- 555 ± 12	-1550	-1013	-1837
373.15	- 343 ± 15	-1256	- 859	-1401
373.15	- 353 ± 3			-1411

2. R.J. Powell, Ph.D. Thesis, University of Strathclyde (1969).
    Class II.

T	B_{12}	T	B_{12}
365.056	-1498	414.988	- 892
389.721	-1217	444.202	- 715

HEXAFLUOROBENZENE+c-HEXANE C_6F_6 + $CH_2(CH_2)_4CH_2$

1. E.M. Dantzler and C.M. Knobler, J. phys. Chem., Ithaca __73__ 1602 (1969).

    E determined from measurements of pressure changes on mixing.

T	E	B_{11}	B_{22}	B_{12}
348.15	98 ± 10	-1550	-1121	-1238

2. R.J. Powell, Ph.D. Thesis, University of Strathclyde (1969).
    Class II.

T	B_{12}
364.195	-997
389.692	-830
414.856	-690

HEXAFLUOROBENZENE + NITROGEN C_6F_6 + N_2

1. B.W. Gainey, Ph.D. Thesis, University of Bristol (1967).
    Chromatographic method.

T	B_{12}
313.2	-128 ± 6

2. B.W. Gainey and R.L. Pecsok, J. phys. Chem., Ithaca __74__ 2548 (1970).
    Chromatographic method.

T	B_{12}
313.2	-126 ± 6

PERFLUORO-n-HEXANE + n-HEXANE $CF_3(CF_2)_4 \cdot CF_3$ $+CH_3 \cdot (CH_2)_4 \cdot CH_3$

1. E.M. Dantzler Siebert and C.M. Knobler, J. phys. Chem., Ithaca 75 3863 (1971).

   E determined from measurements of pressure changes on mixing.

T	E	B_{11}	B_{22}	B_{12}
323.15	315 ± 6	-1712	-1513	-1297
373.15	211 ± 3	-1128	-1035	$- 871$

BROMOBENZENE + BENZENE $C_6H_5Br + C_6H_6$

1. M.P. Khosla, B.S. Mahl, S.L. Chopra, and P.P. Singh, Ind. J. Chem. 10 1098 (1972).

   Boyle's apparatus. Error in B_{12} estimated to be 3.5%.

T	B_{12}
303.2	-2166

CHLOROBENZENE + BENZENE $C_6H_5Cl + C_6H_6$

1. M.P. Khosla, B.S. Mahl, S.L. Chopra, and P.P. Singh, Ind. J. Chem. 10 1098 (1972).

   Boyle's apparatus. Error in B_{12} estimated to be 3.5%.

T	B_{12}
303.2	-2025

FLUOROBENZENE + BENZENE $C_6H_5F + C_6H_6$

1. Sh. D. Zaalishvili, Z.S. Belousova and V.P. Verkhova, Russ. J. phys. Chem. 46 291 (1972).

   Constant volume piezometer: P 6 → 12 atm.

T	B_{11}	B_{22}	B_{12}
453.2	-627	-584	-598
463.2	-593	-555	-563
473.2	-560	-532	-540

483.2	-530		-506	-507
493.2	-502		-483	-471

FLUOROBENZENE + TOLUENE $C_6H_5F + C_6H_5.CH_3$

1. Z.S. Belousova and T.D. Sulimova, Russ. J. phys. Chem. $\underline{50}$ 1272
    (1976);   Zh. fiz. Khim. $\underline{50}$ 2122 (1976).

    Constant volume piezometer.

T	B_{12}	T	B_{12}
443.2	-795	473.2	-655
453.2	-774	483.2	-576
463.2	-709	493.2	-519

FLUOROBENZENE + n-HEPTANE $C_6H_5F + CH_3.(CH_2)_5.CH_3$

1. Z.S. Belousova and T.D. Sulimova, Russ. J. phys. Chem. $\underline{50}$ 585 (1976);
    Zh. fiz. Khim. $\underline{50}$ 981 (1976).

    Constant volume piezometer.

    Class III.

T	B_{12}	T	B_{12}
433.2	-772	473.2	-564
443.2	-706	483.2	-536
453.2	-648	493.2	-515
463.2	-600		

FLUOROBENZENE + n-OCTANE $C_6H_5F + CH_3.(CH_2)_6.CH_3$

1. T.D. Sulimova and Z.S. Belousova, Russ. J. phys. Chem. $\underline{48}$ 620 (1974);
    Zh. fiz. Khim. $\underline{48}$ 1061 (1974).

    Constant volume piezometer.

T	B_{12}	T	B_{12}
443.2	-993	473.2	-855
453.2	-944	483.2	-816
463.2	-897	493.2	-782

BENZENE + c-HEXANE $C_6H_6 + \underline{CH_2(CH_2)_4CH_2}$

1. F.G. Waelbroeck, J. chem. Phys. $\underline{23}$ 749 (1955).

T	B_{11}	B_{22}	B_{12}
328.2	-	-1355 ± 16	-1227 ± 30
333.2	-1117 ± 11	-1268 ± 13	-1203 ± 23
338.2	-1088 ± 2	-1236 ± 11	-1158 ± 12
343.2	-1035 ± 11	-1180 ± 10	-1109 ± 20
348.2	-1011 ± 9	-1171 ± 3	-1075 ± 9

2. J.D. Cox and D. Stubley, Trans. Faraday Soc. 56 484 (1960).

T	B_{11}	B_{22}	B_{12}
373.2	-814	-910	-918

3. G.A. Bottomley and I.H. Coopes, Nature 193 268 (1962).
    Low pressure differential method.

T	B_{11}	B_{22}	B_{12}
308.2	-1338	-1457	-1346
323.2	-1186	-1309	-1215
343.2	-1028	-1121	-1041

4. Z.S. Belousova and T.D. Sulimova, Russ. J. phys. Chem. 50 292 (1976);
    Zh. fiz. Khim. 50 507 (1976).
    Constant volume piezometer.
    Class III.

T	B_{12}	T	B_{12}
433.2	-614	473.2	-517
443.2	-567	483.2	-492
453.2	-552	493.2	-478
463.2	-535		

BENZENE + n-HEXANE $C_6H_6 + CH_3 \cdot (CH_2)_4 \, CH_3$

1. Sh. D. Zaalishvili, Z.S. Belousova and V.P. Verkhova, Russ. J. phys.
    Chem. 45 149 (1971); Zh. fiz. Khim 45 268 (1971).
    B_M for mixtures with mole fraction   n-hexane 0.25 and 0.49,
    P range 5 - 12 atm.

T	B_{11}	B_{22}	B_{12}
433.2	-632	-714	-761
443.2	-602	-677	-738
453.2	-573	-644	-674
463.2	-545	-613	-610
473.2	-520	-583	-557
478.2	-508	-570	-545

2. Z.S. Belousova and V.P. Verkhova, Russ. J. phys. Chem. 47 236 (1973); Zh. fiz. Khim. 47 424 (1973).

T	B_{12}
483.2	-498
493.2	-435

BENZENE + TOLUENE $C_6H_6 + C_6H_5.CH_3$

1. Z.S. Belousova and T.D. Sulimova, Russ. J. phys. Chem. 48 254 (1974); Zh. fiz. Khim. 48 439 (1974).

Constant volume piezometer. Maximum absolute error in B given as ± 10.

T	B_{12}	T	B_{12}
453.2	-676	483.2	-559
463.2	-637	493.2	-530
473.2	-594		

BENZENE + n-HEPTANE $C_6H_6 + CH_3.(CH_2)_5.CH_3$

1. Z.S. Belousova and Sh. D. Zaalishvili, Russ. J. phys. Chem. 41 1290 (1967); Zh. fiz. Khim. 41 2388 (1967).

B_M for 11 mixtures : P range 4 - 10 atm.

T	B_{11}	B_{22}	B_{12}
463.2	-555	-886	-668
473.2	-532	-825	-621
483.2	-506	-766	-581
493.2	-483	-703	-546
503.2	-458	-655	-523

BENZENE + n-OCTANE $C_6H_6 \div CH_3 \cdot (CH_2)_6 \, CH_3$

1. Sh. D. Zaalishvili, Z.S. Belousova and V.P. Verkhova, Russ. J.
    phys. Chem. <u>45</u> 902 (1971); Zh. fiz. Khim. <u>45</u> 1589 (1971).

    B_M for mixtures with mole fraction n-octane 0.47, density range
    0.16 - 0.23 mole/ℓ , T range 473.2 - 498.2 K.

T	B_{11}	B_{22}	B_{12}
478.2	-508	-1022	-666
488.2	-486	- 960	-642
498.2	-468	- 900	-595

BENZENE + HYDROGEN $C_6H_6 + H_2$

1. J.F. Connolly, Physics Fluids <u>4</u> 1494 (1961).

    Solubility measurements of benzene in the compressed gas; P → 50
    atm.

T	B_{12}
323.15	-7.9
323.15	-7.2
373.15	+1.6

Compressibility measurements on three mixtures.

T	B_{12}	C_{112}	C_{122}
473.2	16	8000	1000
493.2	19		
513.2	21		
533.2	23	8000	1000
553.2	25		
573.2	27	8000	1000

2. D.H. Desty, A. Goldup, G.R. Luckhurst, and W.T. Swanton, Gas
    Chromatography, p.76, table 3, Butterworths, London, 1962. (Ed.
    M. van Swaay).

    Gas chromatographic measurements using squalane; P → 5 atm.

    Maximum error in B_{12} estimated to be ± 30.

T	B_{12}
298.2	-13

3. B.W. Gainey and C.L. Young, Trans. Faraday Soc. <u>64</u> 349 (1968).
    Gas chromatographic measurements.

T	B_{12}
313.15	-9 ± 8

4. D.H. Everett, B.W. Gainey and C.L. Young, Trans. Faraday Soc. <u>64</u> 2667 (1968).
    Gas chromatographic measurements, using squalane.

T	B_{12}
323.15	-5 ± 8

5. C.R. Coan and A.D. King, Jr., J. Chromat. <u>44</u> 429 (1969).
    B_{12} determined from measurements of solubility of benzene in the compressed gas: P → 65 atm.

T	B_{12}
323.2	4 ± 3

BENZENE + HELIUM C_6H_6 + He

1. D.H. Everett, B.W. Gainey and C.L. Young, Trans. Faraday Soc. <u>64</u> 2667 (1968).
    Gas chromatographic measurements, using n-octadecane.

T	B_{12}
323.15	+49 ± 8

    Gas chromatographic measurements, using squalane.

T	B_{12}
323.15	-57 ± 8

2. C.R. Coan and A.D. King, Jr., J. Chromat. <u>44</u> 429 (1969).
    B_{12} determined from measurements of solubility of benzene in the compressed gas: P → 65 atm.

T	B_{12}
323.2	+67 ± 4

BENZENE + NITROGEN $C_6H_6 + N_2$

1. D.H. Desty, A. Goldup, G.R. Luckhurst, and W.T. Swanton, Gas
   Chromatography, p.76, table 3, Butterworths, London, 1962. (Ed.
   M. van Swaay).

   Gas chromatographic measurements using squalane; $P \to 5$ atm.

   Maximum error in B_{12} estimated to be ± 30.

T	B_{12}
298.2	-117

2. A.J.B. Cruikshank, M.L. Windsor and C.L. Young, Proc. R. Soc. A295
   271 (1966).

   Gas chromatographic measurements.

T	B_{12}
303.2	-121 ± 6
313.2	-108 ± 6

3. A.J.B. Cruikshank, B.W. Gainey and C.L. Young, Trans. Faraday Soc.
   64 337 (1968).

   Gas chromatographic measurements.

T	B_{12}
308.15	-104 ± 9

4. D.H. Everett, B.W. Gainey and C.L. Young, Trans. Faraday Soc. 64
   2667 (1968).

   Gas chromatographic measurements, using squalane.

T	B_{12}
323.15	-87 ± 8

5. B.W. Gainey and C.L. Young, Trans. Faraday Soc. 64 348 (1968).

   Gas chromatographic measurements.

T	B_{12}	T	B_{12}
293.15	-120 ± 12	333.15	- 89 ± 10
298.15	-109 ± 10	338.15	- 92 ± 40
303.15	-107 ± 10	338.15	- 79 ± 10
308.15	-104 ± 10	340.65	- 86 ± 10
323.15	- 94 ± 10	343.15	- 74 ± 10
328.15	- 86 ± 10	348.15	- 70 ± 10
333.15	- 93 ± 10	348.15	- 79 ± 10

6. C.R. Coan and A.D. King, Jr., J. Chromat. <u>44</u> 429 (1969).

B_{12} determined from measurements of solubility of benzene in the compressed gas: P → 65 atm.

T	B_{12}
308.2	-97 ± 3
323.2	-85 ± 3

7. A.J.B. Cruikshank, B.W. Gainey, C.P. Hicks, T.M. Letcher, R.W. Moody and C.L. Young, Trans. Faraday Soc. <u>65</u> 1014 (1969).
Gas chromatographic measurements.

T	B_{12}
323.2	-98 ± 9

8. B.W. Gainey and R.L. Pecsok, J. phys. Chem., Ithaca <u>74</u> 2548 (1970).
Chromatographic method.

T	B_{12}
313.2	-100 ± 6

9. M. Vigdergauz and V. Semkin, J. Chromat. <u>58</u> 95 (1971).
Chromatographic measurements using dinonyl phthalate.
Class III.

T	B_{12}
353.2	-74

10. B.K. Kaul, A.P. Kudchadker and D. Devaprabhakara, J. chem. Soc. Faraday 1 <u>69</u> 1821 (1973).
Gas-liquid chromatography: P → 6 atm. Values of B_{12} read from graph; quoted experimental uncertainty in B ± 12.

T	B	T	B
303.15	-128	333.15	- 83
313.15	-112	343.15	- 77
323.15	- 92		

11. Y.-K. Leung and B.E. Eichinger   J. phys. Chem., Ithaca <u>78</u> 60 (1974).
Gas chromatographic measurements using polyisobutylene. Nitrogen pressures up to 9 atm.

T	B_{12}
298.16	-126 ± 15

BENZENE + OXYGEN $C_6H_6 + O_2$

1. D.H. Desty, A. Goldup, G.R. Luckhurst, and W.T. Swanton, Gas
    Chromatography, p.76, table 3, Butterworths, London, 1962. (Ed.
    M. van Swaay).

    Gas chromatographic measurements using squalane; $P \rightarrow 5$ atm.

    Maximum error in B_{12} estimated to be ± 30.

T	B_{12}
298.2	-157

2. H. Bradley, Jr., and A.D. King, Jr., J. chem. Phys. <u>47</u> 1189 (1967).
    B_{12} determined from measurements of solubility of benzene in the
    compressed gas: $P \rightarrow 160$ atm.

T	B_{12}
298.2	-104 ± 6

3. D.H. Everett, B.W. Gainey and C.L. Young, Trans. Faraday Soc. <u>64</u>
    2667 (1968).

    Gas chromatographic measurements, using squalane.

T	B_{12}
323.15	-79 ± 8

Cis-HEXA-1,3,5-TRIENE + NITROGEN $CH_2{:}CH.CH{:}CH.CH{:}CH_2 + N_2$

1. T.M. Letcher and F. Marsicano, J. chem. Thermodyn. <u>6</u> 501 (1974).
    Chromatographic measurements using octadecane.

T	B_{12}
308.15	-122 ± 20
323.15	-105 ± 20

Chromatographic measurements using octadec-1-ene.

T	B_{12}
308.15	-132 ± 20
323.15	-113 ± 20

Chromatographic measurements using hexadecane.

T	B_{12}
308.15	-119 ± 20

Chromatographic measurements using hexadec-1-ene.

T	B_{12}
308.15	-148 ± 20

TRANS-HEXA-1,3-DIENE + NITROGEN $CH_2:CH.CH:CH.CH_2.CH_3 + N_2$

1. T.M. Letcher and F. Marsicano, J. chem. Thermodyn. <u>6</u> 501 (1974).
    Chromatographic measurements using octadecane.

T	B_{12}
308.15	-100 ± 20
323.15	- 96 ± 20

Chromatographic measurements using octadec-1-ene.

T	B_{12}
308.15	-122 ± 20
323.15	-109 ± 20

Chromatographic measurements using hexadecane.

T	B_{12}
308.15	-114 ± 20

Chromatographic measurements using hexadec-1-ene.

T	B_{12}
308.15	-114 ± 20

TRANS-HEXA-1,4-DIENE + NITROGEN $CH_2:CH.CH_2.CH:CH.CH_3 + N_2$

1. T.M. Letcher and F. Marsicano, J. chem. Thermodyn. <u>6</u> 501 (1974).
    Chromatographic measurements using octadecane.

T	B_{12}
308.15	-114 ± 20

Chromatographic measurements using octadec-1-ene.

T	B_{12}
308.15	$- 95 \pm 20$

HEXA-1,5-DIENE +NITROGEN $CH_2:CH.CH_2.CH_2.CH:CH_2 + N_2$

1. T.M. Letcher and F. Marsicano, J. chem. Thermodyn. <u>6</u> 501 (1974).
    Chromatographic measurements using octadecane.

T	B_{12}
323.15	$- 91 \pm 20$

Chromatographic measurements using octadec-1-ene.

T	B_{12}
308.15	-107 ± 20
323.15	$- 98 \pm 20$

TRANS,TRANS-HEXA-2,4-DIENE + NITROGEN $CH_3.CH:CH.CH:CH.CH_3 + N_2$

1. T.M. Letcher and F. Marsicano, J. chem. Thermodyn. <u>6</u> 501 (1974).

    Chromatographic measurements using octadecane.

T	B_{12}
308.15	-134 ± 20

Chromatographic measurements using octadec-1-ene.

T	B_{12}
308.15	-113 ± 20

Chromatographic measurements using hexadecane.

T	B_{12}
308.15	-118 ± 20

Chromatographic measurements using hexadec-1-ene.

T	B_{12}
308.15	-142 ± 20

CYCLOHEXANE + HYDROGEN $\quad CH_2 \cdot (CH_2)_4 \cdot CH_2 + H_2$

1. D.H. Desty, A. Goldup, G.R. Luckhurst, and W.T. Swanton, Gas Chromatography, p.76, table 3, Butterworths, London, 1962. (Ed. M. van Swaay).

   Gas chromatographic measurements using squalane; $P \to 5$ atm.

   Maximum error in B_{12} estimated to be ± 30.

T	B_{12}
298.2	-18

CYCLOHEXANE + NITROGEN $\quad CH_2 \cdot (CH_2)_4 \cdot CH_2 + N_2$

1. D.H. Desty, A. Goldup, G.R. Luckhurst, and W.T. Swanton, Gas Chromatography, p.76, table 3, Butterworths, London, 1962. (Ed. M. van Swaay).

   Gas chromatographic measurements using squalane: $P \to 5$ atm.

   Maximum error in B_{12} estimated to be ± 30.

T	B_{12}
298.2	-120

2. A.J.B. Cruikshank, B.W. Gainey and C.L. Young, Trans. Faraday Soc. 64 337 (1968).

   Gas chromatographic measurements.

T	B_{12}
308.15	-122 ± 13

3. C.L. Young, Trans. Faraday Soc. 64 1537 (1968).

   Gas chromatographic measurements.

T	B_{12}
333.2	-80 ± 10

4. B.W. Gainey and R.L. Pecsok, J. phys. Chem., Ithaca 74 2548 (1970).

   Chromatographic method.

T	B_{12}
313.2	-116 ± 6

5. M. Vigdergauz and V. Semkin, J. Chromat. 58 95 (1971).

Chromatographic measurements using dinonyl phthalate.
Class III.

T	B_{12}
353.2	-63

6. B.K. Kaul, A.P. Kudchadker and D. Devaprabhakara, J. chem. Soc. Faraday 1 <u>69</u> 1821 (1973).

Gas-liquid chromatography: $P \to 6$ atm. Values of B_{12} read from graph; quoted uncertainty in $B \pm 12$.

T	B_{12}	T	B_{12}
303.15	-142	323.15	-110
313.15	- 98	333.15	-100

7. Y.-K. Leung and B.E. Eichinger, J. phys. Chem., Ithaca <u>78</u> 60 (1974).

Gas chromatographic measurements using polyisobutylene. Nitrogen pressures up to 9 atm.

T	B_{12}
298.16	-130 ± 15

CYCLOHEXANE + OXYGEN $CH_2 \cdot (CH_2)_4 \cdot CH_2 + O_2$

1. D.H. Desty, A. Goldup, G.R. Luckhurst, and W.T. Swanton, Gas Chromatography, p.76, table 3, Butterworths, London, 1962. (Ed. M. van Swaay).

Gas chromatographic measurements using squalane; $P \to 5$ atm.

Maximum error in B_{12} estimated to be ± 30.

T	B_{12}
298.2	-163

METHYLCYCLOPENTANE + HYDROGEN $CH(CH_3) \cdot (CH_2)_3 \cdot CH_2 + H_2$

1. D.H. Desty, A. Goldup, G.R. Luckhurst, and W.T. Swanton, Gas Chromatography, p.76, table 3, Butterworths, London, 1962. (Ed. M. van Swaay).

Gas chromatographic measurements using squalane; $P \to 5$ atm.

Maximum error in B_{12} estimated to be ± 30.

T	B_{12}
298.2	-12

METHYLCYCLOPENTANE + NITROGEN $\underline{CH(CH_3).(CH_2)_3.CH_2}$ + N_2

1. D.H. Desty, A. Goldup, G.R. Luckhurst, and W.T. Swanton, Gas Chromatography, p.76, table 3, Butterworths, London, 1962. (Ed. M. van Swaay).

   Gas chromatographic measurements using squalane; $P \rightarrow 5$ atm.

   Maximum error in B_{12} estimated to be ± 30.

T	B_{12}
298.2	-128

n-HEX-1-ENE + HYDROGEN $CH_2:CH(CH_2)_3CH_3$ + H_2

1. A.J.B. Cruikshank, B.W. Gainey and C.L. Young, Trans. Faraday Soc. $\underline{64}$ 337 (1968).

   Gas chromatographic measurements.

T	B_{12}
308.15	- 13 $\pm$ 9

n-HEX-1-ENE + NITROGEN $CH_2:CH(CH_2)_3.CH_3$ + N_2

1. A.J.B. Cruikshank, B.W. Gainey and C.L. Young, Trans. Faraday Soc. $\underline{64}$ 337 (1968).

   Gas chromatographic measurements.

T	B_{12}
308.15	-110 $\pm$ 9

2. T.M. Letcher and F. Marsicano, J. chem. Thermodyn. $\underline{6}$ 501 (1974).

   Chromatographic measurements using octadecane.

T	B_{12}
308.15	-122 $\pm$ 20
323.15	- 99 $\pm$ 20

   Chromatographic measurements using octadec-1-ene.

T	B_{12}
308.15	-107 $\pm$ 20
323.15	-110 $\pm$ 20

TRANS-HEX-2-ENE + NITROGEN $CH_3 \cdot CH:CH.(CH_2)_2 \cdot CH_3 + N_2$

1. T.M. Letcher and F. Marsicano, J. chem. Thermodyn. <u>6</u> 501 (1974).
    Chromatographic measurements using octadecane.

T	B_{12}
308.15	-127 ± 20

Chromatographic measurements using octadec-1-ene.

T	B_{12}
308.15	-105 ± 20

TRANS-HEX-3-ENE + NITROGEN $CH_3 \cdot CH_2 \cdot CH:CH.CH_2 \cdot CH_3 + N_2$

1. T.M. Letcher and F. Marsicano, J. chem. Thermodyn. <u>6</u> 501 (1974).
    Chromatographic measurements using octadecane.

T	B_{12}
308.15	-108 ± 20

Chromatographic measurements using octadec-1-ene.

T	B_{12}
308.15	-108 ± 20

Chromatographic measurements using hexadecane.

T	B_{12}
308.15	-120 ± 20

Chromatographic measurements using hexadec-1-ene.

T	B_{12}
308.15	-121 ± 20

BUTYL ACETATE + NITROGEN $CH_3 \cdot COO.(CH_2)_3 CH_3 + N_2$

1. M. Vigdergauz and V. Semkin, J. Chromat. <u>58</u> 95 (1971).
    Chromatographic measurements using dinonyl phthalate.
    Class III.

T	B_{12}
353.2	-79

n-HEXANE + n-OCTANE $CH_3 \cdot (CH_2)_4 \cdot CH_3 + CH_3 \cdot (CH_2)_6 \cdot CH_3$

1. C.J. Wormald, E.J. Lewis, and D.J. Hutchings, J. chem. Thermodyn.
   <u>11</u> 1 (1979).

   B_{12} values derived from measured excess enthalpies.

T	B_{12}
403.2	-1173 ± 26
413.2	-1105 ± 25

n-HEXANE + HYDROGEN $CH_3 \cdot (CH_2)_4 \cdot CH_3 + H_2$

1. D.H. Desty, A. Goldup, G.R. Luckhurst, and W.T. Swanton, Gas
   Chromatography, p.76, table 3, Butterworths, London, 1962. (Ed.
   M. van Swaay).

   Gas chromatographic measurements using squalane; P → 5 atm.

   Maximum error in B_{12} estimated to be ± 30.

T	B_{12}
298.2	0

2. A.J.B. Cruikshank, M.L. Windsor and C.L. Young, Proc. R. Soc. <u>A295</u>
   271 (1966).

   Gas chromatographic measurements.

T	B_{12}
298.2	+ 6 ± 6

n-HEXANE + NITROGEN $CH_3 \cdot (CH_2)_4 \cdot CH_3 + N_2$

1. D.H. Desty, A. Goldup, G.R. Luckhurst, and W.T. Swanton, Gas
   Chromatography, p.76, table 3, Butterworths, London, 1962. (Ed.
   M. van Swaay).

   Gas chromatographic measurements using squalane; P → 5 atm.

   Maximum error in B_{12} estimated to be ± 30.

T	B_{12}
298.2	-128

2. A.J.B. Cruikshank, M.L. Windsor and C.L. Young, Proc. R. Soc. <u>A295</u>
   271 (1966).

   Gas chromatographic measurements.

T	B_{12}
298.2	-112 ± 6
303.2	-103 ± 6
323.2	$- 79 \pm 6$

3. B.W. Gainey, Ph.D. Thesis, University of Bristol (1967).
Chromatographic method.

T	B_{12}
313.2	-107 ± 6

4. A.J.B. Cruikshank, B.W. Gainey and C.L. Young, Trans. Faraday Soc. 64 337 (1968).
Gas chromatographic measurements.

T	B_{12}
308.15	-113 ± 9

5. C.P. Hicks and C.L. Young, Trans. Faraday Soc. 64 2675 (1968).
Gas chromatographic measurements.

T	B_{12}
303.2	-108 ± 15
313.2	-107 ± 10
333.2	$- 93 \pm 10$

6. C.L. Young, Trans. Faraday Soc. 64 1537 (1968).
Gas chromatographic measurements.

T	B_{12}
328.2	$- 90 \pm 15$
333.2	$- 98 \pm 10$
338.2	-104 ± 15

7. B.W. Gainey and R.L. Pecsok, J. phys. Chem., Ithaca 74 2548 (1970).
Chromatographic method.

T	B_{12}
313.2	-110 ± 6

8. M. Vigdergauz and V. Semkin, J. Chromat. <u>58</u> 95 (1971).
Chromatographic measurements using dinonyl phthalate.
Class III.

T	B_{12}
353.2	-69

9. B.K. Kaul, A.P. Kudchadker and D. Devaprabhakara, J. chem. Soc.
Faraday 1 <u>69</u> 1821 (1973).

Gas-liquid chromatography: $P \rightarrow 6$ atm. Values of B_{12} read from graph;
quoted uncertainty in B_{12} ± 12.

T	B_{12}
303.15	-132
313.15	-103
323.15	- 57

10. T.M. Letcher and F. Marsicano, J. chem. Thermodyn. <u>6</u> 501 (1974).
Chromatographic measurements using hexadecane.

T	B_{12}
308.15	-121 ± 20

Chromatographic measurements using octadecane.

T	B_{12}
308.15	-115 ± 20
323.15	- 96 ± 20

Chromatographic measurements using octadec-1-ene.

T	B_{12}
308.15	-103 ± 20
323.15	- 95 ± 20

11. Y.-K. Leung and B.E. Eichinger, J. phys. Chem., Ithaca <u>78</u> 60 (1974).
Gas chromatographic measurements using polyisobutylene. Nitrogen
pressure up to 9 atm.

T	B_{12}
298.16	-118 ± 10

n-HEXANE + OXYGEN $CH_3 \cdot (CH_2)_4 \cdot CH_3 + O_2$

1. D.H. Desty, A. Goldup, G.R. Luckhurst, and W.T. Swanton, Gas
    Chromatography, p.76, table 3, Butterworths, London, 1962. (Ed.
    M. van Swaay).

    Gas chromatographic measurements using squalane; P → 5 atm.

    Maximum error in B_{12} estimated to be ± 30.

T	B_{12}
298.2	-163

2-METHYLPENTANE + HYDROGEN $(CH_3)_2 \cdot CH \cdot (CH_2)_2 \cdot CH_3 + H_2$

1. D.H. Desty, A. Goldup, G.R. Luckhurst, and W.T. Swanton, Gas
    Chromatography, p.76, table 3, Butterworths, London, 1962. (Ed.
    M. van Swaay).

    Gas chromatographic measurements using squalane; P → 5 atm.

    Maximum error in B_{12} estimated to be ± 30.

T	B_{12}
298.2	- 7

2-METHYLPENTANE + NITROGEN $(CH_3)_2 \cdot CH \cdot (CH_2)_2 \cdot CH_3 + N_2$

1. D.H. Desty, A. Goldup, G.R. Luckhurst, and W.T. Swanton, Gas
    Chromatography, p.76, table 3, Butterworths, London, 1962. (Ed.
    M. van Swaay).

    Gas chromatographic measurements using squalane; P → 5 atm.

    Maximum error in B_{12} estimated to be ± 30.

T	B_{12}
298.2	-127

2. C.L. Young, Ph.D. Thesis, University of Bristol (1967).
    Chromatographic method.

T	B_{12}
303.2	-93 ± 6

3. A.J.B. Cruikshank, B.W. Gainey and C.L. Young, Trans. Faraday Soc.
    <u>64</u> 337 (1968).

    Gas chromatographic measurements.

T	B_{12}
308.15	- 91 ± 9

4. B.W. Gainey and R.L. Pecsok, J. phys. Chem., Ithaca <u>74</u> 2548 (1970). Chromatographic method.

T	B_{12}
313.2	-106 ± 6

2-METHYLPENTANE + OXYGEN $(CH_3)_2.CH.(CH_2)_2.CH_3 + O_2$

1. D.H. Desty, A. Goldup, G.R. Luckhurst, and W.T. Swanton, Gas Chromatography, p.76, table 3, Butterworths, London, 1962. (Ed. M. van Swaay).

   Gas chromatographic measurements using squalane; P → 5 atm.

   Maximum error in B_{12} estimated to be ± 30.

T	B_{12}
298.2	-155

3-METHYLPENTANE + HYDROGEN $CH_3.CH_2.CH(CH_3).CH_2.CH_3 + H_2$

1. D.H. Desty, A. Goldup, G.R. Luckhurst, and W.T. Swanton, Gas Chromatography, p.76, table 3, Butterworths, London, 1962. (Ed. M. van Swaay).

   Gas chromatographic measurements using squalane; P → 5 atm.

   Maximum error in B_{12} estimated to be ± 30.

T	B_{12}
298.2	+3

3-METHYLPENTANE + NITROGEN $CH_3.CH_2.CH(CH_3).CH_2.CH_3 + N_2$

1. D.H. Desty, A. Goldup, G.R. Luckhurst, and W.T. Swanton, Gas Chromatography, p.76, table 3, Butterworths, London, 1962. (Ed. M. van Swaay).

   Gas chromatographic measurements using squalane; P → 5 atm.

   Maximum error in B_{12} estimated to be ± 30.

T	B_{12}
298.2	-117

2. B.W. Gainey and R.L. Pecsok, J. phys. Chem., Ithaca <u>74</u> 2548 (1970).
    Chromatographic method.

T	B_{12}
313.2	-96 ± 6

3-METHYLPENTANE + OXYGEN $CH_3.CH_2.CH(CH_3).CH_2.CH_3 + O_2$

1. D.H. Desty, A. Goldup, G.R. Luckhurst, and W.T. Swanton, Gas
    Chromatography, p.76, table 3, Butterworths, London, 1962. (Ed.
    M. van Swaay).

    Gas chromatographic measurements using squalane; P → 5 atm.

    Maximum error in B_{12} estimated to be ± 30.

T	B_{12}
298.2	-163

2,2-DIMETHYLBUTANE + HYDROGEN $(CH_3)_3.C.CH_2.CH_3 + H_2$

1. D.H. Desty, A. Goldup, G.R. Luckhurst, and W.T. Swanton, Gas
    Chromatography, p.76, table 3, Butterworths, London, 1962. (Ed.
    M. van Swaay).

    Gas chromatographic measurements using squalane; P → 5 atm.

    Maximum error in B_{12} estimated to be ± 30.

T	B_{12}
298.2	$+26$

2. A.J.B. Cruikshank, M.L. Windsor and C.L. Young, Proc. R. Soc. <u>A295</u>
    271 (1966).

    Gas chromatographic measurements.

T	B_{12}
298.2	$+ 9 \pm 6$

2,2-DIMETHYLBUTANE + NITROGEN $(CH_3)_3.C.CH_2.CH_3 + N_2$

1. D.H. Desty, A. Goldup, G.R. Luckhurst, and W.T. Swanton, Gas
    Chromatography, p.76, table 3, Butterworths, London, 1962. (Ed.
    M. van Swaay).

    Gas chromatographic measurements using squalane; P → 5 atm.

    Maximum error in B_{12} estimated to be ± 30.

T	B_{12}
298.2	-105

2. D.H. Everett, Trans. Faraday Soc. __61__ 1637 (1965).
   Chromatographic method using squalane; P → 12 atm.

T	B_{12}
298.2	-88 ± 3

3. A.J.B. Cruikshank, M.L. Windsor and C.L. Young, Proc. R. Soc. __A295__ 271 (1966).
   Gas chromatographic measurements.

T	B_{12}
298.2	-94 ± 6

4. A.J.B. Cruikshank, B.W. Gainey and C.L. Young, Trans. Faraday Soc. __64__ 337 (1968).
   Gas chromatographic measurements.

T	B_{12}
308.15	-60 ± 20

5. B.W. Gainey and R.L. Pecsok, J. phys. Chem., Ithaca __74__ 2548 (1970).
   Chromatographic method.

T	B_{12}
313.2	-83 ± 6

2,2-DIMETHYLBUTANE + OXYGEN $(CH_3)_3 \cdot C.CH_2.CH_3 + O_2$

1. D.H. Desty, A. Goldup, G.R. Luckhurst, and W.T. Swanton, Gas Chromatography, p.76, table 3, Butterworths, London, 1962. (Ed. M. van Swaay).
   Gas chromatographic measurements using squalane; P → 5 atm.
   Maximum error in B_{12} estimated to be ± 30.

T	B_{12}
298.2	-145

2,3-DIMETHYLBUTANE + HYDROGEN $(CH_3)_2 \cdot CH \cdot CH \cdot (CH_3)_2 + H_2$

1. D.H. Desty, A. Goldup, G.R. Luckhurst, and W.T. Swanton, Gas
    Chromatography, p.76, table 3, Butterworths, London, 1962.   (Ed.
    M. van Swaay).

    Gas chromatographic measurements using squalane; $P \to 5$ atm.

    Maximum error in B_{12} estimated to be $\pm$ 30.

T	B_{12}
298.2	+ 9

2,3-DIMETHYLBUTANE + NITROGEN $(CH_3)_2 \cdot CH \cdot CH \cdot (CH_3)_2 + N_2$

1. D.H. Desty, A. Goldup, G.R. Luckhurst, and W.T. Swanton, Gas
    Chromatography, p.76, table 3, Butterworths, London, 1962.   (Ed.
    M. van Swaay).

    Gas chromatographic measurements using squalane; $P \to 5$ atm.

    Maximum error in B_{12} estimated to be $\pm$ 30.

T	B_{12}
298.2	-112

2. B.W. Gainey and R.L. Pecsok, J. phys. Chem., Ithaca 74 2548 (1970).
    Chromatographic method.

T	B_{12}
313.2	-99 $\pm$ 6

2,3-DIMETHYLBUTANE + OXYGEN $(CH_3)_2 \cdot CH \cdot CH \cdot (CH_3)_2 + O_2$

1. D.H. Desty, A. Goldup, G.R. Luckhurst, and W.T. Swanton, Gas
    Chromatography, p.76, table 3, Butterworths, London, 1962.   (Ed.
    M. van Swaay).

    Gas chromatographic measurements using squalane; $P \to 5$ atm.

    Maximum error in B_{12} estimated to be $\pm$ 30.

T	B_{12}
298.2	-153

o-FLUOROTOLUENE + TOLUENE $C_6H_4 \cdot (F)(CH_3) + C_6H_5 \cdot CH_3$

1. L.V. Mozhginskaya and L.E. Kolysko, Russ. J. phys. Chem. 49 505
    (1975); Zh. fiz. Khim. 49 860 (1975).

Constant volume piezometer.

T	B_{12}	T	B_{12}
373.2	-1593	423.2	-1133
388.2	-1445	443.2	- 959
403.2	-1301		

m-FLUOROTOLUENE + TOLUENE $C_6H_4 \cdot (F)(CH_3) + C_6H_5 \cdot CH_3$

1. L.V. Mozhginskaya and L.E. Kolysko, Russ. J. phys. Chem. <u>48</u> 1094 (1974);   Zh. fiz. Khim. <u>48</u> 1849 (1974).

   Constant volume piezometer.

T	B_{12}	T	B_{12}
373.2	-1218	423.2	- 802
388.2	-1069	443.2	- 640
403.2	- 967	458.2	- 516

p-FLUOROTOLUENE + TOLUENE $C_6H_4 \cdot (F)(CH_3) + C_6H_5 \cdot CH_3$

1. L.V. Mozhginskaya and L.E. Kolysko, Russ. J. phys. Chem. <u>49</u> 983 (1975);   Zh. fiz, Khim. <u>49</u> 1666 (1975).

   Constant volume piezometer.

T	B_{12}	T	B_{12}
373.2	-1199	423.2	- 868
388.2	-1072	443.2	- 779
403.2	- 975		

TOLUENE + HYDROGEN $C_6H_5 \cdot CH_3 + H_2$

1. J.M. Prausnitz and P.R. Benson, A.I. Ch. E. Jl <u>5</u> 161 (1959).

   Vapour phase solubility measurements.

T	B_{12}
323.15	19.0 ± 5.9
348.15	37.2 ± 6.0

TOLUENE + NITROGEN $C_6H_5 \cdot CH_3 + N_2$

1. J.M. Prausnitz and P.R. Benson, A. I. Ch. E. Jl <u>5</u> 161 (1959).

Vapour phase solubility measurements.

T	B_{12}
323.15	-98.1 ± 6.2
348.15	-88.1 ± 7.5

2. M. Vigdergauz and V. Semkin, J. Chromat. 58 95 (1971).
    Chromatographic measurements using dinonyl phthalate.
    Class III.

T	B_{12}
353.2	-94

n-HEPT-1-ENE + HYDROGEN $CH_2:CH.(CH_2)_4CH_3 + H_2$

1. A.J.B. Cruikshank, B.W. Gainey and C.L. Young, Trans. Faraday Soc.
    64 337 (1968).
    Gas chromatographic measurements.

T	B_{12}
308.15	-2 ± 9

n-HEPT-1-ENE + NITROGEN $CH_2:CH(CH_2)_4CH_3 + N_2$

1. A.J.B. Cruikshank, B.W. Gainey and C.L. Young, Trans. Faraday Soc.
    64 337 (1968).
    Gas chromatographic measurements.

T	B_{12}
308.15	-142 ± 9

2. C.L. Young, Trans. Faraday Soc. 64 1537 (1968).
    Gas chromatographic measurements.

T	B_{12}
333.2	-95 ± 10

n-HEPTANE + n-OCTANE $CH_3.(CH_2)_5.CH_3 + CH_3.(CH_2)_6.CH_3$

1. C.J. Wormald, E.J. Lewis, and D.J. Hutchings, J. chem. Thermodyn.
    11 1 (1979).

440

B_{12} values derived from measured excess enthalpies.

T	B_{12}
403.2	-1407 ± 30
413.2	-1322 ± 28

n-HEPTANE + HYDROGEN $CH_3 \cdot (CH_2)_5 \cdot CH_3 + H_2$

1. D.H. Desty, A. Goldup, G.R. Luckhurst, and W.T. Swanton, Gas
    Chromatography, p.76, table 3, Butterworths, London, 1962. (Ed.
    M. van Swaay).

    Gas chromatographic measurements using squalane; P → 5 atm.

T	B_{12}
298.2	-36

n-HEPTANE + NITROGEN $CH_3 \cdot (CH_2)_5 \cdot CH_3 + N_2$

1. D.H. Desty, A. Goldup, G.R. Luckhurst, and W.T. Swanton, Gas
    Chromatography, p.76, table 3, Butterworths, London, 1962. (Ed.
    M. van Swaay).

    Gas chromatographic measurements using squalane; P → 5 atm.
    Maximum error in B_{12} estimated to be ± 30.

T	B_{12}
298.2	-154

2. B.W. Gainey, Ph.D. Thesis, University of Bristol (1967).
    Chromatographic method.

T	B_{12}
313.2	-111 ± 6

3. A.J.B. Cruikshank, B.W. Gainey and C.L. Young, Trans. Faraday Soc.
    64 337 (1968).
    Gas chromatographic measurements.

T	B_{12}
308.15	-132 ± 9

4. C.P. Hicks and C.L. Young, Trans. Faraday Soc. **64** 2675 (1968).
    Gas chromatographic measurements.

T B_{12}

313.2   -111 ± 10

5. C.L. Young, Trans. Faraday Soc. 64 1537 (1968).
    Gas chromatographic measurements.

T B_{12}

333.2   -101 ± 10
348.2   - 87 ± 10

6. B.W. Gainey and R.L. Pecsok, J. phys. Chem., Ithaca 74 2548 (1970).
    Chromatographic method.

T B_{12}

313.2   -110 ± 6

7. M. Vigdergauz and V. Semkin, J. Chromat. 58 95 (1971).
    Chromatographic measurements using dinonyl phthalate.
    Class III.

T B_{12}

353.2   -81

8. Y.-K. Leung and B.E. Eichinger, J. phys. Chem., Ithaca 78 60 (1974).
    Gas chromatographic measurements using polyisobutylene. Nitrogen
    pressure up to 9 atm.

T B_{12}

298.16   -142 ± 15

n-HEPTANE + OXYGEN $CH_3 \cdot (CH_2)_5 \cdot CH_3 + O_2$

1. D.H. Desty, A. Goldup, G.R. Luckhurst, and W.T. Swanton, Gas
    Chromatography, p.76, table 3, Butterworths, London, 1962. (Ed.
    M. van Swaay).
    Gas chromatographic measurements using squalane; P → 5 atm.
    Maximum error in B_{12} estimated to be ± 30.

T B_{12}

298.2   -183

2-METHYLHEXANE + HYDROGEN $(CH_3)_2.CH.(CH_2)_3.CH_3 + H_2$

1. D.H. Desty, A. Goldup, G.R. Luckhurst, and W.T. Swanton, Gas
    Chromatography, p.76, table 3, Butterworths, London, 1962. (Ed.
    M. van Swaay).

    Gas chromatographic measurements using squalane; $P \rightarrow 5$ atm.

    Maximum error in B_{12} estimated to be ± 30.

T	B_{12}
298.2	-35

2-METHYLHEXANE + NITROGEN $(CH_3)_2.CH.(CH_2)_3.CH_3 + N_2$

1. D.H. Desty, A. Goldup, G.R. Luckhurst, and W.T. Swanton, Gas
    Chromatography, p.76, table 3, Butterworths, London, 1962. (Ed.
    M. van Swaay).

    Gas chromatographic measurements using squalane; $P \rightarrow 5$ atm.

    Maximum error in B_{12} estimated to be ± 30.

T	B_{12}
298.2	-135

2. B.W. Gainey and R.L. Pecsok, J. phys. Chem., Ithaca 74 2548 (1970).
    Chromatographic method.

T	B_{12}
313.2	-115 ± 6

2-METHYLHEXANE + OXYGEN $(CH_3)_2.CH.(CH_2)_3.CH_3 + O_2$

1. D.H. Desty, A. Goldup, G.R. Luckhurst, and W.T. Swanton, Gas
    Chromatography, p.76, table 3, Butterworths, London, 1962. (Ed.
    M. van Swaay).

    Gas chromatographic measurements using squalane; $P \rightarrow 5$ atm.

    Maximum error in B_{12} estimated to be ± 30.

T	B_{12}
298.2	-175

3-METHYLHEXANE + HYDROGEN $CH_3.CH_2.CH(CH_3).(CH_2)_2.CH_3 + H_2$

1. D.H. Desty, A. Goldup, G.R. Luckhurst, and W.T. Swanton, Gas
    Chromatography, p.76, table 3, Butterworths, London, 1962.

(Ed. M. van Swaay).

Gas chromatographic measurements using squalane; $P \to 5$ atm.

Maximum error in B_{12} estimated to be ± 30 .

T	B_{12}
298.2	-32

3-METHYLHEXANE + NITROGEN $CH_3 \cdot CH_2 \cdot CH(CH_3) \cdot (CH_2)_2 \cdot CH_3 + N_2$

1. D.H. Desty, A. Goldup, G.R. Luckhurst, and W.T. Swanton, Gas Chromatography, p.76, table 3, Butterworths, London, 1962. (Ed. M. van Swaay).

   Gas chromatographic measurements using squalane; $P \to 5$ atm.

   Maximum error in B_{12} estimated to be ± 30 .

T	B_{12}
298.2	-136

2. B.W. Gainey and R.L. Pecsok, J. phys. Chem., Ithaca <u>74</u> 2548 (1970).
   Chromatographic method.

T	B_{12}
313.2	-105 ± 6

3-METHYLHEXANE + OXYGEN $CH_3 \cdot CH_2 \cdot CH(CH_3) \cdot (CH_2)_2 \cdot CH_3 + O_2$

1. D.H. Desty, A. Goldup, G.R. Luckhurst, and W.T. Swanton, Gas Chromatography, p.76, table 3, Butterworths, London, 1962. (Ed. M. van Swaay).

   Gas chromatographic measurements using squalane; $P \to 5$ atm.

   Maximum error in B_{12} estimated to be ± 30 .

T	B_{12}
298.2	-180

3-ETHYLPENTANE + HYDROGEN $(CH_3 \cdot CH_2)_3 \cdot CH + H_2$

1. D.H. Desty, A. Goldup, G.R. Luckhurst, and W.T. Swanton, Gas Chromatography, p.76, table 3, Butterworths, London, 1962. (Ed. M. van Swaay).

   Gas chromatographic measurements using squalane; $P \to 5$ atm.

   Maximum error in B_{12} estimated to be ± 30 .

T	B_{12}
298.2	-36

3-ETHYLPENTANE + NITROGEN $(CH_3 \cdot CH_2)_3 CH + N_2$

1. D.H. Desty, A. Goldup, G.R. Luckhurst, and W.T. Swanton, Gas
    Chromatography, p.76, table 3, Butterworths, London, 1962. (Ed.
    M. van Swaay).

    Gas chromatographic measurements using squalane; $P \to 5$ atm.

    Maximum error in B_{12} estimated to be $\pm$ 30.

T	B_{12}
298.2	-144

3-ETHYLPENTANE + OXYGEN $(CH_3 \cdot CH_2)_3 CH + O_2$

1. D.H. Desty, A. Goldup, G.R. Luckhurst, and W.T. Swanton, Gas
    Chromatography, p.76, table 3, Butterworths, London, 1962. (Ed.
    M. van Swaay).

    Gas chromatographic measurements using squalane; $P \to 5$ atm.

    Maximum error in B_{12} estimated to be $\pm$ 30.

T	B_{12}
298.2	-176

2,2-DIMETHYLPENTANE + HYDROGEN $(CH_3)_3 \cdot C \cdot (CH_2)_2 \cdot CH_3 + H_2$

1. D.H. Desty, A. Goldup, G.R. Luckhurst, and W.T. Swanton, Gas
    Chromatography, p.76, table 3, Butterworths, London, 1962. (Ed.
    M. van Swaay).

    Gas chromatographic measurements using squalane; $P \to 5$ atm.

    Maximum error in B_{12} estimated to be $\pm$ 30.

T	B_{12}
298.2	-14

2,2-DIMETHYLPENTANE + NITROGEN $(CH_3)_3 C \cdot (CH_2)_2 \cdot CH_3 + N_2$

1. D.H. Desty, A. Goldup, G.R. Luckhurst, and W.T. Swanton, Gas
    Chromatography, p.76, table 3, Butterworths, London, 1962. (Ed.
    M. van Swaay).

    Gas chromatographic measurements using squalane; $P \to 5$ atm.

Maximum error in B_{12} estimated to be ± 30.

T	B_{12}
298.2	-133

2. B.W. Gainey and R.L. Pecsok, J. phys. Chem., Ithaca <u>74</u> 2548 (1970).
    Chromatographic method.

T	B_{12}
313.2	-177 ± 6

2,3-DIMETHYLPENTANE + HYDROGEN $(CH_3)_2.CH.CH(CH_3).CH_2.CH_3 + H_2$

1. D.H. Desty, A. Goldup, G.R. Luckhurst, and W.T. Swanton, Gas
    Chromatography, p.76, table 3, Butterworths, London, 1962. (Ed.
    M. van Swaay).

    Gas chromatographic measurements using squalane; $P \to 5$ atm.

    Maximum error in B_{12} estimated to be ± 30.

T	B_{12}
298.2	- 6

2,3-DIMETHYLPENTANE + NITROGEN $(CH_3)_2.CH.CH(CH_3).CH_2.CH_3 + N_2$

1. D.H. Desty, A. Goldup, G.R. Luckhurst, and W.T. Swanton, Gas
    Chromatography, p.76, table 3, Butterworths, London, 1962. (Ed.
    M. van Swaay).

    Gas chromatographic measurements using squalane; $P \to 5$ atm.

    Maximum error in B_{12} estimated to be ± 30.

T	B_{12}
298.2	-133

2. B.W. Gainey and R.L. Pecsok, J. phys. Chem., Ithaca <u>74</u> 2548 (1970).
    Chromatographic method.

T	B_{12}
313.2	-108 ± 6

2,3-DIMETHYLPENTANE + OXYGEN $(CH_3)_2.CH.CH(CH_3).CH_2.CH_3 + O_2$

1. D.H. Desty, A. Goldup, G.R. Luckhurst, and W.T. Swanton, Gas

Chromatography, p.76, table 3, Butterworths, London, 1962. (Ed. M. van Swaay).

Gas chromatographic measurements using squalane; $P \rightarrow 5$ atm.

Maximum error in B_{12} estimated to be ± 30.

T	B_{12}
298.2	-167

2,4-DIMETHYLPENTANE + HYDROGEN $(CH_3)_2 \cdot CH \cdot CH_2 \cdot CH \cdot (CH_3)_2 + H_2$

1. D.H. Desty, A. Goldup, G.R. Luckhurst, and W.T. Swanton, Gas Chromatography, p.76, table 3, Butterworths, London, 1962. (Ed. M. van Swaay).

    Gas chromatographic measurements using squalane; $P \rightarrow 5$ atm.

    Maximum error in B_{12} estimated to be ± 30.

T	B_{12}
298.2	- 5

2,4-DIMETHYLPENTANE + NITROGEN $(CH_3)_2 \cdot CH \cdot CH_2 \cdot CH \cdot (CH_3)_2 + N_2$

1. D.H. Desty, A. Goldup, G.R. Luckhurst, and W.T. Swanton, Gas Chromatography, p.76, table 3, Butterworths, London, 1962. (Ed. M. van Swaay).

    Gas chromatographic measurements using squalane; $P \rightarrow 5$ atm.

    Maximum error in B_{12} estimated to be ± 30.

T	B_{12}
298.2	-130

2. C.L. Young, Ph.D. Thesis, University of Bristol (1967). Chromatographic method.

T	B_{12}
303.2	-109 ± 6

3. A.J.B. Cruikshank, B.W. Gainey and C.L. Young, Trans. Faraday Soc. <u>64</u> 337 (1968). Gas chromatographic measurements.

T	B_{12}
308.15	-109 ± 9

4. B.W. Gainey and R.L. Pecsok, J. phys. Chem., Ithaca $\underline{74}$ 2548 (1970).
Chromatographic method.

T	B_{12}
313.2	-115 ± 6

2,4-DIMETHYLPENTANE + OXYGEN $(CH_3)_2.CH.CH_2.CH.(CH_3)_2 + O_2$

1. D.H. Desty, A. Goldup, G.R. Luckhurst, and W.T. Swanton, Gas Chromatography, p.76, table 3, Butterworths, London, 1962. (Ed. M. van Swaay).

Gas chromatographic measurements using squalane; P → 5 atm.

Maximum error in B_{12} estimated to be ± 30.

T	B_{12}
298.2	-165

3,3-DIMETHYLPENTANE + HYDROGEN $(CH_3.CH_2)_2C(CH_3)_2 + H_2$

1. D.H. Desty, A. Goldup, G.R. Luckhurst, and W.T. Swanton, Gas Chromatography, p.76, table 3, Butterworths, London, 1962. (Ed. M. van Swaay).

Gas chromatographic measurements using squalane; P → 5 atm.

Maximum error in B_{12} estimated to be ± 30.

T	B_{12}
298.2	-12

3,3-DIMETHYLPENTANE + NITROGEN $(CH_3.CH_2)_2C(CH_3)_2 + N_2$

1. D.H. Desty, A. Goldup, G.R. Luckhurst, and W.T. Swanton, Gas Chromatography, p.76, table 3, Butterworths, London, 1962. (Ed. M. van Swaay).

Gas chromatographic measurements using squalane; P → 5 atm.

Maximum error in B_{12} estimated to be ± 30.

T	B_{12}
298.2	-131

2. B.W. Gainey and R.L. Pecsok, J. phys. Chem., Ithaca $\underline{74}$ 2548 (1970).
Chromatographic method.

T	B_{12}
313.2	-117 ± 6

3,3-DIMETHYLPENTANE + OXYGEN $(CH_3.CH_2)_2C(CH_3)_2 + O_2$

1. D.H. Desty, A. Goldup, G.R. Luckhurst, and W.T. Swanton, Gas Chromatography, p.76, table 3, Butterworths, London, 1962. (Ed. M. van Swaay).

   Gas chromatographic measurements using squalane; $P \to 5$ atm.

   Maximum error in B_{12} estimated to be ± 30.

T	B_{12}
298.2	-166

2,2,3-TRIMETHYLBUTANE + HYDROGEN $(CH_3)_3C.CH(CH_3)_2 + H_2$

1. D.H. Desty, A. Goldup, G.R. Luckhurst, and W.T. Swanton, Gas Chromatography, p.76, table 3, Butterworths, London, 1962. (Ed. M. van Swaay).

   Gas chromatographic measurements using squalane; $P \to 5$ atm.

   Maximum error in B_{12} estimated to be ± 30.

T	B_{12}
298.2	-6

2,2,3-TRIMETHYLBUTANE + NITROGEN $(CH_3)_3C.CH(CH_3)_2 + N_2$

1. D.H. Desty, A. Goldup, G.R. Luckhurst, and W.T. Swanton, Gas Chromatography, p.76, table 3, Butterworths, London, 1962. (Ed. M. van Swaay).

   Gas chromatographic measurements using squalane; $P \to 5$ atm.

   Maximum error in B_{12} estimated to be ± 30.

T	B_{12}
298.2	-123

2. B.W. Gainey and R.L. Pecsok, J. phys. Chem., Ithaca 74 2548 (1970). Chromatographic method.

T	B_{12}
313.2	-100 ± 6

2,2,3-TRIMETHYLBUTANE + OXYGEN $(CH_3)_3 \cdot C \cdot CH(CH_3)_2 + O_2$

1. D.H. Desty, A. Goldup, G.R. Luckhurst, and W.T. Swanton, Gas
    Chromatography, p.76, table 3, Butterworths, London, 1962. (Ed.
    M. van Swaay).

    Gas chromatographic measurements using squalane; P → 5 atm,
    Maximum error in B_{12} estimated to be ± 30.

T	B_{12}
298.2	-160

STYRENE + NITROGEN $C_6H_5 \cdot CH:CH_2 + N_2$

1. M. Vigdergauz and V. Semkin, J. Chromat. 58 95 (1971).
    Chromatographic measurements using dinonyl phthalate.
    Class III.

T	B_{12}
353.2	-117

n-OCTANE + HYDROGEN $CH_3 \cdot (CH_2)_6 \cdot CH_3 + H_2$

1. J.F. Connolly, Physics Fluids 4 1494 (1961).
    Solubility measurements of n-octane in the compressed gas;
    P → 50 atm.

T	B_{12}
373.15	+18.4
373.15	18.2
373.15	17.7

    Compressibility measurements on four mixtures.

T	B_{12}	T	B_{12}
473.2	41	533.2	48
493.2	42	553.2	50
513.2	46	573.2	51

2. D.H. Everett, B.W. Gainey and C.L. Young, Trans. Faraday Soc. 64
    2667 (1968).

    Gas chromatographic measurements.

T	B_{12}
313.15	9 ± 8
338.20	22 ± 8

n-OCTANE + NITROGEN $CH_3 \cdot (CH_2)_6 \cdot CH_3 + N_2$

1. C.L. Young, Ph.D. Thesis, University of Bristol (1967).
   Chromatographic method.

T	B_{12}
303.2	-143 ± 6

2. A.J.B. Cruikshank, B.W. Gainey and C.L. Young, Trans. Faraday Soc.
   64 337 (1968).
   Gas chromatographic measurements.

T	B_{12}
308.15	-143 ± 9

3. C.L. Young, Trans. Faraday Soc. 64 1537 (1968).
   Gas chromatographic measurements.

T	B_{12}
333.2	-119 ± 10
348.2	- 92 ± 10

4. B.W. Gainey and R.L. Pecsok, J. phys. Chem., Ithaca 74 2548 (1970).
   Chromatographic method.

T	B_{12}
313.2	-134 ± 6

5. M. Vigdergauz and V. Semkin, J. Chromat. 58 95 (1971).
   Chromatographic measurements using dinonyl phthalate.
   Class III.

T	B_{12}
353.2	-98

6. Y.-K. Leung and B.E. Eichinger, J. phys. Chem., Ithaca 78 60 (1974).
   Gas chromatographic measurements using polyisobutylene. Nitrogen
   pressures up to 9 atm.

T	B_{12}
298.16	-146 ± 15

2,2-DIMETHYLHEXANE + NITROGEN $(CH_3)_3.C.(CH_2)_3.CH_3 + N_2$

1. B.W. Gainey and R.L. Pecsok, J. phys. Chem., Ithaca 74 2548 (1970).
    Chromatographic method.

T	B_{12}
313.2	-150 ± 6

ISOOCTANE + HYDROGEN $(CH_3)_3.C.CH_2.CH(CH_3)_2 + H_2$

1. J.M. Prausnitz and P.R. Benson, A.I. Ch. E. Jl 5 161 (1959).
    Vapour phase solubility measurements.

T	B_{12}
323.15	37.6 ± 5.9
348.15	55.7 ± 6.0

2. J.F. Connolly, Physics Fluids 4 1494 (1961).
    Solubility measurements of the liquid in the compressed gas;
    P → 50 atm.

T	B_{12}
348.15	19.0
348.15	18.4

3. D.H. Everett, B.W. Gainey and C.L. Young, Trans. Faraday Soc. 64
    2667 (1968).
    Gas chromatographic measurements.

T	B_{12}
313.15	4 ± 8
338.20	6 ± 8

ISOOCTANE + NITROGEN $(CH_3)_3C.CH_2.CH(CH_3)_2 + N_2$

1. J.M. Prausnitz and P.R. Benson, A. I. Ch. E. Jl 5 161 (1959).
    Vapour phase solubility measurements.

452

T	B_{12}
323.15	-82.3 ± 5.7
348.15	-60.3 ± 6.5

2. M. Vigdergauz and V. Semkin, J. Chromat. 58 95 (1971).
    Chromatographic measurements using dinonyl phthalate.
    Class III.

T	B_{12}
353.2	-47

2,2,3,3-TETRAMETHYLBUTANE + NITROGEN $(CH_3)_3C.C(CH_3)_3 + N_2$

1. B.W. Gainey and R.L. Pecsok, J. phys. Chem., Ithaca 74 2548 (1970).
    Chromatographic method.

T	B_{12}
313.2	-128 ± 6

n-NONANE + NITROGEN $CH_3.(CH_2)_7.CH_3 + N_2$

1. M. Vigdergauz and V. Semkin, J. Chromat. 58 95 (1971).
    Chromatographic measurements using dinonyl phthalate.
    Class III.

T	B_{12}
353.2	-117

2. Y.-K. Leung and B.E. Eichinger, J. phys. Chem., Ithaca 78 60 (1974).
    Gas chromatographic measurements using polyisobutylene. Nitrogen
    pressures up to 9 atm.

T	B_{12}
298.16	-160 ± 15

2,2,5-TRIMETHYLHEXANE + NITROGEN $(CH_3)_3.C.(CH_2)_2.CH(CH_3)_2 + N_2$

1. S.G. D'Avila, B.K. Kaul, and J.M. Prausnitz, J. chem. Engng Data 21
    488 (1976).

    B_{12} calculated from solubility measurements of the hydrocarbon in the compressed gas. Maximum uncertainty in B_{12} given as ± 10.

T	B_{12}	T	B_{12}
298.2	-115	348.2	- 73
323.2	- 93	373.2	- 52

NAPHTHALENE + HYDROGEN $C_{10}H_8 + H_2$

1. A.D. King, Jr. and W.W. Robertson, J. chem. Phys. <u>37</u> 1453 (1962).
    Naphthalene solubility measurements.

T	B_{12}
295.2	-25.0 $\pm$ 3.7
343.2	-12.5 $\pm$ 1.4

NAPHTHALENE + HELIUM $C_{10}H_8 + He$

1. A.D. King, Jr. and W.W. Robertson, J. chem. Phys. <u>37</u> 1453 (1962).
    Naphthalene solubility measurements.

T	B_{12}
305.2	67.5 $\pm$ 2.8
347.2	74.7 $\pm$ 2.8

NAPHTHALENE + NITROGEN $C_{10}H_8 + N_2$

1. A.D. King, Jr. and W.W. Robertson, J. chem. Phys. <u>37</u> 1453 (1962).
    Naphthalene solubility measurements.

T	B_{12}
295.2	-176 $\pm$ 10
345.2	-113 $\pm$ 7

NAPHTHALENE + NITROUS OXIDE $C_{10}H_8 + N_2O$

1. A.D. King, Jr., J. chem. Phys. <u>49</u> 4083 (1968).

    B_{12} determined from measurements of solubility of naphthalene in
    the compressed gas: $P \rightarrow 60$ atm.

T	B_{12}	T	B_{12}
293	-592 ± 8	319.5	-466 ± 9
298	-579 ± 8	324	-443 ± 7
299	-542 ± 10	327	-442 ± 8
300	-587 ± 8	332	-426 ± 8
308	-519 ± 8	339	-414 ± 8
309.5	-521 ± 12	340.5	-405 ± 8
316	-491 ± 13		

NAPHTHALENE+OXYGEN $C_{10}H_8 + O_2$

1. H. Bradley, Jr., and A.D. King, Jr., J. chem. Phys. **47** 1189 (1967).
    B_{12} determined from measurements of solubility of naphthalene in
    the compressed gas: P → 160 atm.

T	B_{12}
297.2	-170 ± 12

tert-BUTYLBENZENE + NITROGEN $C_6H_5 \cdot C(CH_3)_3 + N_2$

1. S.G. D'Avila, B.K. Kaul, and J.M. Prausnitz, J. chem. Engng Data **21**
    488 (1976).
    B_{12} calculated from solubility measurements of the hydrocarbon in
    the compressed gas. Maximum uncertainty in B_{12} given as ± 10.

T	B_{12}	T	B_{12}
323.2	-111	373.2	-70
348.2	-83	398.2	-49

n-DECANE + HYDROGEN $CH_3 \cdot (CH_2)_8 \cdot CH_3 + H_2$

1. J.M. Prausnitz and P.R. Benson, A. I. Ch. E. Jl **5** 161 (1959).
    Vapour phase solubility measurements.

T	B_{12}
323.15	81.2 ± 6.9
348.15	95.3 ± 7.0

n-DECANE + NITROGEN $CH_3 \cdot (CH_2)_8 \cdot CH_3 + N_2$

1. J.M. Prausnitz and P.R. Benson, A. I. Ch. E. Jl **5** 161 (1959).

Vapour phase solubility measurements.

T	B_{12}
323.15	-141 ± 5
348.15	-112 ± 7

2. M. Vigdergauz and V. Semkin, J. Chromat. 58 95 (1971).
Chromatographic measurements using dinonyl phthalate.
Class III.

T	B_{12}
353.2	-130

3. S.G. D'Avila, B.K. Kaul, and J.M. Prausnitz, J. chem. Engng Data 21 488 (1976).
B_{12} determined from solubility measurements of n-decane in the compressed gas. Maximum uncertainty in B_{12} is $\pm$ 10.

T	B_{12}	T	B_{12}
323.2	-148	373.2	-82
348.2	-111	398.2	-74

n-DODECANE + NITROGEN $CH_3 \cdot (CH_2)_{10} \cdot CH_3 + N_2$

1. S.G. D'Avila, B.K. Kaul, and J.M. Prausnitz, J. chem. Engng Data 21 488 (1976).
B_{12} calculated from solubility measurements of the hydrocarbon in the compressed gas. Maximum uncertainty in B_{12} given as $\pm$ 10.

T	B_{12}	T	B_{12}
348.2	-134	398.2	-74
373.2	-104	423.2	-53

ANTHRACENE+OXYGEN $C_6H_4(CH)_2C_6H_4 + O_2$

1. H. Bradley, Jr., and A.D. King, Jr., J. chem. Phys. 47 1189 (1967).
B_{12} determined from measurements of solubility of anthracene in the compressed gas: $P \rightarrow 160$ atm.

T	B_{12}
348.2	-186 ± 6

PHENANTHRENE+NITROUS OXIDE $C_{14}H_{10} + N_2O$

1. H. Bradley, Jr., and A.D. King, Jr., J. chem. Phys. <u>52</u> 2851 (1970).
    B_{12} determined from measurements of solubility of phenanthrene in
    the compressed gas: $P \to 60$ atm. Probable error in $B \pm 24$.

T	B_{12}	T	B_{12}
326	-668	356	-547
330	-643	364	-504
337	-652	380	-455
341	-606	397	-398
346	-564	411	-345
349	-567		

DEUTERIUM + HELIUM D_2 + He

1. J.J.M. Beenakker, F.H. Varekamp, and A. Van Itterbeek, Physica,'s
    Grav. <u>25</u> 9 (1959).

    Low pressure PV measurements.

T	B_{12}
20.4	-19 ± 1

2. F.H. Varekamp and J.J.M. Beenakker, Physica,'s Grav. <u>25</u> 889 (1959).

T	B_{12}	T	B_{12}
18	-25	20	-20
19	-23	21	-18

3. H.F.P. Knaap, M. Knoester, F.H. Varekamp, and J.J.M. Beenakker,
    Physica,'s Grav. <u>26</u> 633 (1960).

T	$B_{12} - 0.5\,(B_{11} + B_{22})$
20.4	65.0

HYDROGEN DEUTERIDE + HELIUM HD + He

1. J.J.M. Beenakker, F.H. Varekamp, and A. Van Itterbeek Physica,'s
    Grav. <u>25</u> 9 (1959).

    Low pressure PV measurements.

T	B_{12}
20.4	-18 ± 1

2. F.H. Varekamp and J.J.M. Beenakker, Physica, 's Grav. <u>25</u> 889 (1959).

T	B_{12}	T	B_{12}
16	-30	19	-21
17	-27	20	-19
18	-24	21	-17

3. H.F.P. Knaap, M. Knoester, F.H. Varekamp, and J.J.M. Beenakker, Physica, 's Grav. <u>26</u> 633 (1960).

T	$B_{12} - 0.5 \, (B_{11} + B_{22})$
20.4	60.6

SULPHUR HEXAFLUORIDE + KRYPTON SF_6 + Kr

1. J. Santafe, J.S. Urieta and C.G. Losa, Chem. phys. <u>18</u> 341 (1974).
    Compressibility measurements. Accuracy of B_{12} estimated to be $\pm$ 3.

T	B_{12}	T	B_{12}
273.2	-99.0	303.2	-70.0
283.2	-91.0	313.2	-64.1
293.2	-80.8	323.2	-57.5

HYDROGEN CHLORIDE + KRYPTON HCl + Kr

1. G. Clockler, C.P. Roe, and D.L. Fuller, J. chem. Phys. <u>1</u> 703 (1933) (*).
    PVT data given; P → 100 atm.

T	B_{12}
328.7	-21.7
368.7	-11.4

HYDROGEN + WATER H_2 + H_2O

1. E.P. Bartlett, J. Am. chem. Soc. <u>49</u> 65 (1927).

Data for water solubility in compressed hydrogen: P range → 1000 atm. T range 298-323 K.

HYDROGEN + HELIUM H_2 + He

1. C.W. Gibby, C.C. Tanner, and I. Masson, Proc. R. Soc. <u>A122</u> 283 (1929) (*).

Class I.

T	B_{12}	T	B_{12}
298.2	15.6	398.6	14.4
323.2	15.6	423.5	15.8
348.2	15.2	448.2	14.6
373.6	15.6		

2. C.C. Tanner and I. Masson, Proc. R. Soc. <u>A126</u> 268 (1930) (*).

PVTx data for x_{He} range 0.16 → 0.84; P → 125 atm.

T	B_{12}
298.2	17.2

3. J.J.M. Beenakker, F.H. Varekamp, and A. Van Itterbeek, Physica, 's Grav. <u>25</u> 9 (1959).

Low pressure PV measurements.

T	B_{12}
20.4	-16 ± 1

4. C.M. Knobler, J.J.M. Beenakker, and H.F.P. Knaap, Physica, 's Grav. <u>25</u> 909 (1959).

Low pressure differential method. Precision ± 3.

T	B_{12}
90	12.8

5. F.H. Varekamp and J.J.M. Beenakker, Physica, 's Grav. <u>25</u> 889 (1959).

T	B_{12}	T	B_{12}
14	-38	18	-22
15	-33	19	-19
16	-28	20	-17
17	-25	21	-15

6. H.F.P. Knaap, M. Knoester, F.H. Varekamp, and J.J.M. Beenakker,
   Physica, 's Grav. <u>26</u> 633 (1960).

T	$B_{12} - 0.5(B_{11} + B_{22})$
20.4	54.7

7. J. Brewer and G.W. Vaughn, J. chem. Phys. <u>50</u> 2960 (1969).
   Measurements of pressure changes on mixing gases at constant volume.

T	B_{11}	B_{22}	B_{12}
148.15	6.78	11.75	15.92
173.15	9.15	11.91	16.20
198.15	10.79	11.98	16.43
223.15	12.10	11.93	16.20
248.15	13.14	11.84	16.43
273.15	13.98	11.76	16.37
298.15	14.64	11.73	16.01
323.15	15.15	11.74	15.82

HYDROGEN + NITROGEN $H_2 + N_2$

1. T.T.H. Verschoyle, Proc. R. Soc. <u>A111</u> 552 (1926) (*).
   PVTx data given for x_{H_2} = 0.25, 0.5 and 0.75; P → 205 atm.

T	B_{12}
273.2	12.0, 12.3, 13.5
293.2	13.0, 13.7, 14.3

2. E.P. Bartlett, J. Am. chem. Soc. <u>49</u> 687, 1955 (1927).
   PVT data given: P range 1 - 1000 atm, T is 273.2 K.

3. E.P. Bartlett, H.L. Cupples and T.H. Tremearne, J. Am. chem. Soc.
   <u>50</u> 1275 (1928).
   PVT data given: P range 50-1000 atm, T range 273-573 K, X_2 is
   0.25.

4. E.P. Bartlett, H.C. Hetherington, H.M. Kvalnes and T.H. Tremearne,
   J. Am. chem. Soc. <u>52</u> 1363 (1930).
   PVT data given: P range 25-1000 atm, T range 203-293 K, X_2 is
   0.25.

5. R. Wiebe and V.L. Gaddy, J. Am. chem. Soc. <u>60</u> 2300 (1938).

PVT data given: P range 25-1000 atm. T range 273-573 K, X_2 is 0.1256, 0.2444, 0.4826 and 0.7388.

6. I. Kritschewsky and V. Markov, Acta phys.-chim. URSS <u>12</u> 59 (1940).

    PVT data given for X_{N_2} = 0.505; P → 500 atm for T range 273 - 473K.

7. A.E. Edwards and W.E. Roseveare, J. Am. chem. Soc. <u>64</u> 2816 (1942) (*).

    Measurement of volume changes on mixing gases at constant pressure of 1 atm, and 0.5 atm.

T	B_{11}	B_{22}	B_{12}
298.2	14.8	-4.5	14.1

8. Ya. S. Kazarnovskii and I.P. Sidorov, Zh. fiz. Khim. <u>21</u> 1363 (1947).

    PVT data for 1:3 mixture; P → 800 atm, at 273.2 K.

9. A. Michels and T. Wassenaar, Appl. sci. Res. <u>A1</u> 258 (1948).

    PVT data given for X_{N_2} = 0.25; P → 340 atm for T range 273 - 423K.

10. B.H. Sage, R.H. Olds, and W.N. Lacey, Ind. Engng Chem. ind. Edn <u>40</u> 1453 (1948).

    PVT data given; $X_{HYDROGEN}$ = 0.76, P → 1000 atm, T range 277.7 - 510.9 K.

11. A. Van Itterbeek and W. Van Doninck, Proc. phys. Soc. <u>B62</u> 62 (1949).

    Velocity of sound measurements.

12. C.O. Bennett and B.F. Dodge, Ind. Engng Chem. ind. Edn <u>44</u> 180 (1952).

    Compressibility of three mixtures from 298K - 398K; P range 70 - 200 atm.

13. Z. Dokoupil, G. Van Soest, and M.D.P. Swenker, Appl. sci. Res. <u>A5</u> 182 (1955).

    Solid-vapour equilibrium measurements; P → 50 atm for T range 25 - 70K.

14. J. Reuss and J.J.M. Beenakker, Physica, 's Grav. <u>22</u> 869 (1956).

    B_{12} determined from influence of pressure on vapour-solid equilibrium.

T	B_{12}	T	B_{12}
36	-227	52	-106
40	-184	56	- 95
44	-150	60	- 92
48	-123		

15. M.G. Ostronov, P.E. Bol'shakov, L.L. Gel'perin, and A.A. Orlova, Russ. J. phys. Chem. <u>41</u> 1171 (1967).

Measurements of isothermal Joule-Thomson effect; P → 50 atm.

T	B_{12}	T	B_{12}
198.15	-7.6	248.15	7.6
223.15	+2.0	273.15	11.0

16. P. Zandbergen and J.J.M. Beenakker, Physica, 's Grav. <u>33</u> 343 (1967).

Measurement of volume changes on mixing gases at constant pressure.

T	B_{12}	T	B_{12}
170	-5.04	230	5.29
180	-2.63	240	6.25
190	-0.55	250	7.24
200	+1.15	260	8.24
210	2.69	270	9.07
220	4.08		

17. J. Brewer and G.W. Vaughn, J. chem. Phys. <u>50</u> 2960 (1969).

Measurements of pressure changes on mixing gases at constant volume.

T	B_{11}	B_{22}	B_{12}
148.15	6.78	-73.54	-10.32
173.15	9.15	-51.86	- 3.39
198.15	10.79	-37.48	1.20
223.15	12.10	-26.38	4.85
248.15	13.14	-17.59	7.96
273.15	13.98	-10.57	10.66
298.15	14.64	- 4.86	12.98
323.15	15.15	- 0.26	14.58

HYDROGEN + NEON H_2 + Ne

1. C.M. Knobler, J.J.M. Beenakker, and H.F.P. Knaap, Physica, 's
    Grav. <u>25</u> 909 (1959).

    Low pressure differential method.   Precision ± 3.

T	B_{12}
90	-1.9

2. J. Brewer and G.W. Vaughn, J. chem. Phys. <u>50</u> 2960 (1969).

    Measurements of pressure changes on mixing gases at constant volume.

T	B_{11}	B_{22}	B_{12}
148.15	6.78	4.36	8.20
173.15	9.15	6.45	9.98
198.15	10.79	7.90	11.23
223.15	12.10	9.12	12.29
248.15	13.14	10.14	13.03
273.15	13.98	10.94	13.69
298.15	14.64	11.54	14.23
323.15	15.15	12.32	14.81

HYDROGEN + OXYGEN H_2 + O_2

1. A. Van Itterbeek and W. Van Doninck, Proc. phys. Soc. <u>B62</u> 62 (1949).
    Velocity of sound measurements.

HYDROGEN + XENON H_2 + Xe

1. A.I. Doroshenko, Kh.-M. A. Sarov, M.B. Iomtev, L.S. Kushner and
    L.T. Kalinichenko, Russ. J. phys. Chem. <u>51</u> 761 (1977);   Zh. fiz.
    Khim. <u>51</u> 1283 (1977).

    B_{12} determined from measurements of the solubility of xenon in the
    compressed gas at pressures → 100 atm.   Standard error in B_{12} is 6.

T	B_{12}	T	B_{12}
80	-125	105	- 89
85	-110	115	- 71
95	- 96	125	- 58

2. B. Schramm and H. Schmiedel (1979) (†).

Estimated error in B_{12} is ± 6.

T	B_{12}	T	B_{12}
295	2.1	400	10.7
350	7.6	450	13.4

WATER + NITROGEN $H_2O + N_2$

1. E.P. Bartlett, J. Am. chem. Soc. **49** 65 (1927).

   Data for water solubility in compressed nitrogen: P range → 1000 atm. T range 298-323 K.

2. A.W. Saddington and N.W. Krase, J. Am. chem. Soc. **56** 353 (1934).

   Measurements of the solubility of water in the compressed gas; P → 300 atm.

3. M. Rigby and J.M. Prausnitz, J. phys. Chem., Ithaca **72** 330 (1968).

   B_{12} derived from measurements of solubility of water in the compressed gas.

T	B_{12}
298.15	-40 ± 6
323.15	-28 ± 5
348.15	-20 ± 4
373.15	-15.5 ± 3

WATER+NITROUS OXIDE $H_2O + N_2O$

1. C.R. Coan and A.D. King, Jr., J. Am. chem. Soc. **93** 1857 (1971).

   B_{12} derived from measurements of solubility of water in the compressed gas. P → 50 atm.

T	B_{12}	T	B_{12}
298.15	-188 ± 9	348.15	-119 ± 6
323.15	-152 ± 9	373.15	- 94 ± 7

HYDROGEN SULPHIDE + NITROGEN $H_2S + N_2$

1. D.B. Robinson, G.P. Hamaluik, T.R. Krishnan and P.R. Bishnoi, J. chem. Engng Data **20** 153 (1975).

   Burnett method, P → 230 atm. Values given are mixture virial coefficients B_M for (a) 8.6 mole % H_2S, (b) 22.2 mole % H_2S.

T	B_M	
	(a)	(b)
292.8	-9.5	-26.41
317.3	-4.9	-21.54
333.1	-2.6	-18.00
352.9	-0.275	-13.3
372.1	-1.917	- 8.84

HELIUM + KRYPTON   He + Kr

1. F.H. Kate, Jr. and R.L. Robinson, Jr., J. chem. Thermodyn. 5 259 (1973).

    Solubility measurements of krypton in the compressed gas; P → 120 atm.

T	B_{12}	C_{112}
90	1.73 ± 0.61	573 ± 57
95	2.97 ± 1.13	572 ± 113
100	4.06 ± 1.03	611 ± 106
105	5.66 ± 0.99	517 ± 106
110	6.99 ± 1.34	439 ± 147
115	8.52 ± 0.66	606 ± 75

2. D.D. Dillard, M. Waxman, and R.L. Robinson, Jr., J. chem. Engng Data 23 269 (1978).

    Burnett method.

T	B_{12}
223.15	19.74
273.15	21.21
323.15	22.05

HELIUM + NITROGEN   He + N_2

1. A.E. Edwards and W.E. Roseveare, J. Am. chem. Soc. 64 2816 (1942) (*).

    Measurement of volume change on mixing gases at constant pressure of 1 atm, and 0.5 atm.

T	B_{11}	B_{22}	B_{12}
298.2	11.8	-4.5	12.5

2. A. Van Itterbeek and W. Van Doninck, Proc. phys. Soc. **B62** 62 (1949).
   Velocity of sound measurements.

3. W.C. Pfefferle, Jr., J.A. Goff and J.G. Miller, J. chem. Phys. **23**
   509 (1955) (*).
   Burnett method: maximum pressure 120 atm.

T	B_{11}	B_{22}	B_{12}
303.2	11.84	-4.17	21.03

4. G.M. Kramer and J.G. Miller, J. phys. Chem.,Ithaca **61** 785 (1957).
   Burnett method. Maximum pressure 130 atm.

T	B_{11}	B_{22}	B_{12}
303.2	11.52	-3.52	21.84

5. C.M. Knobler, J.J.M. Beenakker, and H.F.P. Knaap, Physica, 's
   Grav. **25** 909 (1959).
   Low pressure differential method. Precision ± 3.

T	B_{12}
90	12.4

6. F.B. Canfield, T.W. Leland, Jr., and R. Kobayashi, Adv. cryogen.
   Engng **8** 146 (1963).

T	B_{12}	T	B_{12}
133.15	13.80	183.15	17.85
143.14	15.31	223.13	20.16
158.15	16.50	273.15	21.66

7. R.J. Witonsky and J.G. Miller, J. Am. chem. Soc. **85** 282 (1963).
   Burnett method: maximum pressure 100 atm.

T	B_{11}	B_{22}	B_{12}
448.2	10.89	14.26	22.92 ± 0.38
523.2	10.69	18.32	22.41 ± 0.19
598.2	10.75	20.80	21.73 ± 0.42
673.2	9.67	23.41	21.20 ± 0.35
748.2	9.70	24.73	20.33 ± 0.34

8. P.S. Ku and B.F. Dodge, J. chem. Engng Data <u>12</u> 158 (1967).
    3-term fit of PVT data; P → 300 atm.

T	B_{12}
311.7	22.02
373.2	22.17

9. J. Brewer and G.W. Vaughn, J. chem. Phys. <u>50</u> 2960 (1969).
    Measurements of pressure changes on mixing gases at constant volume.

T	B_{11}	B_{22}	B_{12}
148.15	11.75	-73.54	15.32
173.15	11.91	-51.86	17.25
198.15	11.98	-37.48	17.75
223.15	11.93	-26.38	18.67
248.15	11.84	-17.59	19.85
273.15	11.76	-10.57	20.19
298.15	11.73	- 4.86	21.19
323.15	11.74	- 0.26	21.56

10. K.R. Hall and F.B. Canfield, Physica, 's Grav. <u>47</u> 219 (1969).
    Burnett method.

T	B_{11}	B_{22}	B_{12}
103.15	11.57	-148.5	6.32
113.15	11.77	-117.8	10.43

HELIUM + NEON   He + Ne

1. C.M. Knobler, J.J.M. Beenakker, and H.F.P. Knaap, Physica, 's Grav.
    <u>25</u> 909 (1959).
    Low pressure differential method. Precision ± 3.

T	B_{12}
90	7.4

2. A. Gladun, Cryogen. <u>7</u> 286 (1967).
    Measurements of Joule-Thomson effect for three mixtures.

3. J. Brewer and G.W. Vaughn, J. chem. Phys. <u>50</u> 2960 (1969).
    Measurements of pressure changes on mixing gases at constant volume.

T	B_{11}	B_{22}	B_{12}
148.15	11.75	4.36	10.26
173.15	11.91	6.45	10.91
198.15	11.98	7.90	11.37
223.15	11.93	9.12	11.74
248.15	11.84	10.14	12.00
273.15	11.76	10.94	12.11
298.15	11.73	11.54	12.39
323.15	11.74	12.32	12.71

4. M.B. Iomtev, A.I. Doroshenko, L.S. Kushner, Kh.-M. A. Sarov, and L.T. Kalinichenko, Russ. J. phys. Chem. 51 808 (1977); Zh. fiz. Khim. 51 1373 (1977).

B_{12} and C_{112} determined from measurements of the solubility of neon in the compressed gas at pressures → 120 atm.

T	B_{12}	C_{112}
15	-65 ± 3	600 ± 110
16	-59 ± 3	550 ± 110
17	-57.5 ± 2.5	610 ± 130
18	-50 ± 3	570 ± 110
19	-46 ± 3	550 ± 100
20	-43 ± 2	520 ± 70

HELIUM + OXYGEN He + O_2

1. A. Van Itterbeek and W. Van Doninck, Proc. phys. Soc. B62 62 (1949).
    Velocity of sound measurements.

2. C.M. Knobler, J.J.M. Beenakker, and H.F.P. Knaap, Physica, 's Grav. 25 909 (1959).
    Low pressure differential method. Precision ± 3.

T	B_{12}
90	-4.4

HELIUM + XENON He + Xe

1. F.H. Kate, Jr. and R.L. Robinson, Jr., J. chem. Thermodyn. 5 273 (1973).

Solubility measurements of xenon in the compressed gas; $P \rightarrow 120$ atm.

T	B_{12}	C_{112}
120	9.38 ± 0.96	516 ± 115
130	13.05 ± 0.85	564 ± 114
140	16.42 ± 0.85	683 ± 114
155	20.47 ± 0.82	296 ± 118

KRYPTON + NEON   Kr + Ne

1. G. Thomaes, R. Van Steenwinkel and W. Stone, Molec. Phys. $\underline{5}$ 301 (1962).

    Volume expansion, relative to hydrogen, of mixtures with X_{Ne} = .5767 at 120 and 294.2K, otherwise X_{Ne} = .5919.

T	B_{11}	B_{22}	B_{12}
120	-317.0	-0.5	-23.7
155.8	-187.3	+4.8	-10.5, -9.9
198.8	-122.8	7.9	- 3.4, -2.9
241.2	- 83.9	9.7	+0.04, +2.4
294.2	- 53	11.1	- 1.9, +1.5, +11.4

2. R.C. Miller, A.J. Kidnay and M.J. Hiza, J. chem. Thermodyn. $\underline{4}$ 807 (1972).

    Solubility measurements of krypton in the compressed gas; $P \rightarrow 95$ atm.

T	B_{12}	T	B_{12}
100.00	-34.5	130.00	-19.9
110.00	-31.0	140.00	-14.1
115.00	-27.8	150.00	- 5.7
120.00	-25.5		

3. B. Schramm and R. Gehrmann (1979) (+).

    Estimated error in B_{12} is ± 5.

T	B_{12}	T	B_{12}
213	1.5	262	7.4
223	3.6	276	8.3
242	5.3		

4. B. Schramm and H. Schmiedel (1979) (†).
    Estimated error in B_{12} is ± 5.

T	B_{12}	T	B_{12}
295	10.2	425	16.4
330	12.2	450	15.7
365	14.7	475	15.0
400	14.9		

KRYPTON + XENON Kr + Xe

1. C.A. Pollard and G. Saville (unpublished results). See also C.A.
    Pollard, Ph.D. thesis, University of London (1971).
    Estimated uncertainty in B_{12} is ± 4.

T	B_{12}	T	B_{12}
160.00	-261.3	240.00	-120.1
170.00	-235.0	260.00	-102.9
180.00	-209.9	280.00	- 88.8
190.00	-188.5	300.00	- 77.5
200.00	-170.7	320.00	- 67.7
220.00	-142.1		

2. B. Schramm, H. Schmiedel, R. Gehrmann, and R. Bartl, Ber. (dtsch.)
    Bunsenges. phys. Chem. 81 316 (1977).
    Maximum error in B_{12} estimated to be ± 6.

T	B_{12}	T	B_{12}
202	-164.8	335	- 61.7
213	-150.0	370	- 48.6
233	-130.6	400	- 39.6
253	-112.4	435	- 31.0
278	- 95.8	465	- 25.0
295	- 79.0	500	- 19.0

3. H.-P. Rentschler and B. Schramm, Ber. (dtsch.) Bunsenges. phys.
    Chem. 81 319 (1977).
    Maximum error in B_{12} estimated to be ± 6.

T	B_{12}	T	B_{12}
343	-59.9	553	-12.3

| 411 | -37.0 | 623 | - 4.1 |
| 481 | -21.2 | 700 | + 1.8 |

NITRIC OXIDE + NITROGEN DIOXIDE   NO + NO_2

1. F.T. Selleck, H.H. Reamer, and B.H. Sage, Ind. Engng Chem. ind. Edn 45 814 (1953).

   Volumetric behaviour of fifteen mixtures from 280K to 440K; P → 47 atm.

NITROGEN + NITROUS OXIDE N_2 + N_2O

1. A.E. Markham and K.A. Kobe, J. chem. Phys. 9 438 (1941).

   Percentage volume change measured on mixing gases at 298 K and 1 atm. pressure.

2. A. Charnley, J.S. Rowlinson, J.R. Sutton and J.R. Townley, Proc. R. Soc. A230 354 (1955).

   Values of isothermal Joule-Thomson coefficient at zero pressure given, T range 298 - 318 K.

NITROGEN + NEON N_2 + Ne

1. C.M. Knobler, J.J.M. Beenakker, and H.F.P. Knaap, Physica, 's Grav. 25 909 (1959).

   Low pressure differential method. Precision ± 3.

T	B_{12}
90	-31.0

2. J. Brewer and G.W. Vaughn, J. chem. Phys. 50 2960 (1969).

   Measurement of pressure changes on mixing gases at constant volume.

T	B_{11}	B_{22}	B_{12}
148.15	-73.54	4.36	- 6.54
173.15	-51.86	6.45	- 0.54
198.15	-37.48	7.90	3.27
223.15	-26.38	9.12	6.39
248.15	-17.59	10.14	9.13
273.15	-10.57	10.94	11.27
298.15	- 4.86	11.54	13.28
323.15	- 0.26	12.32	14.96

3. C.M. Knobler, J.J.M. Beenakker, and H.F.P. Knaap, Physica, 's
    Grav. <u>25</u> 909 (1959).

    Low pressure differential method. Precision ± 3.

T	B_{12}
90	-209

NITROGEN + OXYGEN $N_2 + O_2$

1. A.E. Markham and K.A. Kobe, J. chem. Phys. <u>9</u> 438 (1941).

    Percentage volume change measured on mixing gases at 298 K and
    1 atm. pressure.

2. R.A. Gorski and J.G. Miller, J. Am. chem. Soc. <u>75</u> 550 (1953).

    Measurement of volume change on mixing gases at constant pressure,
    below 1 atm.

T	B_{11}	B_{22}	B_{12}
303.15	- 4.34	-15.96	-9.7 ± 0.3

NITROUS OXIDE + OXYGEN $N_2O + O_2$

1. A.E. Markham and K.A. Kobe, J. chem. Phys. <u>9</u> 438 (1941).

    Percentage volume change measured on mixing gases at 298 K and
    1 atm. pressure.

NEON + OXYGEN $Ne + O_2$

1. C.M. Knobler, J.J.M. Beenakker, and H.F.P. Knaap, Physica, 's Grav.
    <u>25</u> 909 (1959).

    Low pressure differential method. Precision ± 3.

T	B_{12}
90	-40.2

NEON + XENON

1. B. Schramm and R. Gehrmann (1979) (†).

    Estimated error in B_{12} is ± 6.

T	B_{12}	T	B_{12}
213	-2.2	262	7.5
223	+0.7	276	9.0
242	5.4		

2. B. Schramm and H. Schmiedel (1979) (†).
   Estimated error in B_{12} is ± 6.

T	B_{12}	T	B_{12}
295	13.4	425	17.4
330	15.7	450	15.6
365	16.4	475	14.4
400	16.9		

PURE GAS FORMULA INDEX

474

$C_2H_6S_2$	2,3-Dithiabutane (Dimethyl disulphide)	83
C_2H_7N	Dimethylamine	83
C_2H_7N	Ethylamine (Aminoethane)	83
C_2N_2	Cyanogen (Oxalic acid dinitrile)	84
C_3F_8	Perfluoro-propane	84
C_3HF_7	1,Hydroperfluoropropane (1,1,1,2,2,3,3,Heptafluoropropane)	84
C_3H_4	Propadiene (Allene, Dimethylenemethane)	85
C_3H_4	Propyne (Propine, Methylacetylene)	85
C_3H_6	Cyclopropane (Trimethylene)	86
C_3H_6	Propene (Propylene)	87
$C_3H_6Cl_2$	2,2-Dichloropropane (Acetone dichloride, Isopropylidene chloride)	90
C_3H_6O	Propanal	90
C_3H_6O	Acetone (2-propanone)	90
$C_3H_6O_2$	Ethyl formate	93
$C_3H_6O_2$	Methyl acetate	94
C_3H_7Br	n-Propyl bromide (1-Bromopropane)	94
C_3H_7Br	Isopropyl bromide (2-Bromopropane)	94
C_3H_7Cl	n-Propyl chloride (1-Chloropropane)	95
C_3H_7Cl	Isopropyl chloride (2-Chloropropane)	96
C_3H_8	Propane	96
C_3H_8O	n-Propanol (1-Propanol)	100
C_3H_8O	2-Propanol (Iso-propanol)	101
C_3H_8S	1-Propanethiol (n-Propyl mercaptan)	102
C_3H_8S	2-Propane thiol (Isopropyl mercaptan)	102
C_3H_8S	Methyl ethyl sulphide	103
C_3H_9N	Trimethylamine	103
$C_3H_9BO_3$	Methyl borate	104
$C_4F_6O_3$	Trifluoroacetic anhydride	104
C_4F_8	Perfluorocyclobutane	104
C_4F_{10}	Perfluoro n-butane	105
C_4H_3ClS	2-Chlorothiophene	105
C_4H_4O	Furan (1,4-epoxy-1,3-butadiene, furfuran)	105
C_4H_4S	Thiophene (Thiofuran)	106
C_4H_5N	Pyrrole	106
C_4H_6	1-Butyne	107
C_4H_6O	2,5-Dihydrofuran	107

C_4H_8	But-1-ene (1-Butene)	108
C_4H_8	2-Methyl propene (Isobutylene)	108
C_4H_8	Cis-2-butene (Cis-β-butylene)	109
C_4H_8	Trans-2-butene (Trans-β-butylene)	109
C_4H_8O	Tetrahydrofuran	109
C_4H_8O	Methyl ethyl ketone (2-Butanone)	110
$C_4H_8O_2$	Butyric acid (Butanoic acid, Ethylacetic acid)	110
$C_4H_8O_2$	n-Propyl formate	110
$C_4H_8O_2$	Ethyl acetate	111
$C_4H_8O_2$	Methyl propionate	111
C_4H_8S	Thiacyclopentane (Tetrahydrothiophene)	111
C_4H_9Br	2-Bromo butane	112
C_4H_9Br	1-Bromo-2-methyl propane	112
C_4H_9Cl	n-Butyl chloride (1-Chloro butane)	113
C_4H_9Cl	Isobutyl chloride	113
C_4H_9Cl	2-Chloro 2-methyl propane (Tert-butyl chloride)	113
C_4H_9N	Pyrrolidine (1-Azacyclopentane, Tetrahydropyrrole)	114
C_4H_{10}	n-Butane	114
C_4H_{10}	Isobutane (2-Methyl propane)	119
$C_4H_{10}O$	n-Butanol (1-Butanol)	120
$C_4H_{10}O$	Iso-butanol (2-Methyl-1-propanol)	121
$C_4H_{10}O$	Sec-butanol (2-Butanol)	121
$C_4H_{10}O$	Tert-butanol (2-Methyl-2-propanol)	121
$C_4H_{10}O$	Diethyl ether (Ethyl ether, Ethoxy ethane)	122
$C_4H_{10}S$	Diethyl sulphide (Ethyl sulphide)	124
$C_4H_{10}S$	1-Butanethiol (Butyl mercaptan)	124
$C_4H_{10}S$	2-Butanethiol	124
$C_4H_{10}S$	2-Methyl 1-propanethiol (Isobutyl mercaptan)	125
$C_4H_{10}S$	2-Methyl 2-propanethiol	125
$C_4H_{10}S$	3-Methyl 2-thiabutane (Methylisopropyl sulphide)	125
$C_4H_{10}S$	2-Thiapentane (Methyl n-propyl sulphide)	125
$C_4H_{11}N$	Diethylamine	126
$C_4H_{12}Si$	Tetramethylsilane	126
C_5D_{12}	Neopentane-d_{12}	127
C_5F_{12}	Perfluoro-n-pentane	128
$C_5H_3D_9$	Neopentane-d_9	128
C_5H_5N	Pyridine (Azine)	128

$C_5H_6D_6$	Neopentane-d_6	129
C_5H_6O	2-Methyl furan	129
C_5H_6S	2-Methyl thiophene (α-Thiotoluene)	129
C_5H_6S	3-Methyl thiophene (β-Thiotoluene)	130
C_5H_7N	1-Methyl pyrrole	130
C_5H_8	Spiropentane	131
C_5H_8	Cyclopentene	131
$C_5H_9D_3$	Neopentane-d_3	131
C_5H_{10}	Cyclopentane	132
C_5H_{10}	Pent-1-ene (1-Pentylene, Propylethylene, 1-Pentene)	132
C_5H_{10}	2-Methyl-1-butene	133
C_5H_{10}	2-Methyl-2-butene	133
$C_5H_{10}O$	Methyl n-propyl ketone	133
$C_5H_{10}O$	Methyl isopropyl ketone	133
$C_5H_{10}O$	Diethyl ketone	134
$C_5H_{10}O_2$	Trimethylacetic acid (2,2-Dimethyl propanoic acid, Pivalic acid)	134
$C_5H_{10}S$	Thiacyclohexane (Pentamethylenesulphide)	134
$C_5H_{10}S$	Cyclopentanethiol (Cyclopentyl mercaptan)	134
$C_5H_{11}Cl$	2-Chloro-2-methylbutane	135
C_5H_{12}	n-Pentane	135
C_5H_{12}	Isopentane (2-Methylbutane)	138
C_5H_{12}	Neopentane	139
$C_5H_{12}S$	1-Pentanethiol (Amyl mercaptan)	143
C_6ClF_5	Chloropentafluorobenzene	143
C_6F_6	Hexafluorobenzene	143
C_6F_{12}	Perfluoro-c-hexane	145
C_6F_{14}	Perfluoro-n-hexane	145
C_6F_{14}	Perfluoro 2-methylpentane	146
C_6F_{14}	Perfluoro-3-methylpentane	146
C_6F_{14}	Perfluoro-2,3-dimethylbutane	147
C_6HF_5	Pentafluorobenzene	147
$C_6H_4F_2$	1,2-Difluorobenzene	148
C_6H_5F	Fluorobenzene	148
C_6H_6	Benzene	150
C_6H_6O	Phenol	159
C_6H_6S	Benzenethiol (Mercaptobenzene)	159
C_6H_7N	α-Picoline (2-Methylpyridine, 2-Picoline)	160
C_6H_7N	β-Picoline (3-Methylpyridine, 3-Picoline)	160

478

480

PURE GAS NAME INDEX

MIXTURE FORMULA INDEX

486

492

C_2H_4 (Ethylene) + NH_3	377
C_2H_4 (Ethylene) + He	377
C_2H_4 (Ethylene) + N_2	377
C_2H_4 (Ethylene) + N_2O	377
C_2H_4 (Ethylene) + O_2	377
$C_2H_4Cl_2$ (1,1-Dichloroethane) + CO_2	343
$C_2H_4F_2$ (1,1-Difluoroethane) + CCl_2F_2	299
C_2H_4O (Acetaldehyde) + C_2H_3N	373
$C_2H_4O_2$ (Methyl formate) + $CHCl_3$	309
$C_2H_4O_2$ (Acetic acid) + CH_2O_2	312
C_2H_5Br (Ethyl bromide) + CH_3Br	313
C_2H_5Br (Ethyl bromide) + $C_4H_{10}O$	378
C_2H_5Br (Ethyl bromide) + C_5H_{12}	378
C_2H_5Cl (Ethyl chloride) + CH_2Cl_2	312
C_2H_5Cl (Ethyl chloride) + CH_3Br	313
C_2H_5Cl (Ethyl chloride) + C_3H_7Cl	378
C_2H_5Cl (Ethyl chloride) + C_4H_9Cl	378
$C_2H_5NO_2$ (Nitroethane) + Ar	262
$C_2H_5NO_2$ (Nitroethane) + CO_2	343
$C_2H_5NO_2$ (Nitroethane) + N_2	378
C_2H_6 (Ethane) + Ar	262
C_2H_6 (Ethane) + CF_4	304
C_2H_6 (Ethane) + CH_4	318
C_2H_6 (Ethane) + CH_4O	337
C_2H_6 (Ethane) + CO_2	343
C_2H_6 (Ethane) + C_2F_6	371
C_2H_6 (Ethane) + C_2H_6O	379
C_2H_6 (Ethane) + C_3H_6	379
C_2H_6 (Ethane) + C_3H_8	379
C_2H_6 (Ethane) + C_4F_{10}	380
C_2H_6 (Ethane) + C_4H_{10}	380
C_2H_6 (Ethane) + $C_4H_{10}O$	381
C_2H_6 (Ethane) + $C_4H_{10}O$	381
C_2H_6 (Ethane) ° C_5H_{12}	381
C_2H_6 (Ethane) + C_6H_{14}	382
C_2H_6 (Ethane) + C_7H_{16}	382
C_2H_6 (Ethane) + C_8H_{18}	382
C_2H_6 (Ethane) + $C_{10}H_8$	383
C_2H_6 (Ethane) + $C_{11}H_{10}$	383

C_3H_6O (Acetone) + CS_2	370
C_3H_6O (Acetone) + C_4H_{10}	390
C_3H_6O (Acetone) + $C_4H_{10}O$	391
C_3H_6O (Acetone) + C_6H_6	391
C_3H_6O (Acetone) + C_6H_{12}	391
C_3H_6O (Acetone) + Hg	392
$C_3H_6O_2$ (Methyl acetate) + $CHCL_3$	309
C_3H_7Cl (n-Propyl chloride) + C_2H_5Cl	378
C_3H_8 (Propane) + Ar	263
C_3H_8 (Propane) + CH_3Br	313
C_3H_8 (Propane) + CH_4	319
C_3H_8 (Propane) + CO_2	345
C_3H_8 (Propane) + C_2H_6	379
C_3H_8 (Propane) + C_3F_6	388
C_3H_8 (Propane) + C_3H_7	389
C_3H_8 (Propane) + C_5H_{12}	392
C_3H_8 (Propane) + C_5H_{12}	393
C_3H_8 (Propane) + C_6H_{14}	393
C_3H_8 (Propane) + C_7H_{16}	393
C_3H_8 (Propane) + C_8H_{18}	394
C_3H_8 (Propane) + HCl	394
C_3H_8 (Propane) + H_2	395
C_3H_8 (Propane) + He	395
C_3H_8 (Propane) + Hg	395
C_3H_8 (Propane) + N_2	395
C_3H_8O (Isopropanol) + Ar	264
C_3H_8O (Isopropanol) + CO_2	346
C_3H_8O (n-Propanol) + N_2	396
C_4F_{10} (Perfluoro-n-butane) + CF_4	304
C_4F_{10} (Perfluoro-n-butane) + CH_4	321
C_4F_{10} (Perfluoro-n-butane) + C_2F_6	371
C_4F_{10} (Perfluoro-n-butane) + C_2H_6	380
C_4F_{10} (Perfluoro-n-butane) + C_4H_{10}	396
C_4F_{10} (Perfluoro-n-butane) + C_5F_{12}	397
C_4F_{10} (Perfluoro-n-butane) + C_6F_{14}	397
C_4F_{10} (Perfluoro-n-butane) + C_6H_{14}	397
C_4H_6 (1,3-Butadeiene) + CO_2	346
C_4H_8 (1-Butene) + C_3H_6	389
C_4H_8O (Methylethyl ketone) + Ar	264

C_4H_8O (Methylethyl ketone) + CO_2	346
C_4H_8O (Methylethyl ketone) + N_2	397
C_4H_8O (Tetrahydrofuran) + H_2O	397
$C_4H_8O_2$ (Dioxan) + Ar	265
$C_4H_8O_2$ (Dioxan) + CO_2	346
$C_4H_8O_2$ (Dioxan) + N_2	398
$C_4H_8O_2$ (ethyl acetate) + $CHCl_3$	309
$C_4H_8O_2$ (n-Propyl formate) + $CHCl_3$	310
C_4H_9Cl (t-Butylchloride) + C_2H_5Cl	378
C_4H_{10} (n-Butane) + Ar	265
C_4H_{10} (n-Butane) + CF_4	304
C_4H_{10} (n-Butane) + CH_3Br	313
C_4H_{10} (n-Butane) + CH_4	321
C_4H_{10} (n-Butane) + CH_4O	338
C_4H_{10} (n-Butane) + CO_2	347
C_4H_{10} (n-Butane) + C_2F_6	372
C_4H_{10} (n-Butane) + C_2H_6	380
C_4H_{10} (n-Butane) + C_3H_6	389
C_4H_{10} (n-Butane) + C_3H_6O	390
C_4H_{10} (n-Butane) + C_4F_{10}	396
C_4H_{10} (n-Butane) + C_4H_{10}	398
C_4H_{10} (n-Butane) + C_5H_{12}	399
C_4H_{10} (n-Butane) + C_6H_{14}	399
C_4H_{10} (n-Butane) + C_8H_{18}	399
C_4H_{10} (n-Butane) + H_2	400
C_4H_{10} (n-Butane) + He	400
C_4H_{10} (n-Butane) + Hg	400
C_4H_{10} (n-Butane) + N_2	401
C_4H_{10} (Isobutane) + CH_4	322
C_4H_{10} (Isobutane) + C_3H_6	389
C_4H_{10} (Isobutane) + C_4H_{10}	398
$C_4H_{10}O$ (1-Butanol) + Ar	265
$C_4H_{10}O$ (1-Butanol) + CH_4	323
$C_4H_{10}O$ (1-Butanol) + CO_2	347
$C_4H_{10}O$ (1-Butanol) + C_2H_6	381
$C_4H_{10}O$ (1-Butanol) + N_2	401
$C_4H_{10}O$ (Diethyl ether) + Ar	266
$C_4H_{10}O$ (Diethyl ether) + $CHCl_3$	310
$C_4H_{10}O$ (Diethyl ether) + CH_3I	314

C_6H_{12} (Cyclohexane) + CO_2	350
C_6H_{12} (Cyclohexane) + C_2H_3N	373
C_6H_{12} (Cyclohexane) + C_3H_6O	391
C_6H_{12} (Cyclohexane) + $C_4H_{11}N$	403
C_6H_{12} (Cyclohexane) + C_6F_6	414
C_6H_{12} (Cyclohexane) + C_6H_6	416
C_6H_{12} (Cyclohexane) + H_2	426
C_6H_{12} (Cyclohexane) + N_2	426
C_6H_{12} (Cyclohexane) + O_2	427
C_6H_{12} (Methyl cyclopentane) + Ar	271
C_6H_{12} (Methyl cyclopentane) + CO_2	350
C_6H_{12} (Methyl cyclopentane) + H_2	427
C_6H_{12} (Methyl cyclopentane) + N_2	428
C_6H_{12} (n-Hex-1-ene) + Ar	271
C_6H_{12} (n-Hex-1-ene) + H_2	428
C_6H_{12} (n-Hex-1-ene) + N_2	428
C_6H_{12} (Trans-hex-2-ene) + N_2	429
C_6H_{12} (Trans-hex-3-ene) + N_2	429
$C_6H_{12}O_2$ (Butyl acetate) + Ar	272
$C_6H_{12}O_2$ (Butyl acetate) + CO_2	351
$C_6H_{12}O_2$ (Butyl acetate) + N_2	429
C_6H_{14} (n-Hexane) + Ar	272
C_6H_{14} (n-Hexane) + CF_4	305
C_6H_{14} (n-Hexane) + $CHCl_3$	311
C_6H_{14} (n-Hexane) + CH_4	328
C_6H_{14} (n-Hexane) + CO_2	351
C_6H_{14} (n-Hexane) + C_2H_6	382
C_6H_{14} (n-Hexane) + C_3H_8	393
C_6H_{14} (n-Hexane) + C_4F_{10}	397
C_6H_{14} (n-Hexane) + C_4H_{10}	399
C_6H_{14} (n-Hexane) + $C_4H_{10}O$	402
C_6H_{14} (n-Hexane) + C_5H_{12}	407
C_6H_{14} (n-Hexane) + C_6F_{14}	415
C_6H_{14} (n-Hexane) + C_6H_6	417
C_6H_{14} (n-Hexane) + C_8H_{18}	430
C_6H_{14} (n-Hexane) + H_2	430
C_6H_{14} (n-Hexane) + N_2	430
C_6H_{14} (n-Hexane) + O_2	433
C_6H_{14} (2-Methylpentane) + Ar	273

C_7H_{14} (n-Hept-1-ene) + N_2	439
C_7H_{16} (n-Heptane) + Ar	277
C_7H_{10} (n-Heptane) + CH_4	329
C_7H_{16} (n-Heptane) + CO_2	353
C_7H_{16} (n-Heptane) + C_2H_6	382
C_7H_{16} (n-Heptane) + C_3H_8	393
C_7H_{16} (n-Heptane) + C_5H_{12}	407
C_7H_{16} (n-Heptane) + C_6H_5F	416
C_7H_{16} (n-Heptane) + C_6H_6	418
C_7H_{16} (n-Heptane) + C_8H_{18}	439
C_7H_{16} (n-Heptane) + H_2	440
C_7H_{16} (n-Heptane) + N_2	440
C_7H_{16} (n-Heptane) + O_2	441
C_7H_{16} (2-Methylhexane) + Ar	277
C_7H_{16} (2-Methylhexane) + CO_2	354
C_7H_{16} (2-Methylhexane) + H_2	442
C_7H_{16} (2-Methylhexane) + N_2	442
C_7H_{16} (2-Methylhexane) + O_2	442
C_7H_{16} (3-Methylhexane) + Ar	277
C_7H_{16} (3-Methylhexane) + CO_2	354
C_7H_{16} (3-Methylhexane) + H_2	442
C_7H_{16} (3-Methylhexane) + N_2	443
C_7H_{16} (3-Methylhexane) + O_2	443
C_7H_{16} (3-Ethylpentane) + CO_2	354
C_7H_{16} (3-Ethylpentane) + H_2	443
C_7H_{16} (3-Ethylpentane) + N_2	444
C_7H_{16} (3-Ethylpentane) + O_2	444
C_7H_{16} (2,2-Dimethylpentane) + Ar	278
C_7H_{16} (2,2-Dimethylpentane) + CO_2	354
C_7H_{16} (2,2-Dimethylpentane) + H_2	444
C_7H_{16} (2,2-Dimethylpentane) + N_2	444
C_7H_{16} (2,3-Dimethylpentane) + Ar	278
C_7H_{16} (2,3-Dimethylpentane) + CO_2	355
C_7H_{16} (2,3-Dimethylpentane) + H_2	445
C_7H_{16} (2,3-Dimethylpentane) + N_2	445
C_7H_{16} (2,3-Dimethylpentane) + O_2	445
C_7H_{16} (2,4-Dimethylpentane) + Ar	278
C_7H_{16} (2,4-Dimethylpentane) + CO_2	355
C_7H_{16} (2,4-Dimethylpentane) + H_2	446

C_7H_{16} (2,4-Dimethylpentane) + N_2	446
C_7H_{16} (2,4-Dimethylpentane) + O_2	447
C_7H_{16} (3,3-Dimethylpentane) + Ar	279
C_7H_{16} (3,3-Dimethylpentane) + CO_2	355
C_7H_{16} (3,3-Dimethylpentane) + H_2	447
C_7H_{16} (3,3-Dimethylpentane) + N_2	447
C_7H_{16} (3,3-Dimethylpentane) + O_2	448
C_7H_{16} (2,2,3-Trimethylbutane) + Ar	279
C_7H_{16} (2,2,3-Trimethylbutane) + CO_2	355
C_7H_{16} (2,2,3-Trimethylbutane) + H_2	448
C_7H_{16} (2,2,3-Trimethylbutane) + N_2	448
C_7H_{16} (2,2,3-Trimethylbutane) + O_2	449
C_8H_8 (Styrene) + Ar	279
C_8H_8 (Styrene) + CO_2	356
C_8H_8 (Styrene) + N_2	449
C_8H_{10} (Ethylbenzene) + Ar	280
C_8H_{10} (Ethylbenzene) + CO_2	356
C_8H_{10} (m-Xylene) + Ar	280
C_8H_{10} (m-Xylene) + CO_2	356
C_8H_{10} (o-Xylene) + Ar	280
C_8H_{10} (o-Xylene) + CO_2	356
C_8H_{10} (p-Xylene) + Ar	280
C_8H_{10} (p-Xylene) + CO_2	357
C_8H_{16} (n-Oct-1-ene) + Ar	281
C_8H_{18} (n-Octane) + Ar	281
C_8H_{18} (n-Octane) + CH_4	329
C_8H_{18} (n-Octane) + CO	341
C_8H_{18} (n-Octane) + CO_2	357
C_8H_{18} (n-Octane) + C_2H_6	382
C_8H_{18} (n-Octane) + C_3H_8	394
C_8H_{18} (n-Octane) + C_4H_{10}	399
C_8H_{18} (n-Octane) + C_5H_{12}	407
C_8H_{18} (n-Octane) + C_6H_5F	416
C_8H_{18} (n-Octane) + C_6H_6	419
C_8H_{18} (n-Octane) + C_6H_{14}	430
C_8H_{18} (n-Octane) + C_7H_{16}	439
C_8H_{18} (n-Octane) + H_2	449
C_8H_{18} (n-Octane) + N_2	450
C_8H_{18} (2,2-Dimethylhexane) + N_2	451

C_8H_{18} (Isooctane) + H_2		451
C_8H_{18} (Isooctane) + N_2		451
C_8H_{18} (2,2,4-Trimethylpentane) + Ar		281
C_8H_{18} (2,2,4-Trimethylpentane) + CO_2		357
C_8H_{18} (2,2,3,3-Tetramethylbutane) + N_2		452
C_9H_{12} (1,2,3-Trimethylbenzene) + Ar		282
C_9H_{12} (1,2,3-Trimethylbenzene) + CO_2		359
C_9H_{12} (1,2,4-Trimethylbenzene) + Ar		283
C_9H_{12} (1,2,4-Trimethylbenzene) + CO_2		359
C_9H_{12} (1,3,5-Trimethylbenzene) + Ar		283
C_9H_{12} (1,3,5-Trimethylbenzene) + CO_2		359
C_9H_{20} (2,2,5-Trimethylhexane) + CH_4		330
C_9H_{12} (1-Methyl-2-ethylbenzene) + Ar		282
C_9H_{12} (1-Methyl-2-ethylbenzene) + CO_2		358
C_9H_{12} (1-Methyl-3-ethylbenzene) + Ar		282
C_9H_{12} (1-Methyl-3-ethylbenzene) + CO_2		358
C_9H_{12} (n-Propylbenzene) + Ar		281
C_9H_{12} (n-Propylbenzene) + CO_2		358
C_9H_{12} (Isopropylbenzene) + Ar		282
C_9H_{12} (Isopropylbenzene) + CO_2		358
C_9H_{20} (n-Nonane) + Ar		283
C_9H_{20} (n-Nonane) + CO_2		350
C_9H_{20} (n-Nonane) + N_2		452
C_9H_{20} (2,2,5-Trimethylhexane) + N_2		452
$C_{10}H_8$ (Naphthalene) + Ar		284
$C_{10}H_8$ (Naphthalene) + CH_4		330
$C_{10}H_8$ (Naphthalene) + CO_2		360
$C_{10}H_8$ (Naphthalene) + C_2H_4		374
$C_{10}H_8$ (Naphthalene) + C_2H_6		383
$C_{10}H_8$ (Naphthalene) + H_2		453
$C_{10}H_8$ (Naphthalene) + He		453
$C_{10}H_8$ (Naphthalene) + N_2		453
$C_{10}H_8$ (Naphthalene) + N_2O		453
$C_{10}H_8$ (Naphthalene) + O_2		454
$C_{10}H_{14}$ (1,2,3,4-Tetramethylbenzene) + Ar		288
$C_{10}H_{14}$ (1,2,3,4-Tetramethylbenzene) + CO_2		364
$C_{10}H_{14}$ (1,2,3,5-Tetramethylbenzene) + Ar		288
$C_{10}H_{14}$ (1,2,4,5-Tetramethylbenzene) + Ar		288
$C_{10}H_{14}$ (1,2,3,5-Tetramethylbenzene) + CO_2		364

$C_{10}H_{14}$ (1,2,4,5-Tetramethylbenzene) + CO_2	364
$C_{10}H_{14}$ (1,2,Dimethyl-3-ethylbenzene) + Ar	286
$C_{10}H_{14}$ (1,2-Dimethyl-3-ethylbenzene) + CO_2	362
$C_{10}H_{14}$ (1,2-Dimethyl-4-ethylbenzene) + Ar	286
$C_{10}H_{14}$ (1,2-Dimethyl-4-ethylbenzene) + CO_2	362
$C_{10}H_{14}$ (1,3-Dimethyl-2-ethylbenzene) + Ar	287
$C_{10}H_{14}$ (1,3-Dimethyl-2-ethylbenzene) + CO_2	363
$C_{10}H_{14}$ (1,3-Dimethyl-4-ethylbenzene) + Ar	287
$C_{10}H_{14}$ (1,3-Dimethyl-4-ethylbenzene) + $C_{10}H_{14}$	363
$C_{10}H_{14}$ (1,3-Dimethyl-5-ethylbenzene) + Ar	287
$C_{10}H_{14}$ (1,3-Dimethyl-5-ethylbenzene) + CO_2	363
$C_{10}H_{14}$ (1,4-Dimethyl-2-ethylbenzene) + Ar	287
$C_{10}H_{14}$ (1,4-Dimethyl-2-ethylbenzene) + CO_2	363
$C_{10}H_{14}$ (1-Methyl-2-propylbenzene) + Ar	285
$C_{10}H_{14}$ (1-Methyl-2-propylbenzene) + CO_2	361
$C_{10}H_{14}$ (1-Methyl-2-isopropylbenzene) + Ar	285
$C_{10}H_{14}$ (1-Methyl-2-isopropylbenzene) + CO_2	361
$C_{10}H_{14}$ (1-Methyl-3-propylbenzene) + Ar	285
$C_{10}H_{14}$ (1-Methyl-3-propylbenzene) + CO_2	361
$C_{10}H_{14}$ (1-Methyl-4-propylbenzene) + Ar	286
$C_{10}H_{14}$ (1-Methyl-4-propylbenzene) + CO_2	362
$C_{10}H_{14}$ (1,4-Diethylbenzene) + Ar	286
$C_{10}H_{14}$ (1,4-Diethylbenzene) + CO_2	362
$C_{10}H_{14}$ (n-Butylbenzene) + Ar	284
$C_{10}H_{14}$ (n-Butylbenzene) + CO_2	360
$C_{10}H_{14}$ (sec-Butylbenzene) + Ar	284
$C_{10}H_{14}$ (sec-Butylbenzene) + CO_2	360
$C_{10}H_{14}$ (tert-Butylbenzene) + CH_4	330
$C_{10}H_{14}$ (tert-Butylbenzene) + N_2	454
$C_{10}H_{14}$ (tert-Butylbenzene) + Ar	284
$C_{10}H_{18}$ (Bicyclohexyl) + CH_4	331
$C_{10}H_{18}$ (Bicyclohexyl) + C_2H_4	374
$C_{10}H_{18}$ (Bicyclohexyl) + C_2H_6	383
$C_{10}H_{22}$ (n-Decane) + CH_4	331
$C_{10}H_{22}$ (n-Decane) + CO_2	364
$C_{10}H_{22}$ (n-Decane) + H_2	454
$C_{10}H_{22}$ (n-Decane) + N_2	454
$C_{11}H_{10}$ (1-Methyl naphthalene) + CH_4	331
$C_{11}H_{10}$ (1-Methyl Naphthalene) + C_2H_4	374

$C_{11}H_{10}$ (1-Methyl naphthalene) + C_2H_6	383
$C_{12}H_{26}$ (n-Dodecane) + CH_4	332
$C_{12}H_{26}$ (n-Dodecane) + N_2	455
$C_{13}H_{12}$ (Diphenylmethane) + CH_4	332
$C_{13}H_{12}$ (Diphenylmethane) + C_2H_4	375
$C_{13}H_{12}$ (Diphenylmethane) + C_2H_6	384
$C_{14}H_{10}$ (Anthracene) + Ar	289
$C_{14}H_{10}$ (Anthracene) + CH_4	332
$C_{14}H_{10}$ (Anthracene) + CO_2	365
$C_{14}H_{10}$ (Anthracene) + C_2H_4	375
$C_{14}H_{10}$ (Anthracene) + C_2H_6	384
$C_{14}H_{10}$ (Anthracene) + O_2	455
$C_{14}H_{10}$ (Phenanthrene) + CH_4	333
$C_{14}H_{10}$ (Phenanthrene) + CO_2	365
$C_{14}H_{10}$ (Phenanthrene) + C_2H_4	375
$C_{14}H_{10}$ (Phenanthrene) + N_2O	456
$C_{16}H_{34}$ (n-Hexadecane) + CH_4	333
$C_{16}H_{34}$ (n-Hexadecane) + C_2H_4	376
$C_{16}H_{34}$ (n-Hexadecane) + C_2H_6	384
$C_{20}H_{42}$ (Eicosane) + CH_4	333
$C_{20}H_{42}$ (Eicosane) + C_2H_6	384
$C_{30}H_{62}$ (Squalene) + CH_4	333
$C_{30}H_{62}$ (Squalene) + C_2H_6	385
D_2 (Deuterium) + He	456
F_6S (Sulphur hexafluoride) + Ar	289
F_6S (Sulphur hexafluoride) + $CClF_3$	298
F_6S (Sulphur hexafluoride) + CF_4	305
F_6S (Sulphur hexafluoride) + CH_4	334
F_6S (Sulphur hexafluoride) + $C_4H_{12}Si$	403
F_6S (Sulphur hexafluoride) + C_5H_{12}	412
F_6S (Sulphur hexafluoride) + Kr	457
HCl (Hydrogen chloride) + Ar	289
HCl (Hydrogen chloride) + C_3H_8	394
HCl (Hydrogen chloride) + Kr	457
HD (Hydrogen deuteride) + He	456
H_2 (Hydrogen) + Ar	290
H_2 (Hydrogen) + CCl_4	301
H_2 (Hydrogen) + CH_4	334
H_2 (Hydrogen) + CO	341

H_2 (Hydrogen) + Xe	462
H_2O (Water) + Ar	291
H_2O (Water) + CH_4	335
H_2O (Water) + CO_2	366
H_2O (Water) + C_2H_6	385
H_2O (Water) + C_4H_8O	397
H_2O (Water) + H_2	457
H_2O (Water) + N_2	463
H_2O (Water) + N_2O	463
H_2S (Hydrogen sulphide) + CH_4	335
H_2S (Hydrogen sulphide) + C_2H_6	385
H_2S (Hydrogen sulphide) + C_5H_{12}	408
H_2S (Hydrogen sulphide) + N_2	463
H_3N (Ammonia) + C_2H_2	372
H_3N (Ammonia) + C_2H_4	377
H_3N (Ammonia) + C_2H_6	386
He (Helium) + Ar	291
He (Helium) + CCl_4	301
He (Helium) + CF_4	307
He (Helium) + CH_4	335
He (Helium) + CO_2	366
He (Helium) + C_2H_4	377
He (Helium) + C_2H_6O	387
He (Helium) + C_3H_6	390
He (Helium) + C_3H_8	395
He (Helium) + C_4H_{10}	400
He (Helium) + C_5H_{12}	412
He (Helium) + C_6H_6	420
He (Helium) + $C_{10}H_8$	453
He (Helium) + D_2	456
He (Helium) + HD	456
He (Helium) + H_2	458
He (Helium) + Kr	464
He (Helium) + N_2	464
He (Helium) + Ne	466
He (Helium) + O_2	467
He (Helium) + Xe	467
Hg (Mercury) + Ar	292
Hg (Mercury) + CH_4O	338

Hg (Mercury) + C_3H_6O	392
Hg (Mercury) + C_3H_8	395
Hg (Mercury) + C_4H_{10}	400
Kr (Krypton) + Ar	292
Kr (Krypton) + CH_4	336
Kr (Krypton) + F_6S	457
Kr (Krypton) + HCl	457
Kr (Krypton) + He	464
Kr (Krypton) + Ne	468
KR (Krypton) + Xe	469
NO (Nitric oxide) + NO_2	470
NO_2 (Nitrogen dioxide) + NO	470
N_2 (Nitrogen) + Ar	294
N_2 (Nitrogen) + BF_3	297
N_2 (Nitrogen) + CCl_4	302
N_2 (Nitrogen) + $CHCl_3$	311
N_2 (Nitrogen) + CH_3NO_2	316
N_2 (Nitrogen) + CH_4	336
N_2 (Nitrogen) + CH_4O	339
N_2 (Nitrogen) + CO_2	367
N_2 (Nitrogen) + CS_2	371
N_2 (Nitrogen) + C_2H_4	377
N_2 (Nitrogen) + $C_2H_5NO_2$	378
N_2 (Nitrogen) + C_2H_6	386
N_2 (Nitrogen) + C_2H_6O	387
N_2 (Nitrogen) + C_3H_8	395
N_2 (Nitrogen) + C_3H_8O	396
N_2 (Nitrogen) + C_4H_8O	397
N_2 (Nitrogen) + $C_4H_8O_2$	398
N_2 (Nitrogen) + C_4H_{10}	401
N_2 (Nitrogen) + $C_4H_{10}O$	401
N_2 (Nitrogen) + $C_4H_{10}O$	402
N_2 (Nitrogen) + C_5H_5N	404
N_2 (Nitrogen) + C_5H_8	404
N_2 (Nitrogen) + C_5H_{10}	405
N_2 (Nitrogen) + C_5H_{10}	406
N_2 (Nitrogen) + C_5H_{12}	409
N_2 (Nitrogen) + C_5H_{12}	411
N_2 (Nitrogen) + C_5H_{12}	413

512